# Nutrient Interactions

# ift Basic Symposium Series

Edited by
INSTITUTE OF FOOD TECHNOLOGISTS
221 N. LaSalle St.
Chicago, Illinois

Foodborne Microorganisms and Their Toxins:
Developing Methodology   edited by Merle D.
Pierson and Norman J. Stern

Water Activity: Theory and Applications to
Food   edited by Louis B. Rockland and
Larry R. Beuchat

Nutrient Interactions   edited by C. E. Bodwell
and John W. Erdman, Jr.

# Nutrient Interactions

*edited by*

## C. E. Bodwell
United States Department of Agriculture
Agricultural Research Station
Beltsville, Maryland

## John W. Erdman, Jr.
Department of Food Science
University of Illinois
Urbana, Illinois

CRC Press
Taylor & Francis Group
Boca Raton London New York

CRC Press is an imprint of the
Taylor & Francis Group, an **informa** business

First published 1988 by MARCEL DEKKAR, INC.

Published 2018 by CRC Press
Taylor & Francis Group
6000 Broken Sound Parkway NW, Suite 300
Boca Raton, FL 33487-2742

Copyright © 1988 by Taylor & Francis Group LLC
CRC Press is an imprint of Taylor & Francis Group, an Informa business

First issued in paperback 2019

No claim to original U.S. Government works

ISBN 13: 978-0-367-45130-1 (pbk)
ISBN 13: 978-0-8247-7868-2 (hbk)

**Visit the Taylor & Francis Web site at**
**http://www.taylorandfrancis.com**

**and the CRC Press Web site at**
**http://www.crcpress.com**

# PREFACE

This volume contains the proceedings of the Eleventh Annual Basic Symposium sponsored by the Institute of Food Technologists and the International Union of Food Science and Technology, which was held in Las Vegas, Nevada, June 15–16, 1987. Current knowledge of nutrient interactions that are of importance in human nutrition is summarized and critically discussed in 15 chapters by experts active in their specific research areas. In general, emphasis is placed on recent research findings from human and animal studies.

Protein and energy interactions are discussed in Chapter 1 and the general role of energy in nutrient interactions is considered. Chapter 2 includes a detailed discussion of metabolic interrelationships of specific amino acids. Critical overviews of current knowledge about minerals and their interactions in food systems and minerals and their physiological interactions are presented in Chapters 3 and 4. Specific interactions involving dietary protein and iron are evaluated in Chapter 5. Chapters 6 through 8 include in-depth coverage of vitamin–mineral interactions and discussions of factors that may affect these interactions, such as oxidative stability of dietary lipids, dietary protein, specific minerals, and electrolytes.

Carbohydrate–mineral interactions and fiber–mineral and fiber–vitamin interactions are evaluated in Chapters 9 and 10. Interacting effects of carbohydrate and lipid on metabolism, including recent data on metabolic effects of omega-3 fatty acids, are discussed in Chapter 11. Vitamin–vitamin interactions are assessed in detail in Chapter 12. In Chapters 13 through 15, the effects of toxic minerals, food additives, and drugs on nutrient interactions are reviewed.

We believe that this book is unique in that, for the first time within a single volume, a comprehensive discussion is presented of nutrient interactions and their significance in human nutrition. Accordingly, this book provides the most critical up-to-date assessment of nutrient interactions available and should be of particular value to food scientists and technologists, clinical and experimental nutritionists, dieticians, food chemists, and toxicologists.

The syposium program was approved and supported by the IFT Basic Symposium Committee, which included Drs. R. V. Josephson (Chairman), Philip G. Crandall, D. T. Gordon, Joseph A. Maga, Merle D. Pierson, Joe M. Regenstein, and Richard A. Scanlan. Dr. John Powers, 1986–1987 IFT President, welcomed participants and expressed his gratitude to the speakers for sharing their expertise. Calvert L. Wiley, IFT Executive Director, John B. Klis, Director of Publications, Anna May Schenck, IFT Associate Scientific Editor, and other IFT staff members provided the support and coordination necessary for the smooth operation of the symposium.

Primary credit for the success of this symposium belongs to the speakers. They performed professionally, meeting the necessary deadlines to result in the rapid and timely publication of this volume. Their cooperation is deeply appreciated.

C. E. Bodwell
John W. Erdman, Jr.

# Contributors

Carolyn D. Berdanier   Department of Nutrition, University of Georgia, Athens, Georgia

Fergus M. Clydesdale   Department of Food Science and Nutrition, University of Massachusetts, Amherst, Massachusetts

Harold H. Draper   Department of Nutritional Sciences, University of Guelph, Guelph, Ontario, Canada

M. R. Spivey Fox   Division of Nutrition, Food and Drug Administration, Washington, D.C.

Naomi K. Fukagawa   The Clinical Research Center, Massachusetts Institute of Technology, Cambridge, Massachusetts, and Departments of Pediatrics and Medicine, Harvard Medical School, Boston, Massachusetts

Janet L. Greger   Nutritional Sciences Department, University of Wisconsin, Madison, Wisconsin

Steven L. Ink   The Quaker Oats Company, Barrington, Illinois

Lillian Langseth   Nutrition Research Newsletter, Palisades, New York

Bo Lönnerdal   Department of Nutrition, University of California at Davis, Davis, California

Lawrence J. Machlin   Department of Clinical Nutrition, Hoffman-La Roche Inc., Nutley, New Jersey

Elaine R. Monsen   Department of Medicine, University of Washington, Seattle, Washington

Sheldon Reiser   Carbohydrate Nutrition Laboratory, Beltsville Human Nutrition Research Center, United States Department of Agriculture, Agricultural Research Station, Beltsville, Maryland

Daphne A. Roe   Division of Nutritional Sciences, Cornell University, Ithaca, New York

Noel W. Solomons   CeSSIAM, Guatemala National Committee for the Blind and Deaf, Hospital de Ojos y Oidos, Guatemala City, Guatemala

Benjamin Torún   Institute of Nutrition of Central America and Panama (INCAP), Guatemala City, Guatemala

John E. Vanderveen   Division of Nutrition, Center for Food Safety and Applied Nutrition, Food and Drug Administration, Washington, D.C.

Vernon R. Young   Department of Applied Biological Sciences, Massachusetts Institute of Technology, Cambridge, Massachusetts

# Contents

# Nutrient Interactions

# 1

# Energy–Nutrient Interactions

**Benjamin Torún**

Institute of Nutrition of Central America and Panama (INCAP)
Guatemala City, Guatemala

## INTRODUCTION

One of the most important roles of food intake is the provision of energy for mechanical work, maintenance of body temperature, biosynthesis, growth, and other metabolic functions. In addition, the intake and metabolism of energy may alter the requirements for other nutrients or influence their metabolism. This is particularly true of protein, which, after energy sources, is the second largest component of the human diet. There is also evidence that the amount and quality of dietary protein influences energy utilization and may have effects on body weight.

The satisfaction of appetite is generally related to the fulfillment of energy needs. A common practice is to prescribe the amounts of food in a diet in terms of its energy content. The bulk of the diet may be a limiting factor to satisfy the needs for many nutrients. Consequently, the concept of nutrient density is very important and has led to the expression of the nutrient content of foods relative to energy.

1

In this overview we will address these issues.

## ENERGY AND PROTEIN

The relationship of protein and energy metabolism has been studied by many investigators in the past four decades. Various aspects of these interactions have been discussed in several reviews (e.g., Calloway, 1981; Young et al., 1983; Torún, 1986). The vast information available has shown that energy intake and metabolism play an important role in the modulation of amino acid and protein metabolism and that this has influenced the experimental assessment of protein requirements.

The extent to which protein metabolism taxes energy metabolism is illustrated in calculations from studies of Reeds et al. (1982) on the proportion of energy expenditure associated with whole body protein synthesis in young pigs. Based on estimates of whole body leucine flux with diets of different compositions, it can be calculated that about 20% of energy expenditure is associated with protein syntheses (Table 1.1). Calculations from studies on whole body protein synthesis in preterm infants (Duffy et al., 1981), full-term neonates fed orally or intravenously (Duffy and Pencharz, 1986), and 4- to 9-year-old children with cystic fibrosis (Parsons et al., 1985) indicate that the cost of protein synthesis represented 10–12% of total energy intake, using a factor of 1.08 kcal/g protein synthesized (Reeds et al., 1982) or 8–9% with a factor of 0.85 kcal/g protein synthesized (Millward and Garlick, 1976). The significant contribution of protein syn-

**TABLE 1.1** Minimum Contribution of Protein Synthesis to Total Daily Energy Expenditure in Young Pigs (28–36 kg Body Weight)

| Diet | Total energy expenditure (kcal/kg/day) | Proportion of energy for protein synthesis[a] (%) |
|---|---|---|
| Whole diet 1 | 127 | 17 |
| Whole diet 2 | 179 | 18 |
| Whole diet 3 | 194 | 20 |
| High carbohydrate | 209 | 20 |
| High fat | 205 | 20 |
| High protein | 188 | 26 |

[a]Assuming 1.08 kcal/g protein synthesis.
*Source: Reeds et al. (1982)*

thesis to total energy expenditure has been confirmed by other invest-igators (Webster, 1981; Garlick, 1986).

There are still some issues that have not been completely resolved per-taining to the manner and extent of the regulation and interactions of energy and protein in persons with different dietary intakes and lifestyles. We will point out some of those issues and raise questions with important implications, particularly for populations with chronically low intakes of protein, energy, or both, and whose way of life imposes additional or spe-cial demands for those nutrients. In doing so, we will consider the effects that energy intake, dietary energy substrates, and energy expenditure may have on the regulation of protein metabolism and requirements.

## Energy Intake and Protein Metabolism

**Nitrogen balance.** It has long been recognized that an increase in energy intake with constant dietary protein intake can reduce nitrogen excretion and increase nitrogen balance (Forbes et al., 1939; Munro, 1951). This nitrogen-sparing effect of nonprotein dietary energy is most probably related to the preferential use that the body makes of carbohydrates and fats as energy sources. This is more obvious in persons eating protein in amounts below or around the protein requirement level, and it has also been demonstrated when energy intakes exceed energy expenditure (e.g., Inoue et al., 1973; Calloway, 1975; Garza et al., 1976; Kishi et al., 1978; But-terfield and Calloway, 1984). There is, however, a limit to the effect of dietary energy on nitrogen balance, as illustrated Fig 1.1. When the diet does not provide adequate amounts of protein, additional intakes of en-ergy beyond a given amount will not allow further improvement of ni-trogen balance.

It would be expected that when energy intake cannot satisfy energy demands, more protein would be oxidized, with a resultant increase in urinary nitrogen excretion and a decrease in nitrogen balance. That reasoning led to the concept that nitrogen balance might be a sensitive in-dicator of the satisfaction of energy requirements. However, this is not always the case. In a study of the energy requirements of preschool-aged children with a constant intake of vegetable protein equivalent to 1.2 g milk protein/kg/day, two consecutive reductions of 10% in dietary energy at 40-day intervals did not affect nitrogen retention measured 17–20 and 37–40 days after each dietary change (Torún and Viteri, 1981a). As Fig. 1.2 shows, energy expenditure diminished with the first reduction in energy intake, and weight gain decreased after the second reduction. Nitrogen balance, however, did not change, nor did height, and there were no con-sistent changes in urinary creatinine excretion related to the levels of en-

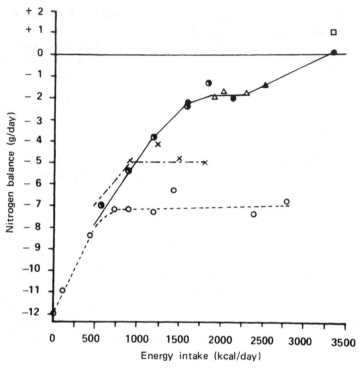

**FIG. 1.1** Nitrogen balance at various levels of energy intake. After certain energy intake, a plateau is reached. The height of the plateau is related to the amount of protein intake. (*From Calloway and Spector, 1954.*)

ergy intake. It seems that the compensatory decrease in energy expenditure, the protein ingestion at a safe level of intake, the good nutritional status of the children, and the hygienic environment in which they lived throughout the study prevented a nitrogen loss due to the reduction in energy intake. This supports Munro's (1951, 1964) contention that the responses of whole body and tissue nitrogen metabolism to altered energy supplies are modulated by the status of protein metabolism and the level of protein intake. It is possible that if the conditions that led to reduced weight gain had continued much longer, if the children had become ill with the frequency that is customary in developing countries, or if their dietary protein had been marginal (i.e., closer to the average requirement for their age), indications of body protein derangement might have become evident.

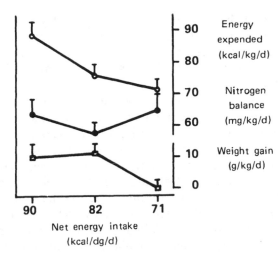

**FIG. 1.2** Energy expenditure, nitrogen balance, and weight gain (mean ± SEM) of six children fed 1.73 g/kg/day of corn and bean proteins (ratio 58:42) with changes in energy intake at 40-day intervals. (*Adapted from data of Torún and Viteri, 1981a.*)

Studies done in India also failed to show a consistent association between energy intake and nitrogen balance. In one study (Iyengar et al., 1979), 4-year-old children ate diets that provided 1, 1.3, 1.6, and 2 g protein and 80 or 100 kcal/kg/day. With three of these levels of protein intake, nitrogen balance fell with the lower energy intake in only 2 of 3 children studied, but these decreases were not related to protein intake level. In another study with adults (Iyengar and Narasinga Rao, 1979), an increase in dietary energy from 44 to 56 kcal/kg/day improved nitrogen retention when the men were fed 1.0 but not 1.2 g protein/kg/day.

These findings indicate that nitrogen balance is more sensitive to changes in dietary energy when protein intake is relatively low and that satisfactory nitrogen balance is a necessary, but not sufficient, criterion of the adequacy of dietary energy supply. They also suggest that when protein intake is adequate, the body is better protected in terms of nitrogen metabolism against a moderate decrease in energy intake.

**Protein synthesis and amino acid oxidation.** The protein-sparing effect of dietary energy has been demonstrated by means other than nitrogen balance. For example, Garlick et al. (1973) and Holliday et al. (1977) showed that dietary energy enhances the rate of muscle protein synthesis in rats. In humans with a protein intake of 0.6 g/kg/day, Motil et al. (1981) used amino acids labeled with stable isotopes. An energy intake 25% in excess of maintenance requirements reduced the rate of leucine oxidation from (mean ± S.D.) 18.0 ± 8.3 to 12.4 ± 8.7 µmol/kg/hr and increased net nitrogen retention.

Studies with $^{15}$N-glycine to measure whole body protein metabolism in children recovering from protein–energy malnutrition showed a positive correlation between food intake and protein turnover and synthesis (Golden et al., 1977). It was assumed that the effect was related to the amount of dietary energy, since similar experiments where dietary protein intake was modified did not affect the rates of protein synthesis (Picou and Taylor-Roberts, 1969). More recent studies by the same group of investigators (Jackson et al., 1983) with children who had recovered or were still recovering from malnutrition failed to corroborate the previous conclusions. Conversely, they found that whole body protein synthesis was greater with 80 than with 90 or 100 kcal/kg/day in children fed either a diet deficient (0.7 g/kg/day) or fairly adequate in protein (1.7 g/kg/day). The large variability of the results, which was greater between individuals on the same diet than within individuals at different levels of energy intake, and the various degrees of nutritional recovery of the children make it difficult to support any conclusions about the energy-protein interactions observed.

**Protein requirements with marginal dietary energy intakes.**    The experimental evidence of energy intake on protein metabolism has raised the question of whether protein requirements have been underestimated for populations prone to have marginal or deficient energy intakes. On the other hand, have such populations adapted to use more efficiently their low energy intakes and preserve body protein? Although a decrease in energy expenditure in response to low energy intake may diminish the risk of body protein loss, it is not known how long such a response can be effectively sustained. Furthermore, this should be regarded as a transient compensation of low intakes and not as a desirable adaptation in view of the physical, emotional, and social costs of diminished energy expenditure, which usually is achieved through a reduction in physical activity (Viteri and Torún, 1975; 1981).

Another important, related question is whether protein requirements are higher for persons with low body stores of energy and whose dietary intakes are sometimes marginal and sometimes adequate, as occurs in groups of migrant workers or persons subject to seasonal changes in food availability.

The most logical answer to these questions would be to raise the dietary protein recommendations. However, this is more expensive and often more difficult than increasing the availability of dietary energy sources. The question, then, that also needs to be answered is whether increasing the recommendations for energy intake in populations with constraints relative to eating more protein will reduce their protein requirements.

## Dietary Energy Substrates

**Nitrogen balance.** The effects of dietary energy on protein metabolism vary with the source of energy, and this might have a bearing on protein requirements. Both carbohydrates and lipids enhance nitrogen and amino acid metabolism, but carbohydrates have specific actions that make them more effective than fats in promoting the utilization of dietary protein, at least transiently (Munro, 1951; 1964). As Fig. 1.3 illustrates, the

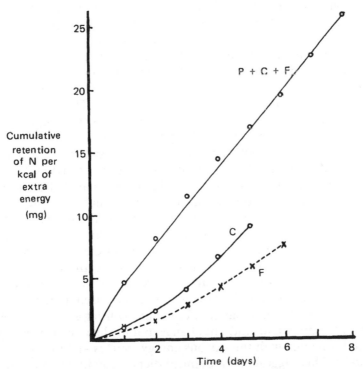

**FIG. 1.3** Cumulative nitrogen retention curves showing the amounts of nitrogen retained by one man after 700 kcal of extra energy were added to his diet as fat (F), carbohydrate (C), or a mixture of protein, carbohydrate, and fat (P + C + F). His basal diet provided 75 g protein and 2850 kcal/day. (*From Munro, 1964.*)

administration of 700 kcal of extra energy as carbohydrate to a man who ate 75 g protein/day reduced his urinary excretion of nitrogen more than fat did and resulted in better nitrogen retention (Munro, 1964). Using isocaloric, isonitrogenous diets with carbohydrate:fat ratios of 1:1 and 2:1, Richardson et al. (1979) showed that the higher nitrogen-sparing effect of carbohydrate relative to fat was more pronounced at lower levels of energy and protein intakes. In another study, these investigators showed that it made no difference whether sucrose or dextrins and maltose were the principal dietary carbohydrates (Richardson et al., 1980).

**Amino acid metabolism.**    The effects of carbohydrates on protein metabolism have also been shown in relation to specific amino acids. The oral or intravenous administration of glucose produces a transient decrease in plasma amino acid concentration, especially the branched-chain amino acids and other large neutral amino acids, such as methionine, phenylalanine, tyrosine, and tryptophan (Munro and Thomson, 1953; Martin-Du Pan et al., 1982). This is accompanied by an increase in synthesis or a reduction in catabolism of muscle protein (Munro et al., 1959; Wassner et al., 1976; Sim et al., 1979). Conversely, dietary carbohydrate restriction, as in high protein, low carbohydrate diets, leads to increased accumulation of plasma branched-chain amino acids after protein feeding or after intravenous infusion of leucine (Gelfand et al., 1979). This is probably due to reduced utilization of these amino acids, which are the major substrates for restoration of muscle tissue after protein feeding.

Using a different model with intravenous infusion of $^{13}$C-leucine and $^{15}$N-lysine, Motil et al. (1981) explored the effects of giving 25% surfeit dietary energy, either as sucrose plus a glucose polymer, as butter fat, or as a combination of both, on the metabolism of whole body leucine and lysine. Those amino acids were selected because the former is metabolized mainly in peripheral tissues and the latter in liver. Net leucine retention was calculated within each individual from the difference between leucine incorporation into protein and its release from protein breakdown, and it was interpreted as an increase in muscle protein. As shown in Table 1.2, in the fed state leucine retention increased more with the carbohydrate or mixed supplementations than with fat alone. Whole body flux of both amino acids did not change, but the leucine flux was greater than the lysine flux with the various energy sources and two levels of energy intake. This suggests that identical dietary conditions may elicit different responses from different essential amino acids.

Glucose administration has also been shown to influence the metabolism of the nonessential amino acid alanine, increasing its de novo syn-

**TABLE 1.2** Net Leucine Retention (i.e., Incorporation into Protein Minus Release from Protein Breakdown) with 25% Excess Energy Intake Provided as Fat, Carbohydrate, or a Mixture of Both[a]

| Condition during isotope infusion | Source of excess energy | Net leucine retention ($\mu$mol/kg/hr) |
|---|---|---|
| 12-hour fast | fat | $-12 \pm 3$ |
| | carbohydrate | $-14 \pm 5$ |
| | mixed | $-9 \pm 2$ |
| fed hourly | fat | $+20 \pm 5$[b] |
| | carbohydrate | $+31 \pm 7$ |
| | mixed | $+30 \pm 2$ |

[a]Mean $\pm$ SEM for 4 men in each group.
[b]Different from carbohydrate and mixed diets, $p < 0.05$.
*Source: Motil et al. (1981).*

thesis, its release to the circulating plasma, and its whole body flux (Felig, 1973; Ben Galim et al., 1980; Shulman et al., 1980; Robert et al., 1982). The significance of these findings with respect to the economy of body protein metabolism is unclear due to the role of alanine as a major precursor of urea, although the rate of alanine-nitrogen formation is not necessarily paralleled by the rate of urea-nitrogen formation (Young et al., 1983).

Munro (1978) has suggested that, in addition to its influence on the metabolism of circulating amino acids, dietary carbohydrate also interacts with amino acids from the same meal. Giving protein and carbohydrates as separate meals can lead to nitrogen loss from the body, which is reversed by giving these two nutrients in the same meal (Cuthbertson and Munro, 1939).

**Mechanisms of action.** The mechanisms by which carbohydrate exerts these effects on protein metabolism are not fully understood. They are partly mediated by the insulin released in response to carbohydrate absorption or infusion (Munro, 1964). This hormone stimulates muscle protein synthesis by increasing amino acid transport and peptide chain formation (Kipnis and Noall, 1958; Jefferson et al., 1974), promoting the uptake of branched-chain amino acids into muscle and decreasing the release of amino acids from muscle in the fasting state (Zinnemann et al., 1966; Pozefsky et al., 1969). The simultaneous infusion of insulin and glucose produces a decrease in plasma urea concentration and urinary nitrogen excretion several times greater than the decrease observed when glucose is infused alone (Fuller et al., 1977).

Another possible mechanism of the effects of glucose may be through the suppression of glucagon release. Glucagon infusion increases the synthesis of urea at the expense of the free amino acid pool and probably also by hydrolysis of visceral protein, and it increases urinary nitrogen excretion (Wolfe et al., 1979). Gelfand et al. (1979) observed a small increase in plasma glucagon concentration when carbohydrates were restricted in the diet, but they concluded that such small hormonal increases did not account for the large increase in circulating branched-chain amino acids after protein ingestion.

**Influence on protein requirements.**   Regardless of the mechanisms involved, the experimental evidence that carbohydrates spare protein more efficiently than fat raises important practical questions, such as: Is there a more efficient use of dietary proteins among populations whose dietary energy is 70–80% from carbohydrates? If so, will their protein requirements be affected by the dietary changes induced by migrations, cultural changes, and access to processed foods?

On the other hand, there may also be adaptive changes in protein utilization with the long-term ingestion of diets rich in fat. Although the stimulation of muscle protein synthesis and reduction of catabolism observed with carbohydrate are not the same when fat is fed as a single meal (Munro, 1978), the studies of Jeejeeboy et al. (1975) with patients on long-term parenteral feeding indicate that the administration of fat as a major energy source on a regular basis over long periods of time may be as effective as carbohydrate in promoting nitrogen retention and net protein synthesis.

## Protein Intake and Energy Metabolism

The influence of protein intake on energy metabolism has long been recognized in animal husbandry, and it is known that animals fed isoenergetic amounts of different diets do not necessarily gain equal weight. Miller and Payne (1962) showed that, in order to maintain weight, baby pigs fed a diet with only 2% protein calories ate five times as much food as their counterparts fed 24% protein calories. However, few studies have looked into the effects of protein intake on energy metabolism in humans, and they usually have dealt with changes in body weight.

**Protein quantity.**   Miller and Mumford (1967) studied 16 adult men and women who were fed, for 4 to 8 weeks, with diets providing about 1400 kcal above their normal daily intake and which contained either about 3% or 15% protein calories. Physical activity in both groups was low and unchanged, food digestibility was normal, and there were no differences in body composition. The group fed the high protein diet gained 3.7 kg in weight, whereas the other group gained only 1.1 kg.

The effects of protein intake on weight gain were also demonstrated in a study designed to assess the effects of isometric exercises on protein requirements of young men (Torún et al., 1977). They ate diets providing either 0.5 or 1 g egg protein/kg/day and habitual dietary energy levels, based on each man's dietary history. In this four-man crossover study, two were fed the low protein diet for 5–7 weeks and then the high protein diet for 4 weeks; the other two men underwent the opposite procedure. Energy intake and physical activity remained constant. Figure 1.4 shows that weight gains were larger with the high protein diet. The men also showed changes in total body potassium (an index of lean body mass) that paralleled the changes in the rates of weight gain.

Dietary protein can also influence the rate of weight gain of infants and young children. MacLean and Graham (1979) evaluated the effects of protein quantity in children recovering from protein–energy malnutrition. They compared the rates of weight gain of six children aged 5 to 17 months who were fed four isoenergetic diets in random order for 14 days each, which provided 125 or 150 kcal/kg/day and 4–8% of their calories from milk protein. The rates of weight gain increased gradually from 2.8 to 6.7 g/kg/day as protein intake increased from 4.0 to 6.4% of dietary energy. No further increments in weight gain were seen with 8% protein calories.

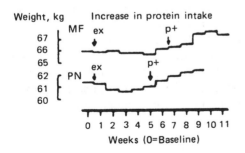 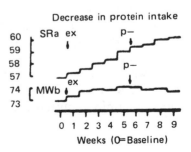

**FIG. 1.4** Changes in rates of weight gain of men who ate isoenergetic diets with either 0.5 or 1 g egg protein/kg/day. p+ indicates doubling, p– halving the dietary protein; ex indicates the beginning of isometric exercise training. (*From Torún et al., 1977.*)

**Protein quality.**    MacLean and Graham (1979) also evaluated the effect
of enriching wheat flour with lysine on the rates of weight gain of five infants aged 3–8 months. The infants were fed a casein diet for 9 days and
then for 15–36 days isoenergetic–isonitrogenous diets in which wheat provided all dietary protein, either by itself or enriched in random order with
0.12, 0.2, or 0.4% lysine. The mean rate of weight gain was lowest with non-
enriched wheat but improved from 2.2–3.5 g/kg/day with the addition of
lysine. There were no differences between casein and wheat with 0.4%
lysine.

In conclusion, the quantity and quality of dietary protein influence the
utilization of dietary energy, at least in terms of weight gain. Further
research is needed to establish whether protein intake affects other aspects of energy metabolism.

## ENERGY AND VITAMINS

Interest in the interactions of energy and micronutrients is based on the
fact that some vitamins serve as cofactors in oxidative reactions by which
the chémical energy present in the molecules of carbohydrates, fats, and
proteins is liberated and made available to the cells of the body for work
or as heat. Consequently, the amount of dietary energy sources that are ingested and the level of physical activity that may modify the cell needs for
energy influence the requirements for several vitamins.

### Thiamin

Thiamin, mainly in its pyrophosphate form, is an important coenzyme
in various reactions of carbohydrate metabolism. Its deficiency can be
accelerated in experimental animals by high carbohydrate intakes and
retarded by substituting fat for carbohydrate. Thus, its requirements máy
vary with the amount of dietary carbohydrates. Since these provide between 55 and 70% of energy in most diets, it has become customary to express thiamin requirements in relation to total energy intake (Table
1.3).

### Niacin

Niacin, mainly as niacinamide, forms part of the coenzymes NAD and
NADP, which are actively involved in carbohydrate and fatty acid

**TABLE 1.3** Examples of Recommended Daily Dietary Intakes of Vitamins Largely Involved in Carbohydrate or Energy Metabolism

| Age (years) | Weight (kg) | Energy[a] (kcal) | Thiamin[b] (mg) 0.5 mg/1000 kcal | Riboflavin[b] (mg) 0.6 mg/1000 kcal | Niacin[b] (mg) 7 mg/1000 kcal |
|---|---|---|---|---|---|
| <1 | 8.5 | 810 | 0.4 | 0.5 | 6 |
| 1–3 | 13 | 1300 | 0.6 | 0.8 | 9 |
| Males | | | | | |
| 7–10 | 27 | 2100 | 1.1 | 1.3 | 15 |
| 14–16 | 56 | 2650 | 1.3 | 1.6 | 19 |
| 18–30 | 70 | 2800 | 1.4 | 1.7 | 20 |
| 30–60 | 70 | 2700 | 1.4 | 1.6 | 19 |
| Females | | | | | |
| 7–10 | 27 | 1800 | 0.9 | 1.1 | 13 |
| 14–16 | 52 | 2150 | 1.1 | 1.3 | 15 |
| 18–30 | 55 | 2100 | 1.1 | 1.3 | 15 |
| 30–60 | 55 | 2100 | 1.1 | 1.3 | 15 |

[a]World Health Organization (1985).
[b]National Research Council (1980).

metabolism. Therefore, it has been postulated that the requirement of humans for niacin is dependent upon energy intake and physical activity. As for thiamim, it is customary to express niacin requirements in relation to total dietay energy intake (Table 1.3).

## Riboflavin

The expression of riboflavin requirements as a function of energy intake is more controversial. Although it participates as part of a coenzyme in a number of oxidative reactions involved in electron transport, mainly as FMN and FAD, the requirements for riboflavin do not seem to be related to energy requirement or to muscular activity. Its urinary excretion is affected by alterations in nitrogen balance, and less is excreted when tissue growth is rapid. This has led to suggestions that riboflavin requirements should be related to protein requirements (Horwitt, 1966). However, it has become customary to express them in terms of dietary energy intake (Table 1.3).

## DIETARY ENERGY DENSITY

Energy sources are the largest nutritional components of a diet. Table 1.4 shows the approximate amounts of macronutrients that a 13-kg child or a 70-kg man must eat every day to satisfy their energy needs. Assuming an average energy density of 1.4 kcal per gram of food, total daily food intake would be of the order of 1 kg for the preschool child and 2 kg for the adult. An increase in energy density to 2 kcal/g reduces the quantity (or bulk) of food needed to meet energy requirements by 30%. This illustrates how the satisfaction of energy requirements can be hindered by bulky diets with low energy density. The younger the person, the more serious the constraint: With the lower density diet, the child in Table 1.4 would have to eat the equivalent of about 7% of his body weight in food, compared to less than 3% of body weight for the adult.

## Energy Requirements

Energy requirements are easily satisfied with the habitual intakes of foods customary in developed countries and among people with middle and high incomes in developing countries. Infants and children of preschool age from low-income families in countries where the diet is mostly

**TABLE 1.4**  Daily Food and Macronutrient Intakes to Satisfy Energy Needs with Mixed Diets (About 10% Protein Energy, 30% Fat Energy, and 60% Carbohydrate Energy)

| | Total energy needs (kcal) | Protein (g) | Fat (g) | CHO (g) | Total macro-nutrients (g) | Daily food intake, g, with energy densities of: | |
|---|---|---|---|---|---|---|---|
| | | | | | | 1.4 kcal/g | 2 kcal/g |
| 13-kg child | 1300 | 35 | 45 | 200 | 280 | 930 | 650 |
| 70-kg man | 2800 | 70 | 100 | 400 | 570 | 2000 | 1400 |

**TABLE 1.5** Average Composition and Frequency of Consumption of Diets Customary for Many Children of Preschool Age in Guatemala

| Food | Intake per day (g) | Frequency of intake (days/week) | Weekly intake Amount (g) | Weekly intake Protein (g) | Weekly intake Energy (kcal) |
|---|---|---|---|---|---|
| Corn tortilla flour | 105 | 7 | 735 | 67.6 | 2,734 |
| Black bean flour | 18 | 7 | 126 | 27.8 | 423 |
| Bread (sweet roll), fresh | 37 | 7 | 259 | 19.7 | 1,000 |
| Vegetables (chayote, squash, or potatoes), raw | 44 | 7 | 308 | 5.8 | 163 |
| Milk products (as fluid milk equivalents) | 100 | 3 | 300 | 9.9 | 195 |
| Fruit (orange, apple, banana), fresh | 30 | 4 | 120 | 0.6 | 60 |
| Egg, fresh | 43 | 2 | 86 | 9.9 | 142 |
| Meat, raw (as beef equivalent) | 40 | 1 | 40 | 7.6 | 97 |
| Sugar | 42 | 7 | 294 | — | 1,176 |
| Oil or lard | 5 | 7 | 35 | — | 315 |
| Total intake per week | | | | 148.9 | 6,305 |
| Mean intake per day | | | | 21.3 | 901 |
| Mean intake/kg/day (assuming weight of 12 kg) | | | | 1.78 | 75 |

Bean:corn ratio = 15:85 by weight and 29:71 by protein contents.
Animal protein = 18% of total.
Energy from fat (including natural fat content of all foods) = 17% of total.

*Source: Torún and Viteri (1981b).*

of plant origin and low in fat may have difficulty in doing so. Table 1.5 shows the foods that constitute the basic diets of low-income preschool children in Guatemala. When cooked and ready to eat, the average daily solid and semisolid foods consumed weigh about 550 g, and their energy density is around 1.6 kcal/g. As Table 1.5 shows, mean energy intake of these children is of the order of 75 kcal/kg/day, or 15–25% below recommended levels (WHO, 1985); stunting and malnutrition are common among them. Addition of vegetable oil and sugar increased energy density to 2.5–2.8 kcal/g in solid and semisolid foods and to 0.8–1 kcal/g in liquid foods and allowed catch-up and adequate growth (Torún and Viteri, 1981b; Torún et al., 1984). Studies by Viteri et al. (1981) with diets based on corn, black beans, and varying amounts of fat further illustrate the importance of energy density to satisfy the energy requirements of preschool children (Table 1.6).

## Protein Requirements

The second largest component of nutritional requirements is protein. The diets in Table 1.4 would satisfy protein needs when eaten in the amounts necessary to satisfy energy requirements. However, if the protein density decreases in relation to total dietary energy, it may not be possible to eat enough food to fulfill protein needs. This has led to the introduction of terms that correlate protein and energy in a diet to assess whether it might be possible to ingest enough of the former when needs for the latter

**TABLE 1.6** Energy Intake and Growth Rate of 2-Year-Old Children Eating Corn-and-Black Bean Diets with Different Energy Densities

| | Energy density | | |
| --- | --- | --- | --- |
| | Low | Middle | High |
| Energy density, kcal/g dry weight | 4.5 | 4.9 | 5.4 |
| Energy from fat, % of total energy | 8 | 22 | 35 |
| Energy intake, kcal/kg/day | 84 ± 6[a] | 96 ± 12 | 102 ± 10 |
| Rate of weight gain, g/day | 6 ± 9[a] | 15 ± 9 | 13 ± 10 |

[a]Lower than with the other two diets, $p < 0.05$.
*Source: Viteri et al. (1981).*

are met. Platt et al. (1961) were largely responsible for the use of the ratio of protein energy to total energy (PE ratio) as a convenient and useful descriptor of this aspect of dietary quality in human nutrition. Since not all dietary protein is absorbed and utilized, the concept of net dietary protein calories as a percentage of total energy (NDPCal%) was introduced to account for both the quality and concentration of protein in the diet (Miller and Payne,1961; Platt et al., 1961).

Most diets worldwide have PE ratios between 8 and 15% (Beaton and Swiss, 1974; UNU, 1979), but their NDPCal% may be markedly lower, particularly in countries where diets are mainly of plant origin (Araya et al., 1981). In such cases, satisfaction of energy needs is not necessarily associated with satisfaction of protein needs, especially for young children weaned from breast-feeding. This can lead to clinical cases of kwashiorkor, which can be treated or prevented by increasing the protein density and quality of the food, thereby raising the NDPCal%. Most diets used to treat protein–energy malnutrition have a PE ratio of 8–13% and NDPCal% about 1% lower.

## Nutritional Quality of Foods and Diets

The amount of food that must be eaten to fulfill the requirements for essential fatty acids, minerals, vitamins, and electrolytes poses less of a problem than do energy and protein requirements. Consequently, it is conceivable that if a person consumes enough food to meet his or her energy needs, the needs for all other nutrients can be met if the diet is adequately balanced in composition. This has led to a practical approach for the assessment of a diet's nutritional quality based on the principle of nutrient density relative to energy. This concept, termed "nutritive ratio" in 1904 by Henry to express the relationship between nutrients and energy in rations for farm animals (Crampton, 1964), was applied to human diets by Rose et al. in 1928 and evolved into the Index of Nutritional Quality (INQ) (Hansen, 1973; Wyse et al., 1976; Hansen et al., 1985). This index is calculated for each nutrient as INQ = Percent of nutrient requirement supplied by a quantity of food/Percent of energy requirement supplied by that quantity of food. Tables and tentative standards have been published in an effort to describe an "adequate" balance of nutrients in a diet, expressing the recommended dietary allowances for each nutrient per 1000 kcal of food intake (Hansen and Wyse, 1980: Hansen et al., 1985; Hegsted, 1985).

## SUMMARY

There is experimental evidence that energy intake influences the metabolism of and nutritional requirements for other nutrients. This is most clearly evident in the case of dietary protein since an important fraction of dietary energy is expended for the synthesis of body protein. The amount of energy intake affects nitrogen balance, particularly when protein intakes are near or below the requirement level, and it also influences the rates of oxidation of specific amino acids. These effects of energy on protein metabolism tend to be greater when carbohydrates, rather than fats, are the energy substrates.

The quality and quantity of dietary protein, in turn, also influences energy metabolism as evaluated through changes in body weight with isoenergetic diets of different protein compositions: Under specific experimental or therapeutic conditions, an increase in protein quality or quantity results in faster weight gain.

These interactions of dietary energy and proteins raise important questions related to currently accepted estimates of requirements and factors that may influence requirements as well as questions about the effects of energy intake on the efficiency of nutrient utilization in general.

Energy metabolism affects the body's need for vitamins that are cofactors in the intracellular oxidation of energy substrates. Thus, the intake of thiamin and niacin is customarily associated with, and recommended in relation to, the amount of energy intake.

Energy density is an important characteristic of foods and diets which suggests whether they will permit the satisfaction of macro- and micronutrient requirements. Although not strictly a nutrient–nutrient interaction, the content of nutrients in a diet relative to its total energy is the basis for practical indices and ratios that give an idea of that diet's nutritional quality, provided that it is consumed in amounts that will fulfill energy requirements. If dietary bulk is too high to allow this, the overall quality of the diet can be improved by increasing its energy density while maintaining an adequate balance of all other nutrients.

## REFERENCES

Araya, H., García, B., and Arroyave, G. 1981. Estudio dietético en embarazadas de Santa María Cauqué. I. Variabilidad de los indicadores proteínicos y su análisis por tiempo de comida. *Arch. Latinoamer. Nutr.* 31: 108.

Beaton, G.H. and Swiss, L.D. 1974. Evaluation of the nutritional quality of food supplies: prediction of "desirable" or "safe" protein-calorie ratios. *Am. J. Clin. Nutr.* 27: 485.

Ben Galim, D., Hruska, K., Bier, D. M., Matthews, D. E., and Haymond, M. W. 1980. Branched chain amino acid nitrogen transfer to alanine in vivo in dogs. *J. Clin. Invest.* 66: 1295.

Butterfield, G. E. and Calloway, D. H. 1984. Physical activity improves protein utilization in young men. *Br. J. Nutr.* 51: 171.

Calloway, D. 1975. Nitrogen balance of men with marginal intakes of protein and energy. *J. Nutr.* 105: 914.

Calloway, D. H. 1981. Energy–protein relationships. In *Protein Quality in Humans.* Bodwell, C. E., Adkins, J. S., and Hopkins, D. T. (Ed.), p. 148. AVI Publishing Co., Westport, CT.

Calloway, D. H. and Spector, H. 1954. Nitrogen balance as related to calorie and protein intake in active young men. *Am. J. Clin. Nutr.* 2: 405.

Crampton, E. W. 1964. Nutrient-to-calorie ratios in applied nutrition. *J. Nutr.* 82: 353.

Cuthbertson, D. P. and Munro, H. N. 1939. The relationship of carbohydrate metabolism to protein metabolism: I. The roles of total dietary carbohydrate and of surfeit carbohydrate in protein metabolism. *Biochem. J.* 33: 128.

Duffy, B. and Pencharz, P. 1986. The effect of feeding route (IV or oral) on the protein metabolism of the neonate. *Am. J. Clin. Nutr.* 43: 108.

Duffy, B., Gunn, T., Collinge, J., and Pencharz, P. 1981. The effect of varying protein quality and energy intake on the nitrogen metabolism of parenterally fed very low birthweight (<1600 g) infants. *Pediatr. Res.* 15: 1040.

Felig, P. 1973. The glucose–alanine cycle. *Metabolism* 22: 179.

Forbes, E. B., Bratzler, J. W., Thacker, E. J., and Marcy, L. F. 1939. Dynamic effects and net energy values of protein, carbohydrate and fat. *J. Nutr.* 18: 57.

Fuller, M. F., Weekes, T. E. C. , Cadenhead, A., and Bruce, J. B. 1977. The protein-sparing effect of carbohydrate. 2. The role of insulin. *Br. J. Nutr.* 38: 489.

Garlick, P. J. 1986. Protein synthesis and energy expenditure in relation to feeding. *Int. J. Vitam. Nutr. Res.* 56: 197.

Garlick, P. J., Millward, D. J., and James, W. P. T. 1973. The diurnal response of muscle and liver protein synthesis in vivo in meal fed rats. *Biochem. J.* 136: 935.

Garza, C., Scrimshaw, N. S., and Young, V. R. 1976. Human protein requirements: The effect of variations in energy intake within the maintenance range. *Am. J. Clin. Nutr.* 29: 280.

Gelfand, R. A., Hendler, R. G., and Sherwin, R. S. 1979. Dietary carbohydrate and metabolism of ingested protein. *Lancet* 2: 65.

Golden, M., Waterlow, J. C., and Picou, D. 1977. The relationship between dietary intake, weight change, nitrogen balance and protein turnover in man. *Am. J. Clin. Nutr.* 30: 1345.

Hansen, R. G. 1973. An index of food quality. *Nutr. Rev.* 31: 1.

Hansen, R. G., Windham, C. T., and Wyse, B. W. 1985. Nutrient density and food labeling. *Clin. Nutr.* 4: 164.

Hansen, R. G. and Wyse, B. W. 1980. Expression of nutrient allowances per 1,000 kilocalories. *J. Amer. Dietet. Assoc.* 76: 223.

Hegsted, D. M. 1985. Dietary standards: Dietary planning and nutrition education. *Clin. Nutr.* 4: 159.

Holliday, M. A., Chantler, C., MacDonnel, R., and Keitges, J. 1977. Effect of uremia on nutritionally induced variations in protein metabolism. *Kidney Int.* 11: 236.

Horwitt, M. K. 1966. Nutritional requirements of man with special reference to riboflavin. *Am. J. Clin. Nutr.* 18: 458.

Inoue, G., Fujita, Y., and Niiyama, Y. 1973. Studies on protein requirements of young men fed egg protein and rice protein with excess and maintenance energy intakes. *J. Nutr.* 103: 1673.

Iyengar, A. and Narasinga Rao, B. S. 1979. Effect of varying energy and protein intake on nitrogen balance in adults engaged in heavy manual labour. *Br. J. Nutr.* 41: 19.

Iyengar, A. K., Narasinga Rao, B. S., and Reddy, V. 1979. Effect of varying protein and energy intakes on nitrogen balance in Indian preschool children. *Br. J. Nutr.* 42: 417.

Jackson, A. A., Golden, M. H., Byfield, R., Jahoor, F., Royes, J., and Soutter, L. 1983. Whole-body protein turnover and nitrogen balance in young children at intakes of protein and energy in the region of maintenance. *Hum. Nutr. Clin. Nutr.* 37C: 433.

Jeejeeboy, K. N., Anderson, G. H., Nakhooda, A. F., Greenberg, G. R., Sanderson, I., and Marliss, E. B. 1975. Metabolic studies in total parenteral nutrition with lipid in man: Comparison with glucose. *J. Clin. Invest.* 57: 125.

Jefferson, L. S., Rannels, D. E., Munger, B. L., and Morgan, H. E. 1974. Insulin in the regulation of protein turnover in heart and skeletal muscle. *Fed. Proc.* 33: 1098.

Kipnis, D. M. and Noall, M. W. 1958. Stimulation of amino acid transport by insulin in the isolated rat diaphragm. *Biochim. Biophys. Acta* 28: 226.

Kishi, K., Miyatani, S., and Inoue, G. 1978. Requirement and utilization of egg protein by Japanese young men with marginal intakes of energy. *J. Nutr.* 108: 658.

MacLean, Jr., W. C. and Graham, G. G. 1979. The effect of level of protein intake in isoenergetic diets on energy utilization. *Am. J. Clin. Nutr.* 32: 1381.

Martin-DuPan, R., Mauron, C., Glaeser, B., and Wurtman, R. J. 1982. Effect of various oral glucose doses on plasma neutral amino acid levels. *Metabolism* 31: 937.

Miller, D. S. and Mumford, P. 1967. Gluttony: 1. An experimental study of overeating low- or high-protein diets. *Am. J. Clin. Nutr.* 20: 1212.

Miller, D. S. and Payne, P. R. 1962. Weight maintenance and food intake. *J. Nutr.* 78: 1.

Miller, D. S. and Payne, P. R. 1961. Problems in the prediction of protein value of diets: Caloric restriction. *J. Nutr.* 75: 225.

Millward, D. J. and Garlick, P. J. (1976). The energy cost of growth. *Proc. Nutr. Soc.* 35: 339.

Motil, K. J., Bier, D. M., Matthews, D. E., Burke, J. F., and Young, V. R. 1981. Whole body leucine and lysine metabolism studied with 1-$^{13}$C-leucine and 1-$^{15}$N-lysine. Response in healthy young men given excess energy intakes. *Metabolism* 30: 783.

Munro, H. N. 1951. Carbohydrate and fat as factors in protein utilization and metabolism. *Physiol. Rev.* 31: 449.

Munro, H. N. 1964. General aspects of the regulation of protein metabolism by diet and hormones. In *Mammalian Protein Metabolism*. Munro H. N. and Allison J. B. (Ed.), Vol. 1, p. 381. Academic Press, New York.

Munro, H. N. 1978. Energy and protein intakes as determinants of nitrogen balance. *Kidney Internat.* 14: 313.

Munro, H. N., Black, J. G., and Thomson, W. S. T. 1959. The mode of action of dietary carbohydrate on protein metabolism. *Br. J. Nutr.* 13: 475.

Munro, H. N. and Thomson, W. S. T. 1953. Influence of glucose on amino acid metabolism. *Metabolism* 2: 354.

National Research Council. 1980. *Recommended Dietary Allowances,* 9th ed. National Academy of Sciences, Washington, DC.

Parsons, H. G., Beaudry, P., and Pencharz, P. B. 1985. The effect of nutritional rehabilitation on whole body protein metabolism of children with cystic fibrosis. *Pediatr. Res.* 19: 189.

Picou D. and Taylor-Roberts, T. 1969. The measurement of total protein synthesis and catabolism and nitrogen turnover in infants in different nutritional states and receiving different amounts of dietary protein. *Clin. Sci.* 36: 283.

Platt, B. S., Miller, D. S., and Payne, P. R. 1961. Protein values in human food. In *Recent Advances in Clinical Nutrition.* Brock, J. F (Ed.), p. 35. Churchill, London.

Pozefsky, T., Felig, P., Tobin, J. D., Soeldner, J. S., and Cahill, G. F. 1969. Amino acid balance across tissues of the forearm in postabsorbtive man. Effect of insulin at two dose levels. *J. Clin. Invest.* 48: 2273.

Reeds, P. J., Wahle, K. W. J., and Haggarty, P. 1982. Energy costs of protein and fatty acid synthesis. *Proc. Nutr. Soc.* 41: 155.

Richardson, D. P., Scrimshaw, N. S., and Young, V. R. 1980. The effect of dietary sucrose on protein utilization in healthy young men. *Am. J. Clin. Nutr.* 33: 264.

Richardson, D. P., Wayler, A. H., Scrimshaw, N. S., and Young, V. R. 1979. Quantitative effect of an isoenergetic exchange of fat for carbohydrate on dietary protein utilization in healthy young men. *Am. J. Clin. Nutr.* 32: 2217.

Robert, J. J., Bier, D. M., Zhao, X. H., Matthews, D. E., and Young, V. R. 1982. Glucose and insulin effects on de novo amino acid synthesis in young men: studies with stable isotope labeled alanine, glycine, leucine and lysine. *Metabolism* 31: 1210.

Rose, M. S., Hessler, M. C., Stiebeling, H. K., and Taylor, C. M. 1928. Visualizing food values. *J. Home Econ.* 20: 781.

Shulman, G. I., Lacy, W. W., Liljenquist, J. E., Keller, U., Williams, P. E., and Cherrington, A. D. 1980. Effect of glucose, independent of changes in insulin and glucagon secretion, on alanine metabolism in the conscious dog. *J. Clin. Invest.* 65: 496.

Sim, A. J., Wolfe, B. M., Young, V. R., Clarke, D., and Moore, F. D. 1979. Glucose promotes whole body synthesis from infused amino acids in fasting man. *Lancet* 1: 68.

Torún, B. 1986. Role of energy metabolism in regulation of protein requirements. In *Proceedings of XIII International Congress of Nutrition.*

Taylor, T. G. and Jenkins, N. K. (Ed.), p. 414. John Libbey, London.

Torún, B., Caballero, B., Flores-Huerta, S., and Viteri, F. 1984. Habitual Guatemalan diets and catch-up growth of children with mild to moderate malnutrition. In *Protein–Energy Requirement Studies in Developing Countries: Results of International Research,* Rand, W. M., Uauy R., and Scrimshaw, N. S. (Ed.), p. 216. United Nations Univ., Tokyo.

Torún, B., Scrimshaw, N. S., and Young, V. R. 1977. Effect of isometric exercises on body potassium and dietary protein requirements of young men. *Am. J. Clin. Nutr.* 30: 1983.

Torún, B. and Viteri, F. E. 1981a. Energy requirements of pre-school children and effects of varying energy intakes on protein metabolism. In *Protein–energy Requirements of Developing Countries: Evaluation of New Data.* Torún, B., Young, V. R. and Rand, W. M. (Ed.), p. 229. United Nations Univ., Tokyo.

Torún, B. and Viteri, F. E. 1981b. Capacity of habitual Guatemalan diets to satisfy protein requirements of pre-school children with adequate dietary energy intakes. In *Protein–Energy Requirements of Developing Countries: Evaluation of New Data.* Torún, B., Young, V. R. and Rand, W. M. (Ed.), p. 210. United Nations Univ., Tokyo.

Torún, B. and Young, V. R. 1981. Interaction of energy and protein intakes in relation to dietary requirements. In *Nutrition in Health and Disease and International Development.* Harper, A. E. and Davis, G. K. (Ed.), p. 47. Alan R. Liss, New York.

United Nations University 1979. *Protein–Energy Requirements under Conditions Prevailing in Developing Countries: Current Knowledge and Research Needs.* Publication WHTR-1/UNUP-18. United Nations Univ., Tokyo.

Viteri, F. E. and Torún, B. 1975. Ingestión calórica y trabajo físico de obreros agrícolas en Guatemala. *Bol. Of. Sanit. Panamer.* 78: 58.

Viteri, F. E. and Torún, B. 1981. Nutrition, physical activity and growth. In *The Biology of Normal Human Growth.* Ritzen, M., Aperia, A., Hall, K., Larsson, Al, Zetterberg, A., and Zetterstrom, R. (Ed.), p. 265. Raven Press, New York.

Viteri, F. E., Torún, B., Arroyave, G., and Pineda, O. 1981. Use of corn-bean mixtures to satisfy protein and energy requirements of pre-school children. In *Protein–Energy Requirements of Developing Countries: Evaluation of New Data,* Torún B., Young, V. R., and Rand, W. M. (Ed.), p. 202. United Nations Univ., Tokyo.

Wassner, S. J., Orloff, S., and Holliday, M. A. 1976. Use of 3-methylhistidine to show cyclic variation of, and effects of starvation on, muscle protein catabolic rats (Abstr.). *Fed. Proc.* 35: 497.

Webster, A. J. F. 1981. The energetic efficiency of metabolism. *Proc. Nutr. Soc.* 40: 121.

Wolfe, B. M., Culebras, J. M., Aoki, T. T., O'Connor, N. E., Finley, R. J., Kaczowka, A., and Moore, F. D. 1979. The effects of glucagon on protein metabolism in normal man. *Surgery* 86: 248.

World Health Organization. 1985. *Energy and Protein Requirements.* Report of a Joint FAO/WHO/UNU Expert Consultation. WHO Tech. Rep. Ser. No. 724. WHO, Geneva.

Wyse, B. W., Sorenson, A. W., Wittwer, A. J., and Hansen, R. G. 1976. Nutritional quality index identifies consumer nutrient needs. *Food Technol.* 30: 22.

Young, V. R., Munro, H. N., Matthews, D. E., and Bier, D. M. 1983. Relationship of energy metabolism to protein metabolism. In *New Aspects of Clinical Nutrition.* Kleinberg, G. and Deutsch, E. (Ed.), p. 43. Karger, Basel.

Zinneman, H. H., Nuttall, F. Q., and Goetz, F. C. 1966. Effect of endogenous insulin on human amino acid metabolism. *Diabetes* 15:5.

# 2

# Amino Acid Interactions: A Selective Review

**Vernon R. Young**

Massachusetts Institute of Technology
Cambridge, Massachusetts

**Naomi K. Fukagawa**

Massachusetts Institute of Technology
Cambridge, Massachusetts
Harvard Medical School
Boston, Massachusetts

## INTRODUCTION

Amino acids serve as the currency of body protein metabolism. Thus, to maintain an adequate state of protein nutriture, an appropriate intake level and balance of indispensable (essential) amino acids and of so-called nonspecific nitrogen [usually from dispensable (nonessential) amino acids] is required. Various factors, including host and other dietary factors, affect the utilization of amino acids and the integration of whole body nitrogen metabolism, and, in turn, these determine the quantitative nature of the diet that is needed to support and maintain the adequate functioning of cells and organs.

In this brief review, we will consider a number of metabolic interactions between amino acids that have potentially important nutritional implications. Our overall concern is with respect to the dietary requirements in healthy individuals and the design of effective formulations intended to support the nutritional needs of those who require specific therapeutic approaches as part of their comprehensive medical treatment. We will not deal here with the interactions between amino acids and other dietary components, particularly the possible consequences of processing procedures on the availability of individual amino acids, such as lysine or methionine. Our focus will be on the physiology of amino acid metabolism, and our attention will be devoted largely to studies conducted in human subjects, since we are concerned ultimately with the question of the maintenance of an adequate state of human protein nutrition under various dietary and environmental circumstances. In any event, the topic of amino acid interactions is too large to cover comprehensively in the space we have available. However, the subject is important to consider, even if covered in a selective way, because interactions among amino acids may, for example, (a) influence the behavior of the individual (Anderson, 1981), (b) have adverse effects on overall utilization of dietary "protein" with inappropriate levels of amino acid supplementation (Scrimshaw et al., 1958), and (c) be altered with disease conditions, which may change the appropriate balance of amino acids that should be consumed to maintain an adequate nutritional status and function (Horowitz et al., 1981; Chawla et al., 1985)

In the following we will consider some interactions between the branched-chain amino acids and the effects of these amino acids on the utilization of other amino acids. We will then discuss, briefly, some interrelationships among the sulfur-containing amino acids before turning to a consideration of interactions between the indispensable and dispensable amino acids. Our purpose, in part, is to take this opportunity to review the results of a number of studies carried out in young adults utilizing stable isotope tracer techniques, since these will provide an important basis for the discussion of each of the above topics.

## FEATURES OF AMINO ACID METABOLISM AND APPROACHES FOR STUDY OF AMINO ACID INTERACTIONS

We will first briefly review the major pathways and metabolic fates of the amino acids. Then attention should be given to the rate at which amino acids flow along these pathways and the factors that modulate this flow in the intact organism. For this  purpose, consideration will be given to measurement of rates of amino acid flux and protein turnover in the body as a whole and how the rates and pattern of turnover are affected by the intake and profile of amino acids entering the body. In this way we may be able to establish a reasonable basis for the assessment of amino acid interactions, their metabolic basis, and possible nutritional significance.

## Major Metabolic Fates of Amino Acids

Following their release from ingested proteins and their absorption from the gastrointestinal tract (or direct entry into the blood stream via an intravenous line), amino acids experience one of three major metabolic reactions (Fig. 2.1): (a) They act as a substrate for the net synthesis of new proteins and maintenance of tissue and organ proteins; (b) they serve as precursors for many metabolically significant, nonprotein nitrogen-containing compounds, such as creatine, epinephrine, serotonin, and the polyamines, as well as providing nitrogen and carbon skeletons for the dispensable amino acids; (c) they may be converted to other amino acids and/or enter catabolic pathways leading to the elimination of the nitrogen moiety, principally in the form of urea and ammonia, with incorporation of the carbon skeletons into pathways of metabolism common to carbohydrates and lipids. Various factors regulate the fate of the amino acids: These include, in addition to the level and pattern of the amino acid supply, (a) the status of energy and hormone balance and (b) availability of cellular factors, such as high energy phosphate compounds (ATP, GTP), trace elements, and various cofactors, including the active forms of vitamins.

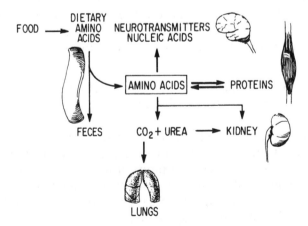

**FIG. 2.1**  A general picture of the major metabolic fates of amino acids.

On the basis of this very general view of the metabolism of amino acids, it should be evident that various approaches might be taken to explore and to assess the significance of amino acid interactions, especially in relation to metabolism and nutrition in the intact individual. These include measurements of amino acids in body fluids and tissues, particularly in blood plasma; the status of the nitrogen economy of the whole organism, as determined with the aid of the nitrogen balance technique; and the dynamic state of amino acid metabolism as evaluated with the aid of isotope tracer techniques. Of course, more mechanistically focused studies are possible with nonhuman models, and these might involve investigations of membrane and interorgan transport (Christensen, 1982, 1984; Oxender et al., 1986) and the status of specific protein synthesis and breakdown in cell-free, isolated cell and organ preparations (Munro et al., 1975). From such studies it appears, for example, that a reduction in the level of a specific indispensable amino acid, which could arise as a consequence of amino acid interactions, results in an inhibition of protein synthesis with polysome disaggregation and reduced formation of the 40S initiation complex (Clemens et al., 1985). Furthermore, there appears to be a defect in the activity of initiation factor eIF-2 and/or eIF-2B (Clemens et al., 1985; Pain, 1986). However, the mechanisms remain obscure, and they may not now be due to changes in the ratio of charged to uncharged tRNA molecules (Austin et al., 1982), as had been earlier proposed (Allen et al., 1969).

We will largely devote our attention here to approaches with which we have had an interest and that have been taken in studies with intact human subjects. Occasional reference will be made to investigations that have been conducted to explore the metabolic and cellular basis for the amino acid interactions that have been chosen for our discussion.

## Rates of Whole Body Amino Acid Flux, Body Protein Synthesis and Breakdown: Some Issues of Methodology

Whole body amino acid and protein turnover is an integration of those major components or pathways of protein and amino acid metabolism discussed above. Assessment of the status of the dynamic state of amino acid and protein metabolism in man is worthwhile in relation to the study of the normal metabolic changes taking place during growth and development, as well as for understanding the changes in metabolism that occur when the level and/or balance of amino acid intake is altered.

Detailed knowledge of the quantitative aspects of amino acid metabolism and of whole body protein turnover in man is still lacking, and, in part, this reflects the difficulty of developing reliable methods for measuring turnover in vivo. In all methods, an isotopically labeled amino acid is given as a tracer, and measurement of the isotope content is made on samples of plasma, urine, or expired $CO_2$ and, in some cases, tissue proteins. Different conceptual and mathematical models are then used to analyze and interpret results and to calculate the components of amino acid and whole body protein turnover. The various models, and assumptions made in applying them, have been described in detail by us (Bier and Young, 1986; Young et al., 1986) and other investigators (Waterlow et al., 1978a), and so this particular topic will not be repeated here, except by way of making two general points.

First, for determination of whole body nitrogen (protein) and amino acid turnover there are two major methods (Table 2.1) : (1) precursor or "plasma" methods, where a tracer amino acid, such as leucine (Matthews et al., 1980) or another indispensable or dispensable amino acid, such as threonine (Zhao et al., 1986) or glycine (Robert et al., 1982), is given by continuous intravenous infusion, often after a priming dose; (2) end-product methods, based on administration of $^{15}N$-labeled amino acids (e.g., glycine) and measurement of $^{15}N$ in products of amino acid metabolism, such as urea or ammonia (Waterlow et al., 1978b; Fern et al., 1981). This latter approach has been used extensively in clinical studies (Waterlow et al., 1978a; Waterlow, 1984; Bier and Young, 1986; Young et al., 1986).

Second, for purposes of assessing the status of protein turnover, and the consequence of ingesting diets of varying amino acid content and balance, labeled lysine, leucine, and tyrosine have been commonly employed as "representative" amino acids. Figure 2.2 illustrates the model employed for this approach. The tracer, in this case $[1\text{-}^{13}C]$ leucine, is introduced (via the plasma) into the leucine pool, where its dilution by dietary or tissue protein-derived leucine occurs. Then, by measuring this dilution and

**TABLE 2.1** Summary of Major Tracer Methods of Measuring Protein Turnover and Amino Acid Kinetics in the Human

1. Precursor or plasma methods
   Amino acid tracer—intravenously or intragastrically
   Measurement of tracer in plasma (or urine) and of isotope in $CO_2$ (or urea)
2. End-product methods
   $^{15}N$-tracer—intravenously or intragastrically
   Measurement of $^{15}N$ in urea and $NH_3$

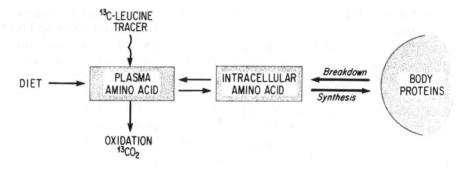

Flux = Synthesis + Oxidation = Breakdown + Intake

**FIG. 2.2** Simplified model for estimating the whole body flux of an amino acid and the rates of body protein synthesis and breakdown, as derived from the disappearance and appearance of an indispensable amino acid in the sampled compartment (plasma) and the oxidation rate of the amino acid. The example shown is for [1-$^{13}$C] leucine as a tracer.

quantifying tracer lost as expired $^{13}CO_2$, it is possible, using mass balance equations, to estimate rates of whole body protein synthesis and breakdown. While generally similar, there are several crucial differences between this approach and the end-product methods. These have been previously discussed (Bier and Young, 1986; Young et al., 1986) and do not need to be described in detail here. Perhaps it will suffice to say that the available methods and approaches for determining the quantitative, dynamic aspects of protein and amino acid metabolism in the intake host each have their limitations. Furthermore, the choice of the particular approach should be made with reference to the aims of the experiment. In the case of studies concerned with amino acid interactions, it is more likely that the precursor or plasma methods would be most useful, as we will illustrate below.

## AMINO ACID FLUX AND PROTEIN TURNOVER IN HEALTHY INDIVIDUALS

Despite the various problems arising from the choice and application of the different approaches to be taken to obtain a reliable and quantitative determination of protein turnover and the the fate of amino acids at the

**TABLE 2.2** Rates of Whole Body Protein Synthesis Compared with Dietary Protein Allowances at Various Ages in Man[a]

| Group | Protein synthesis(A)[b] | Protein allowance(B)[b] | Ratio A/B |
|---|---|---|---|
| Infant (premature) | 11.3, 14 | ~3.0 | 4.5 |
| Child (150 mo) | 6.3 | ~1.3 | 5 |
| Child (2–8 yr) | 3.9 | ~1.1 | 4 |
| Adolescent (~13 yr) | ~5 | ~1.0 | 5 |
| Young adult (~20 yr) | ~4.6 | ~0.75 | |

[a]Values are g protein/kg/day.
[b]Taken from summary by Young et al. (1985).

whole body level, a useful picture is now beginning to emerge concerning the dynamic status of protein and amino acid metabolism in humans under various pathophysiological states. We have summarized some of this information in earlier reviews (Young et al., 1985, 1986) especially in reference to the healthy subject, and this topic has also been considered in detail by others (Waterlow et al., 1978a: Waterlow, 1984). Here, perhaps, it is more useful to make a couple of points based on the summary of some estimates of the rates of whole body protein synthesis at different stages in life, as provided in Table 2.2.

The first point is that whole body protein synthesis and breakdown rates are considerably higher than the estimated dietary requirement for protein and for the total individual indispensable amino acids at each stage in life. This means that most of the new protein formed in tissues and organs arises by the recycling of the amino acids that are liberated during the course of protein breakdown. From the standpoint of amino acid interactions, it might be anticipated that alterations in the level and balance of amino acid intake would be buffered by the considerable and continuous flow of amino acids entering the free amino acid pools of tissues via this extensive protein breakdown. This, however, seems not to be the case in view of the prompt and rapid changes that occur in plasma amino acid levels, and presumably tissue amino acid concentrations, when amino acid intake and balance are altered by even single meals (Harper et al., 1970; Hussein et al., 1971).

Indeed, a different argument might be made: Because of the significant and continued synthesis and turnover of proteins, coupled with a requirement by the protein synthetic machinery for an appropriate balance of amino acids (Munro et al., 1975; Clemens et al., 1985), alterations in the concentration of specific amino acids within tissues might well have a profound effect on protein balance within tissues and organs. This is speculation, and there is little information to support this possibility, at least in relation to the effects of amino acid *interactions* on the nitrogen economy of the intact human host. Unfortunately, we do not have a precise understanding of the quantitative extent to which alterations in tissue amino acid levels, and the relationships among them, must occur before changes in the pattern and rate of organ and whole body protein synthesis and breakdown are noted. However, such effects might occur and, in this case, help to explain why, for example, more wheat gluten is required to maintain body nitrogen balance in young adults (Young, 1975) than would seem to be necessary on theoretical grounds, or at least from a consideration of the FAO/WHO/UNU (1985) estimation of the lysine requirement in adults in relation to the content of this amino acid in wheat gluten.

A second point concerning amino acid interactions in relation to the extensive nature of body protein synthesis and breakdown is that the major flow of amino acids into and from the free amino pools is due to the breakdown and synthesis of tissue proteins, respectively. Although qualitatively and physiologically important, the flow of amino acids via other pathways, such as in relation to tryptophan for serotonin synthesis and glutamine for purine and pyrimidine synthesis, is low compared to that associated with the protein synthesis–protein breakdown cycle. In this sense the organization of organ and whole body amino acid and nitrogen metabolism might be viewed, as proposed earlier (Crabtree and Newsholme, 1985), as a branched metabolic pathway, where the flux through one branch (i.e., protein turnover) is greatly in excess of that in the other (e.g., Fig. 2.3). Furthermore, Crabtree and Newsholme (1985) and Newsholme et al. (1985) propose, according to principles of metabolic control, that the sensitivity of the flux of the low-flux pathway can be markedly increased without decreasing significantly the concentration of amino acids and intermediates in the main pathway. Thus, the extensive movement of amino acids into and from proteins provides the organism with an effective mechanism to meet increased needs for amino acids for various biosynthetic purposes, as would be expected to occur, for example, following exposure to a significant infection or following physical injury. On the other hand, this also means that if the concentration of free amino acids in tissues is altered or reduced, perhaps as a consequence of an interaction between amino acids at the membrane level, then the ability of the cells to use amino acids via the low-flux pathway or, component of overall body

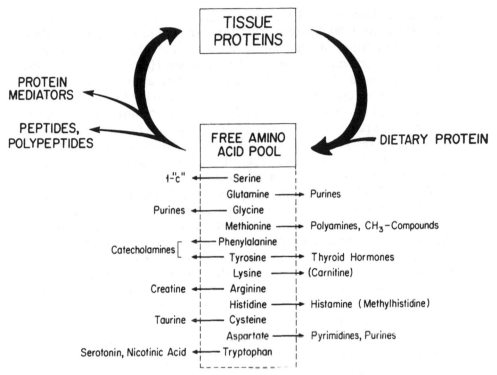

**FIG. 2.3** A schematic indication that the major flux of amino acids in tissue pools is associated with the processes of protein synthesis and breakdown (high-flux pathway). Some examples of the fate of amino acids via other pathways are also indicated (low-flux pathway). The organization of amino acid metabolism is, therefore, viewed in the form of a branched metabolic pathway, according to *logic* of metabolic control proposed by Crabtree and Newsholme (1985).

amino acid metabolism, for various biosynthetic purposes (e.g., purine synthesis, neurotransmitter and peptide synthesis, etc.) would be diminished. Furthermore, this could occur without an immediate or major impact on the high-flux pathway of amino acid metabolism. In this event, amino acid interactions would influence cell function more profoundly via their effect on these other pathways and uses of amino acids, rather than via a major influence on the status of protein synthesis and/or breakdown per se. Thus, in this context it is reasonable to question whether the heightened sensitivity of rats given low-protein diets to amino acid disproportions (Harper et al., 1970) might be due, in part, to a protein turnover lower than would be expected under these conditions (Garlick, 1980; Millward, 1980; Laurent et al., 1984) and the possibly greater effects these disproportions might have on the utilization of amino acids within the low-flux pathways of amino acid metabolism. Harper et al. (1970) have also suggested that the more adverse effects of an amino acid disproportion on animals given low-protein diets are due to the low activity of the enzymes of amino acid catabolism. Indeed, this may also explain the untoward consequences of amino acid disproportions (such as imbalances and antagonisms) under these conditions of protein inadequacy, but it is probably just one of a number of possible mechanisms that could account for this observation (Harper et al., 1970).

With this background concerning amino acid and protein turnover, we now turn to a discussion of a number of specific interactions among amino acids, in order to arrive at some general conclusions about the significance of amino acid interactions in animal and human nutrition. Not all of the biochemically and nutritionally interesting interactions can be discussed here, so we have omitted the specific consideration of the interrelationships and interactions between phenylalanine and tyrosine (Harper et al., 1970) and between lysine and arginine (Jones, 1965; Boorman and Fisher, 1966; Jones et al., 1966; Nesheim, 1968; Austic and Calvert, 1981). Rather, we prefer to concentrate our attention on those interactions with which we have had some direct research interest over the past decade or so.

## BRANCHED-CHAIN AMINO ACID INTERACTIONS

### Interactions Between Branched-Chain Amino Acids (BCAA)

Because leucine, isoleucine, and valine share the same enzyme systems, branched-chain amino acid transferase and branched-chain 2-oxoacid dehydrogenase, for their initial degradative steps, the branched-chain amino acids are commonly thought of as a group in terms of their roles in amino acid homeostasis. For example, Harper et al. (1970) have stated that "nature has dealt with the three branched-chain amino acids as a group." However, unique roles of individual branched-chain amino acids have been uncovered. For example, leucine-specific effects on the regulation of muscle protein metabolism (Buse and Weigand, 1977; Tischler et al., 1982) and adipose tissue valine oxidation (Frick et al., 1981) have now been emphasized. We have also had the opportunity to study some metabolic aspects of the in vivo relationships between leucine and the other branched chain amino acids, especially valine. Thus, we (Hambraeus et al., 1976) observed in earlier studies that subjects fed leucine-deficient diets had pronounced elevations of plasma valine. However, we did not observe the reverse situation, that is, plasma leucine concentrations did not change significantly when individuals were given valine-deficient diets.

The fall in plasma leucine concentration when a low-leucine diet is consumed is probably, at least in part, due to a reduced rate of protein turnover (Fukagawa et al., 1985; Meguid et al., 1986), but, as noted above, there is a simultaneous and marked rise in both plasma valine and isoleucine levels. This response mirrors the reduction in these two branched-chain amino acids when high levels of leucine are given to humans (Swendseid et al., 1965; Hagenfeldt et al., 1980) as well as other species (Harper et al., 1970; Oestemer et al., 1973; Block and Harper, 1984). Furthermore, the effects of dietary leucine on plasma valine and isoleucine levels appear to be leucine-specific, because neither low nor high intakes of either of these latter two branched-chain amino acids have important effects on the plasma concentration of leucine (Ozalp et al.,1972; Oestemer et al., 1973; Block and Harper, 1984). In this context, it has been shown (Block et al., 1985; Aftring et al., 1986) that a dietary excess of leucine or the injection of leucine in rats results in increased branched-chain keto acid dehydrogenase (BCKAD) activity in the liver, mainly associated with activation of the enzyme complex. BCKAD is controlled by a unique phosphorylation (inactivation)–dephosphorylation (activation) mechanism, catalyzed by a specific branched-chain dehydrogenase kinase and phosphatase (Harris and Paxton, 1985; Harris et al., 1986). Further, Harris et al. (1985) observed an inactivation of the hepatic BCKAD complex when rats were given a low-protein diet, probably a consequence of low hepatic levels of branched-chain alpha-keto acids. The latter inhibits branched-chain alpha-keto acid dehydrogenase kinase and, thus, at high leucine or protein intakes, the BCKAD complex is activated. This is further supported by Espinal et al. (1986), who found that in rats fed a 0% casein diet for 10 days there was a fourfold increase in the activity of the branched-chain alpha-keto acid kinase in extracts of liver mitochondria. The changes in plasma valine and isoleucine that are seen when a low-leucine diet is given presumably arise,therefore, as a result of a reduction in the activity of the dehydrogenase in various tissues; the low plasma leucine levels in humans receiving a low leucine intake would be paralleled by low plasma alpha-ketoisocaproate concentrations and, presumably, diminished intracellular levels of this keto acid (Hutson and Harper, 1981; Matthews et al., 1982).

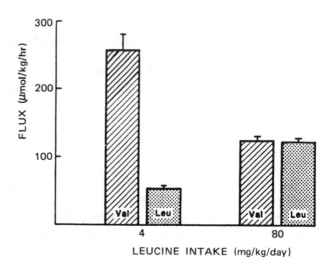

**FIG. 2.4**  Effect of a deficient (Chawla et al., 1985) and adequate (Miller et al., 1985) leucine intake on the plasma flux of leucine and valine in healthy young men. (*From Meguid et al., 1983.*)

    To investigate the dynamics of these interrelationships, we have examined both leucine and valine kinetics under circumstances where intake of one amino acid was reduced from a sufficient or excess level to a deficient level (Meguid et al., 1983). In these studies subjects received for six days a crystalline amino acid mixture patterned after the amino acid composition of protein, and the intake of valine or leucine was adjusted to provide two different levels. On the morning of the seventh day of the dietary periods, and while consuming hourly aliqouts of their daily test diet, the subjects were infused with either L-[1-$^{13}$C] leucine or L-[1-$^{13}$C] valine to determine branched-chain amino acid flux and oxidation. The effects we observed following the alteration of leucine intake from a sufficient to deficient level on plasma leucine and valine flux are shown in Fig. 2.4; valine and leucine oxidation were also altered (Fig. 2.5). Leucine oxidation declined, as expected, with reduced dietary leucine intake. Despite the dramatic increase in valine turnover at the lower dietary leucine intake, the fraction of valine flux oxidized was reduced by only about 27%. In contrast to these responses  of valine metabolism to a low leucine intake, valine turnover and oxidation declined (Figs. 2.6, 2.7) with reduced dietary valine intake, but the turnover of leucine and its oxidation turnover remained unchanged (Figs. 2.6, 2.7).

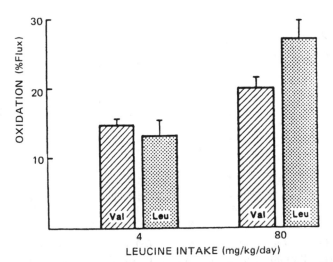

**FIG. 2.5** Effect of a deficient (Chawla et al., 1985) and adequate (Miller et al., 1985) leucine intake on the oxidation of leucine and valine, when expressed as a percent of the corresponding plasma flux (see Fig. 2.4). (*From Meguid et al., 1983.*)

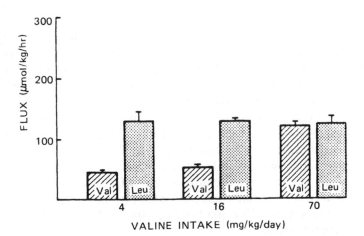

**FIG. 2.6** Effect of a low (Chawla et al., 1985), marginal (Matthews et al., 1980), and generous (Bapurao and Krishnaswamy, 1978) valine intake on the plasma flux of valine and leucine in healthy young men (*From Meguid et al., 1983.*)

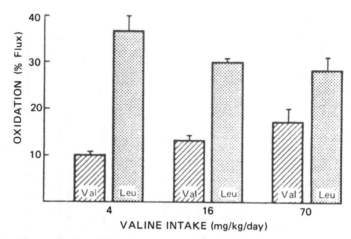

**FIG. 2.7** Effect of a low (Chawla et al., 1985), marginal (Matthews et al., 1980), and generous (Bapurao and Krishnaswamy, 1978) intake of valine on the oxidation of valine and leucine when expressed as a percentage of the corresponding flux (see Fig. 2.6). (*From Meguid et al., 1983.*)

While it is tempting to explain these data in terms of leucine-specific substrate effects on valine metabolism, it is important to remember that in clinical studies the changes observed might be the result of secondary hormonal or other regulatory events due to an altered dietary leucine consumption. For example, Clark et al. (1968) were unable to depress plasma valine in alloxan diabetic rats following ingestion of leucine. They interpreted these results as indicating that leucine-induced insulin release was the mediator of oral leucine's effect on plasma valine in nondiabetic animals. In contrast, Hagenfeldt et al. (1980) suggested that the fall in plasma valine observed during a continuous 2.5-hour infusion of L-leucine (0.3 μmol/min) in humans was independent of insulin action, but peripheral plasma insulin values rose from 17 to 22 μU/ml in their experiment, and it is conceivable that portal insulin levels rose even further. Nevertheless, since valine and isoleucine have less potent effects on the activation of the branched-chain 2-oxo acid dehydrogenase in vivo (Aftring et al., 1986), it seems reasonable to attribute part of the differential effects of leucine and valine on plasma amino acid levels to the regulation of the dehydrogenase.

In summary, these leucine-induced changes in plasma valine and isoleucine levels can be attributed to alterations primarily in either (a) membrane transport, (b) amino acid oxidation, or (c) rates of net tissue protein deposition. Harper et al. (1984) have considered these possibilities in detail and, in summary, they have concluded that changes in protein turnover cannot explain the selective effects of leucine on the concentrations of the other branched-chain amino acids. The findings from our experiments in humans would support this view. Since these amino acids share a common transport system, the L-system (Christensen, 1984), it would be reasonable to find that leucine at high or low levels affects the influx and/or influx of the other BCAA in tissues. Although this possibility has not been examined in detail, it was concluded by Harper et al., (1984) that transport effects are an unlikely basis for the observed changes in plasma amino acid levels. It follows, therefore, that the most likely explanation appears to be that changes in BCAA oxidation are brought about by alterations in the activation of the BCKAD complex. We will return to this point later, with a possible counterargument.

These observations on the effects of either excess or deficient intakes of leucine on valine and isoleucine metabolism are relevant to improving the design of amino acid mixtures intended for treatment of patients suffering from multiple trauma (Blackburn et al., 1981) and those with hepatic encephalopathy (Wahren et al., 1983). Specifically, BCAA-enriched solutions have been proposed as a means for normalizing blood amino acid patterns and for promoting a protein anabolic state under these disease conditions.

There is, however, an important unresolved question as to the effectiveness of such an approach (Wahren et al., 1983; Brennan et al., 1986) and we have not found any favorable effects on the status of amino acid metabolism in burned patients who were given amino acid solutions containing high concentrations of BCAA (Yu et al., 1987), as summarized in Table 2.3. Clearly, it is important for future developments in this area of clinical applications of modified amino acid mixtures to consider the potential for leucine to exert regulatory effects on branched-chain amino acid metabolism in various tissues and organs. These effects may have either a positive or negative consequence, depending upon the specific metabolic condition, for the functioning of the individual. Only additional metabolic research will enable us to maximize the positive, and potentially exciting, clinical effects of a branched-chain amino acid interactions through intervention by nutritional means.

## Interactions Among Other Amino Acids

**Tryptophan-niacin and the BCAA.**   Dietary tryptophan is known to spare the requirement for preformed niacin (Krehl, 1949) and of relevance here

**TABLE 2.3** Summary of Some Parameters of Whole Body Nitrogen and Leucine Metabolism in 12 Severely Burned Patients Receiving Either a Conventional Egg-Protein Based Formulation or an Enriched Branched-Chain Amino Acid (BCAA) Solution

|  | Amino acid intake from: | |
| Parameter | Egg-protein | BCAA |
| --- | --- | --- |
| Leucine intake | 47 ± 3[a] | 112 ± 8 |
| Leucine flux | 223 ± 20 | 317 ± 37 |
| Leucine oxidation | 49 ± 7 | 118 ± 16 |
| Leucine appearance from protein breakdown | 176 ± 20 | 205 ± 34 |
| N balance (mg N/kg/day) | 36 ± 20 | 22 ± 26 |

[a]Values expressed as μmol/kg/hr. Mean ± SEM for 12 patients, whose mean body surface area burn (%) was 36.
*Summarized from Yu et al. (1987).*

is the possibility, raised by various studies, that this sparing effect might be altered by the balance of amino acids in the diet. As summarized in Table 2.4, an earlier series of studies in India (Gopalan and Srikantia, 1960; Belavady et al., 1963; Srikantia et al., 1968; Ghafoorunissa and Narasinga Rao, 1973; Krishnaswamy et al., 1976; Bapurao and Krishnaswamy, 1978; Krishnaswamy and Bapurao, 1978) showed that diets high in leucine could precipitate a niacin deficiency in experimental animals. Furthermore, it was suggested that pellagra, which is endemic among population groups whose stable food is sorghum *(Sorghum vulgare)*, is due to the relatively high leucine content in this millet, possibly coupled with a low vitamin $B_6$ content in the diet. From a metabolic standpoint, an association between tryptophan, niacin, $B_6$, and leucine would be logical since the pathway involving tryptophan conversion to niacin requires $B_6$-dependent enzymes, and the catabolism of leucine is accomplished initially through participation of a transaminase requiring pyridoxal phosphate as a coenzyme. However, as stated by way of a summary in Table 2.5, this work from India was not supported by a series of follow-up studies, published during the mid-to-late 1970s and early 1980s, in rats, dogs, chicks, and human subjects (Nakagawa et al., 1975; Nakagawa and Sasaki, 1977; Manson and Carpenter, 1980a,b). Two relatively recent papers (Bender, 1983; Magboul and Bender, 1983), however, again have suggested that leucine supplements, even in the presence of adequate pyridoxine, alter the metabolism of trypophan by rats in a way that would

**TABLE 2.4**   Relationship of Leucine Intake to Pellagra

| Observation | Reference |
| --- | --- |
| Pellagra endemic where staple is *Sorghum vulgare* | Goplan and Srikantia, 1960 |
| Excess leucine affects tryptophan metabolism | Belavady et al., 1963 |
| | Srikantia et al., 1968 |
| | Krishnaswamy et al., 1976 |
| | Ghafoorunissa and Narasinga Rao, 1973 |
| Possible involvement of B$_6$ status in leucine–tryptophan interaction | Krishnaswamy and Bapurao, 1978 |
| Pellagrins: inadequate B$_6$ status: abnormal leucine tolerance | Bapurao and Krishnaswamy, 1978 |

*As reported in studies at National Institute of Nutrition, Hyderabad, India.*

**TABLE 2.5**   Observations That Do Not Support an Involvement of Leucine in the Etiology of Pellagra

| Reference | Observation |
| --- | --- |
| High leucine in chicks and rats did not precipitate niacin deficit | Manson and Carpenter, 1980a |
| Rats: no evidence of leucine-induced niacin deficit | Nakagawa and Sasaki, 1977 |
| No evidence of leucine-induced niacin deficit in dogs | Manson and Carpenter, 1980b |
| Healthy volunteers—no effect of excess leucine on urinary tryptophan and niacin metabolites | Nakagawa et al., 1975 |
| Rats: large leucine excess in niacin-deficient diet; no major effects on niacin deficiency or tryptophan metabolism | Cook and Carpenter, 1987 |

support the original hypothesis that a diet based on sorghum could result in a metabolic deficiency of niacin. Finally, in a further and careful exploration of this problem, Cook and Carpenter (1987) examined the metabolism of (methylene$^{14}$C)tryptophan in young rats fed a niacin-deficient diet with and without 15 g/kg supplementary L-leucine. These investigators observed that niacin-deficient rats grew slowly, but there were no differences in the severity of the condition that could be due to an effect of a leucine excess. Furthermore, there were no major differences in tryptophan metabolism between leucine-supplemented and -unsupplemented diets. These results do not provide support for the hypothesis that excessive leucine is responsible for the existence of pellagra. In the aggregate, it appears to us that there is little conclusive evidence to believe that relatively high leucine intakes are of etiological significance in the pellagra that exists in areas of India. Additional research will be necessary to resolve these discrepancies.

**Interactions between BCAA and other large neutral amino acids with emphasis on the brain.**   Interactions between BCAA and other amino acids, such as tryptophan and the aromatic amino acids (phenylalanine and tyrosine), that serve as precursors of neurotransmitters are perhaps of more practical importance. Specifically, the availability of these precursors in the brain may be influenced not only by their levels present in the peripheral circulation, but their uptake into the brain may be modulated by the presence of other amino acids that share common transport systems. It is in this way that the branched-chain amino acids may interact importantly with and affect the utilization of tryptophan and the aromatic amino acids.

Briefly, there are three independent transport systems localized in the luminal and antiluminal membranes of brain capillaries for the neutral, basic, and acidic amino acids, respectively (Pardridge and Choi, 1986). The neutral amino acid carrier transports all of the neutral amino acids with the exception of glycine and possibly proline, and this system appears to be analogous to the L-system, identified and characterized by Christensen (1984). Since the system in the brain has a low $K_m$ (Pardridge, 1983; Miller et al., 1985), the transport of tryptophan or phenylalanine into the brain is not only a function of the plasma amino acid concentration of the specific amino acid but also of the concentration of other large neutral amino acids that compete with the same system for membrane transport. Hence, an important parameter for predicting the availabilty of tryptophan and phenylalanine in the brain is the ratio of the plasma tryptophan (or phenylalanine) concentration to that of competing large neutral amino acids. This concept, proposed initially by Fernstrom and Wurtman (1972), indicates that brain tryptophan (or other large neutral amino acid levels) is a direct function of plasma tryptophan concentration and an inverse function of the levels of any or all of the other large neutral amino acids in plasma (Pardridge and Choi, 1986). Furthermore, the conversion of tryptophan, and of tyrosine, to neurotransmitters involves hydroxylation by regulatory enzymes, and these are unsaturated at physiological levels of their substrates. Thus, the transport into the brain and activities of enzymes responsible for the metabolism of tryptophan and phenylalanine (i.e., tryptophan 2, -3 dioxygenase (EC 1.13.11.11), phenylalanine hydroxylase (EC 1.14.16.1), and tyrosine aminotransferase (EC 2.6.1.5)) are factors that determine availability of precursor and the rate of specific neurotransmitter synthesis in the central nervous system (CNS).

Perhaps the control of the uptake and fate of these amino acids is, in reality, more complex than these simple statements might imply, as has been also suggested by Pogson et al. (1986) in their account of the regulation of hepatic amino acid metabolism. Nevertheless, the foregoing helps, for our purpose, to explain the profound consequences of amino acid interactions via changes in plasma amino acid concentrations and their effects on brain function and how selective manipulation of blood amino acid levels can improve the condition of patients suffering from genetic defects in amino acid metabolism, such as phenylketonuria, maple syrup urine disease, and homocystinuria.

In reference to our focus, it is relevant that physiological changes in plasma amino acid concentrations could result in changes in brain function (Pardridge and Choi, 1986). Thus, nutritional conditions experienced in normal, everyday life might well be associated with subtle effects on brain function that might not be clinically evident unless carefully and specifically examined.

For example, ingestion of protein results in more rapid and extensive increases in the levels of branched-chain amino acids in plasma relative to that of tryptophan, and this brings about a decreased ratio of plasma tryptophan to the other neutral amino acids. Furthermore, ingestion of a high carbohydrate meal (Lieberman et al., 1986) or intravenous infusion of glucose (Fukagawa et al., 1987) leads to an increase in the concentration of plasma tryptophan relative to the other large neutral amino acids (valine, isoleucine, leucine, tyrosine, and phenylalanine). It is difficult, at present, to establish causal links among normal variations in diet and alterations in CNS neurochemistry and function. However, as has been pointed out by Anderson (1981), and as based on results from studies in animal models, it seems possible that in humans sleep, food intake, mood, pain sensitivity, and hormone balance might be influenced by nutritionally induced modulations in plasma amino acid levels (Leathwood, 1987). Further research is necessary to establish whether and how physiological changes in blood amino acid levels and their interrelationships affect normal behavior in healthy subjects. At least the profound and prompt depression in food intake when animals are given a high leucine diet is, presumably, causally related to changes in brain serotonin (Ramanamurthy and Srikantia, 1970) and dopamine levels (Harper et al., 1984).

Of course, nutritionally significant interactions of the kind indicated above extend well beyond considerations of branched-chain amino acids, which allowed us to introduce the topic in a relevant way of amino acid interactions and CNS metabolism. Neurotransmitter precursors may be consumed in pure form and separate from food or usual meals, as might be the case when they find pharmacological applications in the treatment of disorders of the CNS (Maher, 1987); or as aspartame, the nonnutritive, dipeptide sweetener (aspartylphenylalanine methylester) (Wurtman and Ritter-Walker, 1987), or where supplements of amino acids such as tryptophan, tyrosine, and phenylalanine are taken without prescription. This makes it particularly important to appreciate the possible consequences of interactions among amino acids with respect to their impact on CNS chemistry and function.

Finally, it should be mentioned here that our understanding of the metabolic consequences of amino acid interactions, due in this case to the excessive accumulation of a given amino acid in body fluids, has been expanded recently by the careful thinking of Christensen (1987a,b). He has emphasized that either the influx or the efflux of an amino acid may be competitively inhibited by the accumulation of another amino acid. Thus, in the case of excess phenylalanine, he proposes that the flux-asymmetry of its inhibition of transport in most tissues acts to cause tissue sequestration of various endogenously derived amino acids. As a direct result of this, coupled with an acceleration of the catabolism of the amino acids in the affected tissues, the plasma levels of various amino acids will decline. In consequence, Christensen says that there will be a lowered mass action of the affected amino acids at the blood-brain barrier and that this leads us to a further effect on the availability of the affected amino acids in the central nervous system.

These refined concepts of Christensen (1987a, b), involving, in part, a consideration of interorgan relationships in amino acid metabolism (Christensen, 1982), are exciting and important to an adequate understanding of the metabolic basis of the responses of the intact individual to changes in the balance of the amino acid supply to tissues and organs. Indeed, it must now be questioned whether the reduced levels of plasma valine and isoleucine that occur with excess leucine intakes, discussed earlier, together with the decline in the levels of methionine, tyrosine, and phenylalanine and the associated increased rates of valine and isoleucine oxidation under these dietary conditions (Harper et al., 1984) are mediated by the influence of high leucine concentrations on the *efflux* of valine and other amino acids from tissues, especially muscle and liver. This will be important to determine, particularly in view of the conclusion by Harper et al., (1984) that changes in the inward transport of amino acids are not likely to explain the alterations in plasma and tissue amino acid concentrations under conditions of low or excess leucine intakes.

## SULFUR AMINO ACIDS

The nutritional and metabolic associations between methionine and cystine were recognized in the earliest studies dealing with methionine, which was the later of the two sulfur amino acids to be identified (Mueller, 1923). Methionine was soon recognized to be the obligatory, indispensable sulfur amino acid (Womack et al., 1937), and since then numerous studies of the ability of dietary cystine to spare the requirements for methionine have been conducted in various animal species and humans (Womack and Rose, 1941; Rose and Wixom, 1955; Reynolds, et al., 1958; Rama et al., 1961). The estimates of the methionine-sparing effect of dietary cystine in humans vary widely, with the extremes being from about 16 to 89% of the total methionine requirements, relative to the methionine need determined with a cystine-free diet. The human studies have been based mainly on the use of the nitrogen balance technique, but this procedure is prone to error and other problems (Hegsted, 1976; Young, 1986). It is not surprising, therefore, that current information about quantitative aspects of the interactions between methionine and cystine in humans is generally inadequate.

The mechanisms that account for the conservation of methionine at low dietary intakes of amino acid and the methionine-sparing action of cystine have been explored using nonhuman models. The effect of cystine on enzymes of methionine catabolism has been studied (Finkelstein and Mudd, 1967; Shannon et al., 1972; Stipanuk and Benevenga, 1977; Finkelstein et al., 1986). In vitro, total activities of methionine catabolic enzymes were determined in livers from rats fed a methionine-cysteine-deficient diet and from a similar group that received cystine (Finkelstein and Mudd, 1967). Cystathionine synthase was found to be inhibited by the addition of cystine in vivo but not when cystine was added to a liver preparation in vitro (Finkelstein and Mudd, 1967). The inhibition appeared within 24 hours of treatment in rats and reached a maximum of about 50% of the activity determined for the control animals receiving a sulfur amino acid–devoid diet. Using an in vivo approach, Stipanuk and Benevenga (1977) gave a carboxyl $^{14}C$-methionine tracer to rats and measured $^{14}CO_2$ production as an index of the rate of transsulfuration. In these investigations it was found that the addition of cystine to a methiionine-deficient diet was associated with a decrease in transsulfuration, in accordance with earlier enzyme studies (Stipanuk and Benevenga, 1977; Finkelstein et al., 1986). In addition, the ability of cystine to spare methionine was then tested, by these investigators, in rats given a diet that was both methionine- and threonine-deficient. Under these circumstances cystine did not decrease the transsulfuration of methionine. Hence, it was concluded (Stipanuk and Benevenga, 1977) that the homocysteine locus (or branching point between homocysteine remethylation to methionine and the conversion of homocysteine to cystathionine (Fig. 2.8)) did not account for the sparing effect of cystine; the sparing mechanism, in this case, appeared to be a cystine-mediated augmentation of protein synthesis, improving the utilization of the limiting supply of methionine for net protein synthesis. Although this in vivo study (Stipanuk and Benevenga, 1977) provides evidence for the involvement of the pathway of protein anabolism in the methionine-sparing action of cystine, it does not rule out the possibility that the homocysteine locus plays a role when methionine is the first limiting amino acid in the diet for protein synthesis and where other amino acids are present in adequate amounts.

Accordingly, to explore various aspects of methionine–cystine interrelationships in humans, we have employed a tracer model of methionine metabolism to quantify the major components of the methionine cycle (Fig. 2.9) and the changes that occur when the intake and balance of methionine and cystine are altered. Briefly, healthy young adult subjects were given a diet providing adequate methionine with no cystine, a methionine-cystine–devoid diet, or a methionine-free cystine-supplemented diet for five days prior to receiving a constant intravenous infusion of $^2H$-methyl-1$^{13}C$-methionine. With this procedure it was possible to determine the role of the methionine and homocysteine loci (Fig. 2.8) in the conservation of methionine when methionine intakes were low and

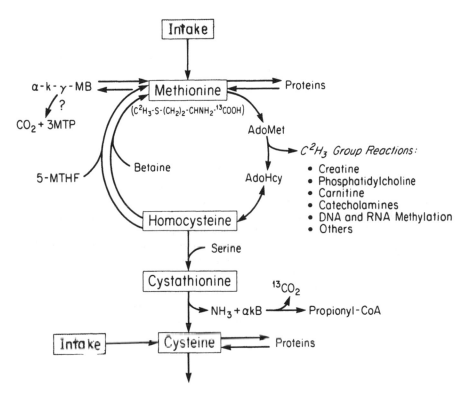

**FIG. 2.8** General scheme of methionine metabolism and also with reference to use of the L-²H-methyl-l-¹³C-methionine tracer (see Fig. 2.9). alpha-k-gamma-MB = alpha-keto-gamma methiolbutyrate; 5-MTHF = 5-methyltetrahydrofolic acid; 3MTP = 3-methylthiobutyrate; as a possible product of the transamination pathway; alpha-kB = alpha-ketobutyrate. The scheme indicates the major sites where the labeled methyl and carboxyl groups of methionine are thought to be removed during metabolism. The *methionine* and *homocysteine loci* are branching points in methionine and homocysteine metabolism, respectively; in the case of methionine this occurs with uptake of methionine into proteins and/or its entry into the transmethylation pathway, with the formation of S-adenosylmethionine (AdoMet). The homocysteine locus is that point in the scheme where homocysteine may either be remethylated to form methionine or converted to cystathionine.

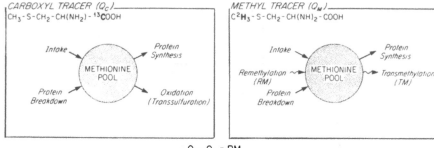

$$Q_M - Q_C = RM$$
$$TM = RM + Oxidation$$

**FIG. 2.9** A schematic outline of a dual isotope-tracer model of methionine kinetics. Dilution of the $^{13}C$-carboxyl moiety (left panel) occurs via methionine entry from diet (intake), and protein breakdown and loss of this moiety occurs principally via protein synthesis and transsulfuration. Dilution of the methyl-$^2H_3$ moiety (right panel) occurs via entry of methionine into pool via diet (intake), protein breakdown and remethylation (RM) of homocysteine. Loss of label occurs via protein synthesis and transmethylation (TM) reactions. Under steady state conditions the difference between the fluxes of the methyl- and carboxyl-labeled methionines is a measure of the rate of remethylation. (*From Storch et al., 1987a.*)

also of their possible involvement in the methionine-sparing effect of cystine (Storch et al., 1987 a, b).

As summarized in Table 2.6, it can be seen that with cystine supplementation of a sulfur amino acid–free diet there was a small but statistically significant decline in the rate of methionine oxidation relative to that observed for the sulfur amino acid–devoid diet. This latter diet, of course, resulted in a markedly lower rate of methionine oxidation (transsulfuration) than that occurring at an adequate level of methionine intake.

In this initial study (Storch et al., 1987a), however, we did not observe any other statistically significant differences between the absolute rates of the various components of methionine cycle for the sulfur amino acid–free diet as compared to the fluxes measured during the cystine-supplemented diet. Nevertheless, the cystine-supplemented diet revealed changes in methionine metabolism that were similar to the changes observed with ingestion of the sulfur amino acid–devoid diet. Hence, there was a tendency for the rate of homocysteine remethylation, in relation to its conversion to cystathionine (revealed by the ratio RM/C), to be higher; the rate of methionine oxidation, relative to the rate of transmethylation

**TABLE 2.6**   Mean Values of the Components of the Methionine Cycle in Young Men Given Various Intakes of Methionine and Cystine

| Component of methionine cycle | Dietary group | | |
|---|---|---|---|
| | Adequate | SAA[a]-devoid | SAA-devoid + cystine |
| Methyl-methionine flux (Qm) | 36.3[b] | 16.2* | 15.5* |
| Carboxyl-methionine flux (Qc) | 31.6 | 13.9* | 13.6* |
| Remethylation of homocysteine (RM) | 4.74 | 2.3 | 1.93* |
| Transsulfuration (C) | 7.62 | 1.1* | 0.65** |
| Transmethylation (TM) | 12.4 | 3.5* | 2.6* |
| Methionine into proteins (S) | 24.0 | 12.7* | 13.0* |
| Methionine from proteins (B) | 18.4 | 13.9* | 13.6* |
| Ratios: | | | |
|   RM/C(%) | 60 | 203* | 336* |
|   C/TM | 0.63 | 0.34* | 0.21* |
|   S/TM | 2.1 | 4.0* | 5.2* |
|   TM/Qm | 0.34 | 0.21* | 0.17* |

[a]SAA = sulfur amino acid.
[b]Values are µmol/kg/hr.
*Different from adequate ($p < 0.05$); **Different from SAA-devoid.
*Extracted from Storch et al. (1987b).*

(C/TM), tended to be lower when cystine was added to the sulfur amino acid–free diet. Similarly, there was a tendency for the ratios S/TM and TM/Qm to be higher and lower, respectively, than those found for the sulfur amino acid–free diet (Table 2.6).

    This study (Storch et al, 1987b) offers a new perspective on the methionine-sparing effect of cystine in the intact human subject and our findings extend those obtained at the in vivo and in vitro levels in rats by Stipanuk and Benevenga (1977) and Finkelstein and co-workers (Finkelstein and Mudd, 1967; Finkelstein et al., 1986). Hence, the sparing of methionine by the addition of cystine to the sulfur amino acid–devoid diet in these human amino acid kinetic studies amounted to, approximately, a 50% reduction in the rate of methionine oxidation. Earlier nutritional studies have shown that, although dietary methionine is essential for the maintenance of protein nutritional status in adults and for normal growth and development in the young, dietary cystine can lower the methionine requirements by about 50% (Rose, 1957). This sparing action of cystine has been shown, with generally comparable quantitative results, in the cat (Teeter et al., 1978), mouse (Leveille et al., 1961), rat (Sowers et al., 1972), chick (Sasse and Baker, 1974), pig (Baker et al., 1969), and trout (Walton et al., 1982)

Apart from these more recent experiments in healthy adult subjects, there is little new information concerning the interrelationships and interactions among the sulfur amino acids except perhaps in relation to the effects of early growth and development on the conversion of methionine to cystine (Gaull et al., 1972) and on the possible importance of an endogenous supply of taurine in the nutritional support of infants requiring parenteral nutrition (Geggle et al., 1985). The area of taurine metabolism and function is an active and indeed exciting one (Kuriyama et al., 1983), but, at its present state of development, a detailed consideration of taurine is really outside of the present area of specific emphasis.

With respect to the interaction between methionine and cystine in the neonate, there is an apparent low hepatic transsulfuration enzyme activity at this stage of life (Gaull et al., 1972), and this has led to the suggestion that cysteine may be an essential dietary constituent for the preterm and newly born infant (Gaull et al., 1972). However direct metabolic studies (Zlotkin et al., 1981; Malloy et al., 1984) have failed to show that addition of cysteine to parenteral amino acid solutions improves growth and nitrogen balance in the newborn, particularly in the preterm infant. Thus, the question of the essentiality of cysteine seems still to be unresolved. Stegink (1986) suggests, based upon considerations of plasma amino acid data, that cysteine should be present in parenteral amino acid solutions. It appears to us that isotope tracer approaches of the kind mentioned above with methionine, coupled with kinetic studies of cystine metabolism, might offer a useful and new approach to the investigation of the metabolic aspects of methionine–cysteine interrelationships and their nutritional significance in the very young (Young et al., 1987).

Finally, it is important to mention methionine toxicity, particularly since this amino acid appears to be the most toxic of all indispensable amino acids (Harper et al., 1970). In addition, it is used as a supplement where soy proteins provide the major source of dietary protein in infant nutrition (Fomon et al., 1986), although there is no reason to add this amino acid to soy protein products when they are given as the sole source of amino acids in the diet of adult subjects (Young et al., 1979, 1984).

As Harper et al. (1970) have reviewed, methionine toxicity in the rat is alleviated by the addition of glycine and serine to high-methionine diets. It is possible, therefore, that the toxicity is related to a significant rise in the homocysteine level in plasma and tissues. The beneficial effects of glycine and/or serine might then be related to an enhanced formation of cystathionine, catalyzed by cystathionine synthase, although there is no direct or unequivocal evidence that this is the mechanism responsible for their effects. The possibility that methionine degradation occurs via a transamination or non-S-adenosylmethionine–dependent pathway (Case and Benevenga, 1976; Mitchell and Benevenga, 1978; Steele and Benevenga, 1978) is of further interest. Perhaps when high methionine intakes are given, with formation via this transamination pathway of methylthiopropionate (Steele and Benevenga, 1978) and methylmercaptan, these intermediates, or others, might be responsible for the symptoms of methionine toxicity. Also, the addition of glycine to a high-methionine diet might reduce the flow of methionine via this transamination pathway. Again, it seems to us that isotope tracer studies with labeled methionine and cystine would help to resolve these intriguing aspects of methionine metabolism. Indeed, investigations of this kind should serve the purpose of establishing the quantitative extent to which methionine is degraded via the transamination pathway under physiological conditions and usual levels of methionine intake. The current data do not allow a judgment as to the importance of this alternative pathway of methionine metabolism in the human subject under any conditions of diet or health.

## DISPENSABLE AMINO ACIDS

The contributions of the dispensable amino acids to the nitrogen economy of the host are qualitatively and quantitatively significant (Table 2.7). Furthermore, some of the dispensable amino acids, such as cystine and tyrosine, might be better classified as conditionally indispensable (Rudman, 1982; Harper, 1983), because they may be required under specific conditions, as in liver disease (Rudman, 1982) or in the biochemically immature, preterm infant, (Gaull et al., 1972). As we have already indicated, this is still debatable. Also, in animal experiments, addition of amino acids, such as serine, glycine, or alanine, to a threonine-limiting diet causes unfavorable effects, as evidenced by an inhibition of growth in rats (Tews et al., 1980). However, relatively little attention has been given to

**TABLE 2.7** Intake, Tissue Content, and Turnover of Indispensable (IDAA) and Dispensable (DAA) Amino Acids in the Adult

| Source of amino acid | Amino acids per 70 kg body wt | | |
| --- | --- | --- | --- |
| | Total (g) | IDAA (g) | DAA (g) |
| Diet: | | | |
|   Minimum needs | 56 | 6 | 50 |
|   Usual diet | 90 | 45 | 45 |
|   Absorbed (and secreted gut proteins) | 150 | 75 | 75 |
| Free amino acid pools: | | | |
|   Plasma | 0.7 | 0.2 | 0.5 |
|   Tissues | 70 | 10 | 60 |
| Daily body protein turnover | 300 | 150 | 150 |

*Kindly provided by Prof. H.N. Munro.*

this class of amino acids with respect to an assessment of the importance of amino acid interactions, specifically in reference to human protein and amino acid requirements and nutritional status.

We have begun to explore the metabolism of dispensable amino acids (Gersovitz et al., 1980) and especially the effects of changes in the total dietary nitrogen intake and in the ratio of indispensable:dispensable amino acids on the kinetics of whole body metabolism of two dispensable amino acids, alanine and glycine, in healthy young adults (Yu et al., 1985). Combined intravenous infusions of indispensable and dispensable amino acids, labeled with stable isotopes (Robert et al., 1982), were given to estimate the rates of whole body de novo synthesis of glycine and alanine under the varying dietary conditions (Fig 2.10). We chose to examine these two amino acids in view of their different responses to changes in dietary protein intake and to meal feeding (Young and Bier, 1981; Yang et al., 1986). As summarized in Fig. 2.11, our results show that glycine synthesis rates changed in response to an alteration in the intake of indispensable amino acids; this response contrasted with that for alanine (Yu et al., 1985), since the rate of de novo alanine nitrogen synthesis was unaffected by the changes in level of dietary nitrogen and amino acid composition examined in this study.

Thus, the metabolism of glycine appears to be responsive to the nitrogen, or amino acid, component of the diet; the rate of glycine synthesis is reduced when the level of total amino acid intake is low and the dis-

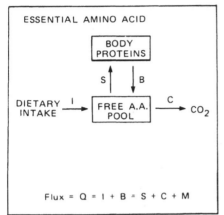
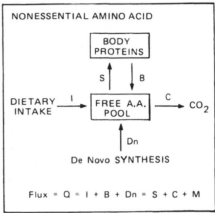

**FIG. 2.10** Simplified model of the metabolism of a nonessential (dispensable) amino acid (right panel) as compared with that of an essential (indispensable) amino acid (left panel) based on a continuous administration of labeled amino acids. For estimation of de novo glycine synthesis ($D_n$), for example, the model involves giving a simultaneous infusion of $^{15}N$-glycine and $1-^{13}C$-leucine. Under steady state conditions, the flux (Q) of the dispensable amino acid is determined by the entry of the amino acid from the diet (I), that released from tissue protein breakdown (B), and that arising via de novo synthesis ($D_n$). The disappearance of the amino acid from the pool occurs via protein synthesis (Su), oxidation (C), and conversion to other metabolites (M). Estimates of S and B are initially derived from the flux estimate for the indispensable amino acid, such as with $1-^{13}C$-leucine and the model in the left panel.

pensable amino acids are replaced by a mixture of indispensable amino acids. However, the physiological or functional significance of the decline in whole body glycine synthesis, especially at the lower intake level of total amino acid intake, is difficult to judge fully at this time. From simultaneous measures of leucine flux in our study (Yu et al., 1985), it was apparent that an absence of dietary dispensable amino acids for the brief dietary periods examined (7 days) did not result in any major change in the status of whole body protein breakdown. Nevertheless, the possible significance of the reduced rate of glycine synthesis might be inferred from estimates of the utilization of glycine for synthesis of creatine, porphyrins, and glutathione; it appears that these compounds require about 60% of the observed rate of new glycine synthesis at the lower amino acid intake level (Yu et al., 1985). Thus, glycine synthesis proceeds at a rate

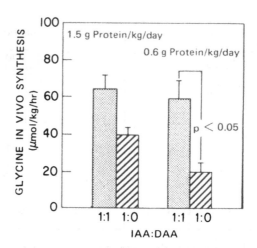

**FIG. 2.11** Estimates of de novo rate of glycine synthesis at two levels of "protein" intake with the dietary ratio of indispensable (IAA) to (DAA) dispensable amino acids being 1:1 or 1:0. (*Drawn from Yu et al., 1985.*)

close to the estimated amount necessary to meet demands for the formation of creatine under basal conditions. Prolonged restriction of dietary nitrogen or of glycine and other dispensable amino acids might be expected to limit the capacity of tissues to form these compounds from glycine. This point is important if we refer back to the earlier discussion concerning the control of branched pathways of metabolism and the regulatory significance of the relationship between the high- and low-flux components of these pathways. Based on the earlier discussion, a reduced rate of de novo glycine synthessis, which accounts for about 60% of the total entry of glycine into the free amino acid pools, due to the combined contributions of both protein breakdown and new glycine synthesis, might compromise the capacity of cells, under times of enhanced metabolic activity, to obtain sufficient glycine for porphyrin and purine synthesis.   This proposition finds some additional support from studies by Beliveau and Brusilow (1987) showing that when rats receiving a glycine-serine–devoid diet were given an injection of sodium benzoate, there was reduced growth and other biochemical changes. Significantly, these effects were reversed when the diet was supplemented with glycine and serine. Hence, these results show that glycine availability can be limiting under certain metabolic conditions, despite the significant capacity for glycine synthesis by tissues. Again, this might mean that a reduced rate of glycine synthesis in the adult human, as discussed above, limits the availability of glycine for meeting various biosynthetic functions that might be increased in times of stress.

Finally, by way of a consideration of dispensable amino acids in relation to the topic of amino acid interactions, we have considered it important to explore aspects of proline metabolism in human subjects with particular reference to the regulation of proline synthesis in vivo. These studies were initiated because it has been shown that there is feedback regulation of proline synthesis in cultured cells (Lodato et al., 1981), and it was important to establish whether this applies to the intact human. Furthermore, it has been suggested that proline is an indispensable amino acid for the young, rapidly growing pig (Ball et al., 1986), and thus we need to better understand the role of proline in the nutrition of humans at various ages.

Again, using a dual tracer amino acid model (Robert et al., 1982) to estimate the in vivo rate of proline synthesis, we (Jaksic et al., 1987) have observed that when proline is given intravenously, at physiologically relevant rates, there is a significant reduction in the de novo rate of proline synthesis (Fig. 2.12). This finding underscores the importance of regulation of dispensable amino acid metabolism and also the fact that nutritional factors can modulate the rate of synthesis of these amino acids.

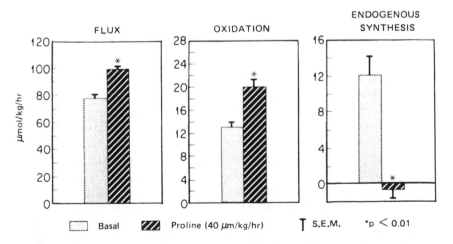

**FIG. 2.12** Effects of an intravenous infusion of proline (40 μmol/kg/hr) on the flux, "oxidation," and rate of de novo proline synthesis in healthy and young men. (*From Jaksic et al., 1987.*)

These various studies emphasize the need to give more critical attention to the so-called nonspecific nitrogen, or dispensable amino acid, component of the total dietary protein need. This is important because such knowledge is necessary for the optimal design of amino acid solutions intended for the long-term nutritional support of patients. It is plausible to suggest, for example, in the severely burned patient or the rapidly growing preterm baby, where rates of collagen synthesis and turnover are likely to be very high, that there might be a specific requirement for an exogenous source of proline, in view of its high concentration in collagens. Furthermore, it seems possible that the balance of other amino acids, such as ornithine and glutamate, which serve as precursors of proline (Jones, 1985), might influence the synthesis and availability of proline in cells and tissues. Amino acid kinetic studies should help to examine these possibilities, and the results of such investigations would add, in a significant way, to the limited data currently available on the nutritional significance of specific dispensable amino acids. Much of the present knowledge about the role of these amino acids and their interactions in human nutrition has been obtained from nitrogen balance studies. In view of the major limitations, both technical and conceptual, of this particular approach, we are forced to conclude that our understanding of the precise metabolic role played by the dispensable amino acids, and the interactions among them, with regard to the long-term maintenance of protein nutritional status, is very incomplete.

## SUMMARY AND CONCLUSIONS

As Lepkovsky (Jukes, 1986) impressed on one of the authors (VRY) of this review a number of years ago, the essence of the study of nutrition is *action* and *interaction,* such as synthesis and breakdown, anabolism and catabolism, influx and efflux, or stimulation and inhibition. We have concentrated our discussion largely on the metabolic aspects of some amino acid interactions, many of which could be predicted from known metabolic pathways and, for example, from knowledge of the kinetics of isolated enzymes and membrane transport systems. However, the nutritional significance and quantitative nature of these interactions in the complex, metabolic system of the intact organism cannot be judged adequately from studies in isolated systems alone because of the biochemically and physiologically important associations among the system of components within and among cells and between different organs. For this reason our focus has been on studies at the whole body level. We have also considered amino acid kinetics, but unfortunately the data available from studies in humans are still quite limited. Indeed, the importance of understanding and quantifying amino acid fluxes within cells and organs has been emphasized by many, including Waterlow (1985) and Porteous (1983a,b), who also said:

> When the final amino acid and nucleotide sequence has been entered into computer records, we shall be vastly better informed than we are now; there will be a huge library to place next to our metabolic maps; but the sequence data, alone or in conjunction with the maps, will not tell us how the living cell works.

This may be taken as a challenge to those of us interested in nutritional biochemistry and the metabolic and physiologic basis of nutrient interactions.

There are numerous interactions between amino acids, as reviewed above and earlier in more extensive detail by Harper et al., (1970), largely from information obtained in studies with experimental animals, that can affect the metabolic and nutritional condition of the organism. It is now becoming increasingly evident that these interactions may have greater physiological significance in the human subject than previously thought. Furthermore, Harper (1974) has also reviewed the literature on the tolerance of adult man to relatively high intakes of individual amino acids, with the conclusion that very large doses have been given with only mild adverse effects (Table 2.8) and doses in considerable excess of the requirement for most of the amino acids are "tolerated well." However, a limitation in our ability to draw definitive conclusions is that the requirements for the specific indispensable amino acids are probably much higher (Young et al., 1985; Young, 1987) than those values accepted in 1974, at the time when Harper (1974) conducted his survey. Furthermore, the more classical methods, including nitrogen balance and growth parameters, as well as the use of crude assessments of behavior are probably too insensitive to be of further major use to uncover the clinical and nutritional significance of the effects of amino acid interactions.

We have highlighted a number of specific interactions dealing with (a) branched-chain amino acids, (b) the sulfur amino acids, and (c) dispensable amino acids, all of which deserve a more in-depth exploration, especially with respect to their significance for normal well-being in human subjects. We have hypothesized that the dynamic status of organ and whole body protein turnover will affect the response of the organism to alterations in the level and balance of amino acids entering the body

**TABLE 2.8** Effects of Administering Various Quantities of Individual Amino Acids to Human Subjects

| Amino acid | Amount (g) | Number of days | Observations |
| --- | --- | --- | --- |
| Methionine | 10–15 | 9,7 | 20 g → behavioral changes and gastric distress |
| Tryptophan | 7,15 | 28,7 | + amine oxidase inhibitor → behavioral changes |
| Lysine | 10–40 | 6 | Highest doses → abdominal cramps and diarrhea |
| Histidine | 20 | 7 | No obvious adverse effects |
|  | 8–64 | 2,2 | Two days with and two days without histidine in scleroderma patients caused fall in serum Zn and taste acuity |
| Phenylalanine | 20 | 7 | No obvious adverse effects |
| Tyrosine | 20 | 7 | No obvious adverse effects |
| Arginine | 42 (i.v.) | 9 | No obvious adverse effects |
|  | 25 | 10 | In 9–12 yr cystic fibrosis patients → beneficial |
| Leucine | 10–20 | Single dose | → 14 g → slight depression in blood glucose concentration |

*Reproduced from Harper (1974) with permission.*

pools. Thus, it would seem important now to more carefully and critically examine the question of amino acid interactions, especially in relation to human aging, and the protein-depleted state that often occurs in hospitalized patients. It also seems, on metabolic grounds, that the clinical significance of amino acid interactions are likely to become more important under conditions of enhanced metabolic demand, as in the injured or infected host. However, initially healthy individuals, or those suffering from genetic disease, and experimental animals have served largely as the models for the study of amino acid interactions in vivo. With a more complete understanding of the metabolic significance of amino acid interactions in the intact organism, under various conditions of nutrition and health, the role of nutrition as a component of treatment and in the maintenance of health is likely to be enhanced.

## ACKNOWLEDGMENTS

The unpublished results from the authors' laboratories have been supported by NIH grants DK15856 and 36616 and a grant from the Shriners Hospital for Crippled Children. We thank our colleagues, Drs. D. Bier, D. Matthews, K. Storch, T. Jacksic, J. Burke, D. Wagner, and Y-M. Yu, for allowing us to refer to their unpublished findings.

## REFERENCES

Aftring, R. P., Block, K. B., and Buse, M. G. 1986. Leucine and isoleucine activate skeletal muscle branched-chain alpha-keto acid dehydrogenase *in vivo*. *Am. J. Physiol.* 250: E599.

Allen, R. E., Raines, P. L. and Regen, D. M. 1969. Regulatory significance of transfer RNA charging levels. I. Measurements of charging levels in livers of chow-fed rats, fasting rats and rats fed balanced or imbalanced mixtures of amino acids. *Biochem. Biophys. Acta* 190: 323.

Anderson, G. H. 1981. Diet, neurotransmitters and brain function. *Brit. Med. Bull.* 37: 95.

Austic, R. E. and Calvert, C. C. 1981. Nutritional interrelationships of electrolytes and amino acids. *Fed. Proc.* 40: 63.

Austin, S. A., Pain, V. M., Lewis, J. A., and Clemens, M. J. 1982. Investigation of the role of uncharged tRNA in the regulation of polypeptide chain initiation by amino acid starvation in cultured mammalian cells; A reappraisal. *Europ. J. Biochem.* 112: 519.

Baker, D. H., Clausing, W. W., Harmon, B. G., Jensen, A. H., and Becker, D. E. 1969. Replacement value of cystine for methionine for the young pig. *Brit. J. Nutr.* 55: 659.

Ball, R. O., Atkinson, J. L., and Bayley, H. S. 1986. Proline is an essential amino acid for the young pig. *Brit. J. Nutr.* 55: 659.

Bapurao, S. and Krishnaswamy, K. 1978. Vitamin $B_6$ nutritional status of pellagrins and their leucine tolerance. *Am. J. Clin. Nutr.* 31: 819.

Belavady, B., Srikantia, S. G., and Gopalan, C. 1963. The effect of the oral administration of leucine on the metabolism of tryptophan. *Biochem. J.* 87: 652.

Beliveau, G. P. and Brusilow, B. W. 1987. Glycine availability limits maximum hippurate synthesis in growing rats. *J. Nutr.* 117: 36.

Bender, D. A. 1983. Effects of a dietary excess of leucine on the metabolism of tryptophan in the rat: a mechanism for pellagragenic action of leucine. *Brit. J. Nurt.* 50: 25.

Bier, D. M. and Young, V. R. 1986. Assessment of whole-body protein-nitrogen kinetics in the human infant. In *Energy and Protein Needs in Infancy,* Fomon, S. J. and Heird, W. C. (Ed.), p. 107. Academic Press, Inc., New York.

Blackburn, G. L., Desai, S. P., Kennan, R. A., Bentley, B. T., Moldawer, L. L., and Bistrian, B. 1981. Clinical use of branched-chain amino acid enriched solutions in the stressed and injured patient. In *Metabolism and Clinical Implications of Branched Chain Amino Acid and Keto Acids.* Walser, M. and Williamson, J. R. (Ed.), p. 521. Elsevier/North Holland, New York.

Block, K. P. and Harper, A. E. 1984. Valine metabolism *in vivo:* Effects of high dietary levels of leucine and isoleucine. *Metabolism* 33: 559.

Block, K. P., Soemitro, S., Heywood, B. M., and Harper, A. E. 1985. Activation of liver branched-chain alpha-keto acid dehydrogenase in rats by excesses of dietary amino acids. *J. Nutr.* 115: 1550.

Boorman, K. N. and Fisher, H. 1966. The arginine-lysine interaction in the chick. *Brit. Poultr. Sci.* 7: 39.

Brennan, M. F., Cerra, F., Daly, J. M., Fischer, J. E., Moldawer, L. L., Smith, R. J., Vinnars, E., Wannemacher, R., and Young, V. R. 1986. Report of a research workshop: Branched-chain amino acids in stress and injury. *JPEN* 10: 446.

Buse, M. G. and Weigand, D. A. 1977. Studies concerning the specificity of the effect of leucine on the turnover of protein in muscles of control and diabetic rats.*Biochim. Biophys. Acta* 475: 81.

Case, G. L. and Benevenga N. J. 1976. Evidence for S-adenosylmethionine independent catabolism of methionine in the rat. *J. Nutr.* 106: 1721.

Chawla, R. K., Berry, C. J., Kutner, M. H., and Rudman, D. 1985. Plasma concentrations of transsulfuration pathway products during nasoenteral and intravenous hyperalimentation of malnourished patients.*Am. J. Clin. Nutr.* 42: 577.

Christensen, H. N. 1982. Interorgan amino acid nutrition. *Physiol. Rev.* 62: 1193.

Christensen, H. N. 1984. Organic ion transport during seven decades. The amino acids. *Biochim. Biophys. Acta* 779: 255.

Christensen, H. N. 1987a. Hypothesis: Where do depleted amino acids go in phenylalaninemia and why? *Persp. Biol. Med.* 30: 186.

Christensen, H. N. 1987b. Role of membrane transport in inter-organ amino acid flows: Where do the depleted amino acids go in phenylketonuria? *J. Cell. Biochem.* 33: (in press).

Clark, A. J., Yamada, C., and Swendseid, M. E. 1968. Effect of L-leucine on amino acid levels in plasma and tissue of normal and diabetic rats. *Am. J. Physiol.* 215: 1324.

Clemens, M. J., Austin, S. A., Galpin, A. R., and Pollard, J. W. 1985. Translational control by amino acids and aminoacyl-tRNA synthetase in cultured mammalian cells. *Biochem. Soc. Trans.* 13: 676.

Cook, N. E. and Carpenter, K. J. 1987. Leucine excess and niacin status in rats. *J. Nutr.* 117: 519.

Crabtree, B. and Newsholme, E. A. 1985. A quantitative approach to metabolic control. *Curr. Topics Regulation* 25: 21.

Espinal, J., Begges, M., Patel, M., and Randle, P. J. 1986. Effects of low-protein diet and starvation on the activity of branched-chain 2-oxo acid dehydrogenase kinase in rat liver and heart. *Biochem. J.* 237: 285.

FAO/WHO/UNU. 1985. *Energy and Protein Requirements.* WHO Tech. Rept. Ser. No. 724. World Health Organization, Geneva, Switzerland.

Fern, E. B., Garlick, P. J., McNurlan, M. A., and Waterlow, J. C. 1981. The excretion of isotope in urea and ammonia for estimating protein turnover in man with $^{15}$N glycine. *Clin. Sci.* 61: 217.

Fernstrom, J. D. and Wurtman, R. J. 1972. Brain serotonin content: physiological regulation by plasma neutral amino acids. *Science* 178: 414.

Finkelstein, J. D., Martin, J. J., and Harris, B. J. 1986. Effect of dietary cystine on methionine metabolism in rat liver. *J. Nutr.* 116: 985.

Finkelstein, J. D. and Mudd, S. H. 1967. Trans-sulfuration in mammals: the methionine-sparing effect of cystine. *J. Biiol. Chem.* 242: 873.

Fomon, S. J., Ziegler, E. E., Nelson, S. E., and Edwards, B. B. 1986. Requirement for sulfur-containing amino acids in infancy. *J. Nutr.* 116: 1405.

Frick, G. P., Tai, L-R., Blinder, L., and Goodman, H. M. 1981. L-leucine activates branched chain alpha-keto acid dehydrogenase in rat adipose tissue. *J. Biol. Chem.* 256: 2618.

Fukagawa, N. K., Minaker, K. L., Rowe, J. W., Goodman, M. N., Matthews, D. E., Bier, D. M., and Young, V. R. 1985. Insulin-mediated reduction of whole body protein breakdown. Dose-response effects on leucine metabolism in post-absorptive men. *J. Clin. Inves.* 76: 2306.

Fukagawa, N., Minaker, K. L., Rowe, J. W., and Young, V. R. 1987. Plasma tryptophan and total neutral amino acid levels in men: Influence of hyperinsulinemia and age. *Metabolism* 36: 683.

Garlick, P. J. 1980. Protein turnover in the whole animal and specific tissues. In *Comprehensive Biochemistry,* Newberger, A. (Ed.), Vol. 19B, Part 1, p. 77. Elsevier Scientific Publ. Co., Amsterdam.

Gaull, G. E., Sturman, J. A., and Raiha, N. C. R. 1972. Development of mammalian sulfur metabolism. Absence of cystathionase in human fetal tissues. *Pediatr. Res.* 6: 538.

Geggle, H. W., Ament, M. E., Heckenlively, J. R., Martin, D. A., and Koppel, J. D. 1985. Nutritional requirements for taurine in patients receiving long-term parenteral nutrition. *New Eng. J. Med.* 312: 142.

Gersovitz, M., Bier, D., Matthews, D., Udall, J., Munro, H. N., and Young, V. R. 1980. Dynamic aspects of whole body glycine metabolism: influence of protein intake in young adult and elderly males. *Metabolism* 29: 1087.

Ghafoorunissa, R. and Narasinga Rao, B.S. 1973. Effect of leucine on enzymes of the tryptophan-niacin metabolic pathway in rat liver and kidney. *Biochem. J.* 134: 425.

Gopalan, C. and Srikantia, S. G. 1960. Leucine and pellagra. *Lancet* 1: 554.

Hagenfeldt, L., Eriksson, S., and Wahren, J. 1980. Influence of leucine on arterial concentrations and regional exchange of amino acids in healthy subjects. *Clin. Sci.* 59: 173.

Hambraeus, L., Bilmazes, C., Dippel, C., Scrimshaw, N., and Young, V. R. 1976. Regulatory role of dietary leucine on plasma branched chain amino acid levels in young men. *J. Nutr.* 106: 230.

Harper, A. E. 1974. Amino acid excess. In *Nutrients in Processed Foods: Proteins.* White P. L. and Fletcher, D. C. (Ed.), p. 49. Publishing Sciences Group, Inc., Acton, MA.

Harper, A. E. 1983. Dispensable and indispensable amino acid interrelationships. In *Amino Acids: Metabolism and Medical Applications,* Blackburn, G., Grant, J. P. and Young, V. R. (Ed.), p. 105. PSG Publishing Co., Inc. Littleton, MA.

Harper, A. E., Benevenga, N. J., and Wohlheuter, R. M. 1970. Effects of ingestion of disproportionate amounts of amino acids. *Physiol. Rev.* 50: 428.

Harper, A. E., Miller, R. H., and Biock, K. P. 1984. Branched-chain amino acid metabolism. *Ann. Rev. Nutr.* 4: 409.

Harris, R. A. and Paxton, R. (Eds.) 1985. Regulation of branched chain alpha-ketoacid dehydrogenase complex by phosphorylation-dephosphorylation. *Fed. Proc.* 44: 305.

Harris, R. A., Paxton, R., Goodwin, G. W., Kuntz, M. J., Shimomura, Y., and Han, A. 1986. Regulation of branched-chain amino acid metabolism. *Biochem. Soc. Trans.* 14: 1005.

Harris, R. A., Powell, S. M., Paxton, R., Gillim, S. E., and Nagae, H. 1985. Physiological covalent regulation of rat tracer branched-chain alpha-ketoacid dehydrogenase. *Arch. Biochem. Biophys.* 243: 542.

Hegsted, D. M. 1976. Balance studies *J. Nutr.* 106: 307.

Horowitz, J. H., Rypins, E. B., Henderson, J. M., Heymsfield, S. B., Moffitt, S. D., Bain, R. P., Chawla, R. K., Bleier, C., and Rudman, D. 1981. Evidence for impairment of transsulfuration pathway in cirrhosis. *Gastroenterology* 81: 668.

Hussein, M. A., Young, V. R., Murray, E., and Scrimshaw, N. S. 1971. Daily fluctuations of plasma amino acid levels in adult men. Effect of dietary tryptophan intake and distribution of meals. *J. Nutr.* 101: 61.

Hutson, S. M. and Harper, A. E. 1981. Blood and tissue branched-chain amino and alpha-keto acid concentrations. Effect of diet, starvation and disease. *Am. J. Clin. Nutr.* 34: 173.

Jaksic, T., Wagner, D. A., Burke, J. F., and Young, V. R. 1987. Plasma proline kinetics and the regulation of proline synthesis in man. *Metabolism* 36: 1040.

Jones, J. D. 1965. Lysine-arginine antagonism in the chick. *J. Nutr.* 84: 313.

Jones, J. D., Wolters, R., and Burnett, P. C. 1966. Lysine-arginine-electrolyte relationships in the rat. *J. Nutr.* 89: 171.

Jones, M. E. 1985. Conversion of glutamate to ornithine and proline: Pyroline-5-carboxylate, a possible modulator of arginine requirements. *J. Nutr.* 115: 509.

Jukes, T. H. 1986. Samuel Lepkovsky (1899–1984).Biographical sketch. *J. Nutr.* 116.

Krehl, W.A. 1949. Niacin in amino acid metabolism. *Vitamins Hormones* 7: 114.

Krishnaswamy, K. and Bapurao, S. 1978. Effect of leucine at different levels of vitamin $B_6$ on hepatic quinolinate phosphoribosyltransferase and leucine amino transferase in rats. *Br. J. Nutr.* 39: 61.

Krishnaswamy, K., Bapurao, S., Raghuram, T. C., and Srikantia, S. G.

1976. Effect of vitamin $B_6$ on leucine induced changes in human subjects. *Am. J. Clin. Nutr.* 29: 177.

Kuriyama, K., Huxtable, R. J., and Iwata H. (Ed.). 1983. *Sulfur Amino Acids. Biochemical and Clinical Aspects,* p. 485. Alan R. Liss, Inc. New York.

Laurent, B. C., Moldawer, L. L., Young, V. R., Bistrian, B. R., and Blackburn, G. L. 1984. Whole body leucine and muscle protein kinetics in rats fed varying protein intakes. *Am. J. Physiol.* 246: E444.

Leathwood, P. D. 1987. Tryptophan availability and serotonin synthesis. *Proc. Nutr. Soc.* 46: 143.

Leveille, G. A., Sauberlich, H. E., and Shockley, J. W. 1961. Sulfur amino acid requirements for growth of mice fed two levels of nitrogen. *J. Nutr.* 75: 445.

Lieberman, H. R., Caballero, B., and Finer, N. 1986. The composition of lunch determines afternoon plasma-tryptophan ratios in humans. *J. Neurol Transmission* 65: 211.

Lodato, R. F., Smith, R. J., Valle, D., Phang, J. M., and Aoki, T. T. 1981. The regulation of proline biosynthesis: The inhibition of pyroline-5-carboxylate synthase activity by ornithine. *Metabolism* 30: 908.

Magboul, B. I. and Bender, D. A. 1983. The effect of a dietary excess of leucine on the synthesis of nicotinamide nucleotides in the rat. *Brit. J. Nutr.* 49: 321.

Maher, T. J. 1987. Natural food constituents and food additives: The pharmacologic connection. *J. Allergy Clin. Immunol.* 79: 413.

Malloy, M. H., Rassin, D. K., and Richardson C. J. 1984. Total parenteral nutrition in sick preterm infants: Effects of cysteine supplementation with nitrogen intakes of 240 and 480 mg/kg/day. *J. Pediatr. Gastro. & Nutr.* 3: 239.

Manson, J. A. and Carpenter, K. J. 1980a. The effect of a high level of dietary leucine on the niacin status of chicks and rats. *J. Nutr.* 108: 1889.

Manson, J. A. and Carpenter, K. J. 1980b. The effect of a high level of dietary leucine on niacin status in dogs. *J. Nutr.* 108: 1889.

Matthews, D. E., Motil, K. J., Rohrbaugh, D. K., Burke, J. F., Young, V. R., and Bier, D. M. 1980. Measurement of leucine metabolism in man from a primed, continuous infusion of L-[1-$^{13}$C] leucine. *Am. J. Physiiol.* 238: E473.

Matthews, D. E., Schwarz, H. P., Yang, R. D., Motil, K. J., Young, V. R., and Bier, D. M. 1982. Relationship of plasma leucine and α-ketoisocaproate during a L-[1-$^{13}$C]leucine infusion in man: a method for measuring human intracellular leucine tracer enrichment. *Metabolism* 81: 1105.

Meguid, M. M., Matthews, D. E., Bier, D. M., Meredith, C., N., Soeldner, J. S., and Young, V. R. 1986. Leucine kinetics at graded leucine intakes in young men. *Am. J. Clin. Nutr.* 43: 770.

Meguid, M. M., Schwarz, H., Matthews, D. E., Karl, I. E., Young, V. R., and Bier, D. M. 1983. *In vivo* and *in vitro* branched-chain amino acid interactions. In *Amino Acids: Metabolism and Medical Applications,* Blackburn, G., Grant, J. P., and Young, V. R. (Eds), p. 147. PSG Publishing Co., Inc., Littleton, MA.

Miller, L. P., Pardridge, W. M., Braun, L. D., and Oldendorf, W. H. 1985. Kinetic constants for blood-brain barrier amino acid transport in conscious rats. *J. Neurochem.* 45: 1427.

Millward, D. J. 1980. Protein degradation in muscle and liver. In *Comprehensive Biochemistry,* Newberger, A. (Ed.), Vol. 19B, Part 1, p. 153. Elsevier Scientific Publ. Co., Amsterdam.

Mitchell, A. D. and Benevenga, N. J. 1978. The role of transamination in methionine oxidation in the rat. *J. Nutr.* 108: 67.

Mueller, J. H. 1923. A new sulfur containing amino acid isolated from the hydrolytic products of protein. *J. Biol. Chem.* 56: 157.

Munro, H. N., Hubert, C., and Baliga, B. S. 1975. Regulation of protein synthesis in relation to amino acid supply—A review. In *Alchohol and Abnormal Protein Biosynthesis. Biochemical and Clinical,* Rothschild, M. A., Oratz, M. and Schreiber, S. S. (Ed.), p. 33. Pergammon Press, Inc., New York.

Nakagawa, I., Ohguri, S., Sasaki, A., et al. 1975. Effects of excess intake of leucine and valine deficiency on tryptophan and niacin metabolites in humans. *J. Nutr.* 105: 1241.

Nakagawa, I. and Sasaki, A. 1977. Effect of an excess intake of leucine, with and without additions of vitamin $B_6$ and/or niacin, on tryptophan and niacin metabolism in rats. *J. Nutr. Sci. Vitaminol.* 23: 535.

Nesheim, M. C. 1968. Genetic variations in arginine and lysine utilization. *Fed. Proc.* 27: 1210.

Newsholme, E. A., Crabtree, B., and Ardawi, M. S. M. 1985. Glutamine metabolism in lymphocytes: Its biochemical, physiological and clinical importance. *Quart. J. Exp. Physiol.* 70: 473.

Oestemer, G. A., Hanson, L.E., and Meade, R. J. 1973. Leucine-isoleucine interrelationships in the young pig. *J. Anim. Sci.* 36: 674.

Oxender, D. L., Collarini, E. J., Shotwell, M. A., Lobaton, C. D., Moreno, A., Campbell, G. S., and El-Gewely, M. R. 1986. Cellular transport in the regulation of amino acid metabolism. *Biochem. Soc. Trans.* 14: 993.

Ozalp, I., Young, V. R., Nagchaudhuri, J., Tontisirin, K., and Scrimshaw,

N. S. 1972. Plasma amino acid response in young men given diets devoid of single essential amino acids. *J. Nutr.* 102: 1147.

Pain, V. M. 1986. Initiation of protein synthesis in mammalian cells. *Biochem. J.* 235: 625

Pardridge, W. M. 1983. Brain metabolism: a perspective from the blood-brain barrier. *Physiol. Rev.* 63: 1481.

Pardridge, W. M. and Choi, T. B. 1986. Neutral amino acid transport at the human blood-brain barrier. *Fed. Proc.* 45: 2073.

Pogson, C. I., Salter, M., and Knowles, R. G. 1986. Regulation of hepatic aromatic amino acid metabolism. *Biochem. Soc. Trans.* 14: 999.

Porteous, J. W. 1983a. Sound practice follows from sound theory—the control analysis of Kacser and Burns evaluated. *Trends Biochem. Sci.* 8: 200.

Porteous, J. W. 1983b. Catalysis and modulation in metabolic systems. The aims and scope of the Colloquium. *Biochem. Soc. Trans.* 11: 29.

Rama Rao, P. B., Norton, H. W., and Johnson, B. C. 1961. The amino acid composition and nutritive value of proteins. IV. Phenylalanine, tyrosine, methionine and cystine requirements of growing rat. *J. Nutr.* 73: 38.

Ramanamurthy, P. S. V. and Srikantia, S. G. 1970. Effects of leucine on brain serotonin. *J. Neurochem.* 17: 27.

Reynolds, M. S., Steel, D. L., Jones, E. M., and Baumann C. A. 1958. Nitrogen balances of women maintained on various levels of methionine and cystine. *J. Nutr.* 64: 99.

Robert, J. J., Bier, D. M., Zhao, X. L., Matthews, D. E., and Young, V. R. 1982. Glucose and insulin effects on *de novo* amino acid synthesis in young men: Studies with stable isotope labeled alanine, glycine, leucine and lysine. *Metabolism* 31: 1210.

Rose, W. C. 1957. The amino acid requirements of adult man. *Nutr. Abstr. Rev.* 27: 631.

Rose, W. C. and Wixom R. L. 1955. The amino acid requirements of man, XII. The sparing effect of cystine on the methionine requirement. *J. Biol. Chem.* 216: 763.

Rudman, D. 1982. Overview: Deficiencies of essential nutrients, clinical and biochemical manifestations. *Am. J. Clin. Nutr.* 35: 1112.

Sasse, C. E. and Baker, D. H. 1974. Sulfur utilization by the chick with emphasis on the effect of inorganic sulfate on the cystine-methionine interrelationship. *J. Nutr.* 104: 244.

Scrimshaw, N. S., Bresanni, R., Behar, M., and Viteri F. 1958. Supplementation of cereal protein with amino acids. I. Effect of amino acid supplementation of corn-masa at high levels of protein intake on the nitrogen retention of young children. *J. Nutr.* 66: 485.

Shannon, B. M., Howe, J. M., and Clark, H. E. 1972. Interrelationships between dietary methionine and cystine as reflected by growth certain hepatic enzymes and liver composition of weaning rats. *J. Nutr.* 102: 557.

Sowers, J. E., Stockland, W. L., and Meade, R. J. 1972. L-methionine and L-cystine requirements of the growing rat. *J. Anin. Sci.* 35: 782.

Srikantia, S. G., Narasinga Rao, B. S., Raghuramulu, N., and Gopalan, C. 1968. Pattern of nicotinamide nucleotides in erythrocytes of pellagrins. *Am. J. Clin. Nutr.* 21: 1306.

Steele, R. D. and Benevenga, N. J. 1978. Identification of 3-methylthiopropionic as an intermcdiate in mammalian methionine metabolism *in vivo. J. Biol. Chem.* 253: 7844.

Steginк, L. D. 1986. Parenteral amino acid requirements: Special problems and possible solutions. In *Energy and Protien Needs during Infancy,* Fomon, S. J. and Heird, W. C. (Ed.), p. 183. Academic Press, New York.

Stipanuk, M. H. and Benevenga, N. J. 1977.Effect of cysteine on the metabolism of methionine in rats. *J. Nutr.* 107: 1455.

Storch, K. J., Wagner, D. A , Burke, J. F., and Young, V. R. 1987a. *In vivo* mechanisms of methionine conservation and methionine-sparing effect of dietary cystine, studied by $^2H_3$-methyl-1-$^{13}$C-methionine kinetics, in man. (Submitted for publication).

Storch, K. J., Wagner, D.A., Burke, J. F., and Young, V. R. 1987b. A quantitative study,*in vivo,* of the methionine cycle in man, using $^2H_3$methyl-1-$^{13}$Cmethionine. *AM. J. Physiol.* (Submitted).

Swendseid, M. E., Villalobos, J., Figueroa, W. S., and Drenick, E. J. 1965. The effect of test doses of leucine, isoleucine or valine on plasma amino acid levels. The unique effect of leucine. *Am. J. Clin. Nutr.* 17: 317-321.

Teeter, R. G., Baker, D. H., and Corbin, J. E. 1978. Methionine and cystine requirements of the cat. *J. Nutr.* 108: 219.

Tews, J. K., Kim, YW. L., and Harper, A. E. 1980. Induction of threonine imbalance by dispensable amino acids: Relationships between tissue amino acids and diet in rats. *J. Nutr.* 110: 394.

Tischler, M. E., Desaultes, M., and Goldberg, A. L. 1982. Does leucine, leucyl-tRNA, or some metabolite of leucine regulate protein synthesis and degradation in skeletal and cardiac muscle? *J. Biol. Chem.* 257: 1613.

Wahren, J., Denis, J., Desurmont, P., Erikkson, L. S., Escopfier, JM, Gauthier, A. P., Hagenfeldt, L., Michel H., Opolon, P., Paris, J-C., and Veyrac, M. 1983. Is intravenous administration of branched chain amino acids effective in the treatment of hepatic encephalopathy? A multi center study. *Hepatology* 3: 475.

Walton, M. J., Cowey, C. B., and Adron, J. W. 1982. Methionine metabolism in rainbow trout fed diets differing in methionine and cystine content. *J. Nutr.* 112: 1525.

Waterlow, J. C. 1984. Protein turnover with special reference to man. *Quart. Rev. Exp. Physiol.* 69: 409.

Waterlow, J. C. 1985. Opening remarks. In *Substrate and Energy Metabolism in Man,* Garrow, J. S. and Halliday, D. (Ed.), p. 1. John Libbey, London.

Waterlow, J. C., Garlick, P. J., and Millward, D. J. 1978a. *Protein Turnover in Mammalian Tissues and in the Whole Body,* p. 804. North-Holland Publishing Co., Amsterdam.

Waterlow, J. C., Golden, M. H. N., and Garlick, P. J. 1978b. Protein turnover in man measured with $^{15}$N: Comparison of end products and dose regimes. *Am. J. Physiol.* 235: E165.

Womack, M., Kremmerer, K. S., and Rose, W. C. 1937. The relation of cysteine and methionine to growth. *J. Biol. Chem. 141: 375.*

Wurtman, R. J. and Ritter-Walker, E. (Ed.) 1987. *Dietary Phenylalanine and Brain Function.* Proc. First Int'l. Meeting on Dietary Phenylalanine and Brain Function, Washington, D.C., May 8–10, 1987, p. 509. Center Brain Sciences and Metabolism Charitable Trust, Cambridge, MA.

Yang, R. D., Matthews, D. E., Bier, D. M., Wen, Z.-M., and Young, V. R. 1986. Response of alanine metabolism in humans to manipulations of dietary protein and energy intakes. *Am. J. Physiol.* 250: E39.

Young, V. R. 1975. *Recent Advances in Evaluation of Protein Quality in Adult Humans.* Proc. 9th. Int'l. Cong. Nutr., Vol. 3, p. 348. Karger, Basel.

Young, V. R. 1986. Nutritional balance studies: Indicators of human requirements or adaptive mechanisms? *J. Nutr.* 116: 700.

Young, V. R. 1987. 1987 McCollum Award Lecture. Kinetics of human amino acid metabolism: nutritional implications and some lessons. *Am. J. Clin. Nutr.* 46: 709.

Young, V. R. and Bier, D. M. 1981. Protein metabolism and nutritional state in man. *Proc. Nutr. Soc.* 40: 343.

Young, V. R., Fukagawa, N., Bier, D. M., and Matthews, D. 1986. Some aspects of *in vivo* human protein and amino acid metabolism, with particular reference to nutritional modulation. In *Verhandlungen der Deutschen Gesellschaft für Innere Medizin, 92 Band,* p. 640. J. F. Bergmann Verlag, München, Germany.

Young, V. R., Fukagawa, N. K., Storch, K. J., Hoerer, R., Jaksic, T., and Bier, D. M. 1987. Stable isotope probes: Potential for application in studies of amino acid utilization in the neonate. In *Perinatal Nutrition,* Vol. 6. Lindblad, B. S. (Ed.), p. 221. Academic Press, San Diego.

Young, V. R., Meredith, C., Hoerr, R., Bier, D. M., and Matthews, D. E. 1985. Amino acid kinetics in relation to protein and amino acid requirements: The primary importance of amino acid oxidation. In *Substrate and Energy Metabolism in Man,* Garrow, J. S. and Halliday, D. (Ed.), p. 119. John Libbey, London.

Young, V. R., Puig, M., Queiroz, E., Scrimshaw, N. S., and Rand, W. M. 1984. Evaluation of the protein quality of an isolated soy protein in young men: relative nitrogen requirements and effect of methionine supplementation. *Am. J. Clin. Nutr.* 39: 16.

Young, V. R., Scrimshaw, N. S., Torun, B., and Viteri F. 1979. Soybean in human nutrition. *J. Am. Oil Chem. Soc.* 56: 110.

Yu, M., Yang, R. D., Matthews, D. E., Burke, J. F., Bier, D. M., and Young, V. R. 1985. Quantitative aspects of glycine and alanine nitrogen metabolism in young men: Effect of level of nitrogen and dispensable amino acid intake. *J. Nutr.* 115: 339.

Yu, Y-M, Wagner, D. A. , Walesrewski, J. C., Burke, J. F., and Young, V. R. 1987. A kinetic study of leucine metabolism in severely burned patients: Comparison between conventional and branched-chain amino acid-enriched nutritional therapy. *Annals Surg.* (submitted).

Zhao, X. L., Wen, Z-M, Meredith, C. N., Matthews, D. E., Bier, D. M., and Young, V. R. 1986. Threonine kinetics at graded threonine intakes in young men. *Am. J. Clin. Nutr.* 43: 795.

Zlotkin, S. H., Bryan, M. H., and Anderson, G. H. 1981. Cysteine supplementation to cysteine free intravenous feeding regimens in newborn infants. *Am. J. Clin. Nutr. 34: 914.*

# 3

# Mineral Interactions in Foods

**Fergus M. Clydesdale**

University of Massachusetts
Amherst, Massachusetts

Great strides have been made in expanding the knowledge base of mineral nutrition in the last ten years. The physiological role of minerals is becoming more fully defined, and a more complete understanding of their absorption is resulting from some excellent studies in both animals and humans. Unfortunately, this increased enthusiasm for mineral nutrition, with some notable exceptions, has not been as evident in studies of the chemical reactions and interactions that minerals undergo in food. Historically this may have been due to a persistent tendency to view the minerals as being rather inert and thus affected only by food processes in which leaching or physical separation occurred. However, it is now well recognized that some foods and/or food components are inhibitors to mineral absorption, whereas others enhance it. Further, with the advent of mineral fortification of foods, it became apparent that often the most bioavailable mineral salts were the most reactive, and, therefore, least functional, thus decreasing quality and, in so doing, being changed to less bioavailable forms. As a result, increased interest in this area has occurred, in part, to explain phenomena not readily apparent from studies on bioavailability alone.

## CHEMICAL REACTIVITY OF MINERALS

The physical and chemical properties of the minerals are governed by the electronic configurations of their electrons, which in turn are defined by the quantum numbers of each electron.

The quantum numbers of the electron are known as n, $l$, $m_l$, and $m_s$. The first or principal quantum number, n, is the most important for defining its energy, increasing with increasing electron energy and average distance of the electron cloud from the nucleus. It is always an integral and cannot be zero. Electrons in a given atom with the same value of n are said to be in the same level or shell.

The second quantum number, $l$, is known as the angular momentum quantum number, which defines the shape or sublevel of the electron cloud. For a given electron, it is related to n as follows:

$$l = 0, 1, 2 \ldots (n - l)$$

The third quantum number, $m_l$, known as the magnetic orbital quantum number, is linked to the directional orientation or orbital of the electron cloud and can have any integral value between $l$ and $-l$, such that $m_l = l, l - 1, l - 2, \ldots, -l$.

The fourth quantum number, $m_s$, the magnetic spin quantum number, is associated with the direction in which the electron moves about its axis. Since there can be only two possible values for $m_s$, and since no two electrons in any atom can have the same set of four quantum numbers, each orbital, defined by $m_l$, is limited to two electrons. Since the value of $m_l$ is dependent on $l$ or n, this means that the number of electrons in each sublevel ($l$) and in each level (n) is also limited.

Electron energies are mainly affected by the values of n and $l$. For this reason, the electronic configuration is therefore expressed by only these two quantum numbers, omitting the values for $m_l$ and $m_s$. Tradition dictates that n be written as an integer but that $l$, representing the sublevel, be defined in terms of letters. The symbols and order of all sublevels ($l$) are as follows:

1s, 2s, 2p, 3s, 3p, 4s, 3d, 4p, 5s, 4d, 5p, 6s, 4f, 5d, 6p, 7s, 5f, 6d

Therefore, the electronic configuration of a mineral such as iron can be defined as:

$$1s^2 2s^2 2p^6 3s^2 3p^6 4s^2 3d^6$$

However, during ionization, the outer $4s^2$ electrons are lost first. The physical properties of a mineral, such as its ionization potential or energy, can be derived from its electron configuration, since it is a measure of the energy required to remove electrons from the mineral. The first electron to be removed, being in the furthest sublevel from the nucleus, would come off more easily than those in sublevels closer to the nucleus, with each sublevel having a particular ionization energy. When a chemical reaction occurs, about 100 kcal per mole of reactant is released. This corresponds to the ionization energy of the electrons in the outermost shell of an atom but is much smaller than those of the inner sublevels. It may be concluded that the electronic rearrangements occurring during chemical reactions involve only the outermost electrons, and therefore the chemical properties of an atom are associated with the electron configuration of its outermost shell. Thus, each of the alkali metals, Li, Na, K, Rb, and Cs, has a single outermost s electron, and all react with water to form a hydroxide of the general formula MOH. Similarly, nearly all of the compounds formed from these elements are water-soluble.

This similarity of properties was the basis for the development of the Periodic Table, which interestingly classified the elements according to atomic weights and other properties prior to an understanding of their electron configuration. However, it is easier to predict properties of the minerals and understand the Table in terms of electron configuration. The elements in the Table are listed in order of increasing atomic number in horizontal rows, such that elements with the same outermost electron configuration fall directly beneath one another. Group 1A, the alkali metals, have an outermost electron configuration of ns, that is, one outermost electron in one of the s sublevels. Group 2A, the alkaline earths, have two outermost electrons in the s sublevels or $ns^2$. Thus, they are not quite as reactive as the ns elements but share some properties.

There are ten additional elements in the fourth period. These are often referred to as the transition elements, where the ten electrons of the 3d sublevel are being filled in an increasing fashion from left to right. The transition elements are known for their ability to form coordination compounds, which can effect their bioavailability dramatically.

Obviously, the minerals represent a wide variety of physically and chemically reactive species. This reactivity must be defined to fully understand the mechanisms involved with bioavailability and to allow physiologically significant fortification of food, which will provide more sound dietary recommendations. A further discussion of mineral chemistry as it relates to food may be found in a recent review by Clydesdale

(1988). The remainder of this chapter will deal with selected reactions and conditions that are either known to, or have the greatest potential of, affecting bioavailability.

## PHYSICOCHEMICAL REACTIONS AS DETERMINANTS OF BIOAVAILABILITY IN FOOD

There are an incredibly large number of reactions in a food material that might effect bioavailability. In order to focus on major reactions, it is first necessary to make some arguable assumptions. The assumptions I would like to make for the purpose of this chapter are that solubility and charge density are two of the most important properties of a mineral or mineral complex in their potential to affect bioavailability. Therefore, we should concern ourselves with the physicochemical reactions that are most likely to affect these properties.

Solubility is of obvious importance, since a mineral must come in contact with all parts of the intestinal mucosa if it is to be efficiently absorbed. The role of charge density is less obvious but is also important for its effect on complex formation as well as membrane permeability.

The solubility of a mineral in a food during preparation, processing, and storage will depend on the effect of the environment on such basic properties as the reduction potential and pH of the food as well as the length of time it is processed and/or stored. Also, it will depend on the type of bonding, complex formation, or chelation the mineral undergoes. If a molecular species in food forms a compound with the mineral that is soluble and can be absorbed by the mucosal cells as such and/or if it undergoes cleavage to release the mineral in a soluble form and/or if it has stability constants that allow the mineral to be transferred to a mucosal or serosal acceptor, it is known as an enhancer. If, on the other hand, the compound so formed is insoluble and/or cannot be absorbed as such and/or cannot be cleaved and/or releases the mineral in an insoluble form and/or has thermodynamic constants that do not allow transfer to a mucosal or serosal acceptor, it is known as an inhibitor.

Therefore, when we speak of solubility in reference to the bioavailability of a mineral, we are referring not only to the solubility of an ion, salt, hydrate, or complex, but also to the type and strength of chemical bonds involved within these species. This explains the anomolous observations often found when an insoluble mineral salt or complex seems bioavailable. It is not due to the fact that "the human intestine has failed to read the physical chemistry textbooks" (Heaney; 1986); it is due to a lack of understanding of the totality of the chemical reactions occurring in that environment.

The importance of solubility as a prerequisite for absorption has been recognized in iron nutrition for many years, as evidenced by the fact that most in vitro techniques rely on solubility as an end point (Motzok et al., 1978; Rao and Prabhavthi, 1978; Miller et al., 1981; Bates et al., 1983), and even the in vivo extrinsic tag technique assumes the solubilization of all nonheme iron in the diet to form a common intraluminal pool. Certainly this fact is recognized in the case of other minerals, but not as much work has been done in the development of assessment techniques for their solubility during digestion.

In a medium such as food, solubility is often obtained only by the formation of mineral complexes, and there is often a correlation between charge density and the ability to form such complexes. Those cations from groups 6B to 2B with a relatively high ionic charge ($+2$, $+3$) and small size form a large number of stable complex ions, whereas the large alkali metal cations, such as $Na^+$ and $K^+$, with a small charge are much less likely to form complexes.

Of importance among the transition metals, which may form more than one cation, is the fact that the ion of higher charge forms complexes more readily. Therefore, complexes formed by $Cr^{+3}$, $Co^{+3}$, and $Fe^{+3}$ are more numerous and more stable than those of their respective $+2$ ions. Therefore, $Fe^{+3}$ would be stabilized more by complexation than $Fe^{+2}$, but $Fe^{+2}$ is inherently more soluble than $Fe^{+3}$ in water due to the solubility of its hydrate. The greater stability of $Fe^{+3}$ complexes has been confirmed with both amino acids and carboxylic acids in several studies (Perrin, 1958; Conrad and Schade, 1968; Flynn and Clydesdale, 1983; Gorman and Clydesdale, 1984).

Charge is also involved with cell permeability as well as solubility prior to entering the cell. However, the rate at which a molecule diffuses across the lipid bilayer that makes up the cell membrane varies enormously depending upon its size and degree of polarity or charge. The lipid bilayer serves as a highly impermeable barrier to most polar molecules, although with enough time any molecule will diffuse across a protein-free lipid bilayer down its concentration gradient. Small nonpolar molecules readily dissolve in lipid bilayers and therefore rapidly diffuse across them. Uncharged polar molecules also will diffuse rapidly if they are small enough, while polar charged molecules are effectively blocked. Interestingly, water, although polar, diffuses rapidly across a lipid bilayer due, in part, to its small size and lack of charge and to the fact that the dipolar structure of water allows it to cross the bilayer containing the lipid head groups unusually rapidly (Alberts et al., 1983).

This biological barrier to charged molecules has a practical significance in nutrition. Emery (1982) pointed out that one of the reasons why ferrichrome is such an effective bacterial siderophore is not only that it binds iron but also that it neutralizes the three positive charges on ferric iron when it binds with it, thus facilitating its passage back through the cell membrane.

## POTENTIAL FOR REACTIONS IN FOOD

There are many reactions that minerals might undergo in foods to effect a change in solubility and/or charge. However, to keep this discussion to a reasonable length, I have chosen those reactions which I think might have the most practical significance in affecting bioavailability. In this section I would like to discuss briefly the characteristics of such reactions and in a subsequent section provide a more detailed description of individual reactions that lead to the enhancement or inhibition of mineral absorption.

## COMPLEX FORMATION AND CHELATION

A complex or complex ion can be formed by any molecule or anion possessing an unshared pair of electrons which are donated to a metal. The molecule that donates the electrons is known as the ligand and the atom within the ligand which provides the electrons is normally one of the more electronegative elements, such as C, N, O, S, F, Cl, Br, or I. The bond so formed is a coordinate covalent bond where the anion acts as a Lewis base and the metal or central atom, a Lewis acid. Therefore, to act as a ligand, a molecule or anion must have an unshared electron pair, which can be donated to the central atom to form a coordinate covalent bond. The number of such coordinate bonds formed by the central atom or mineral is known as its coordination number. Some metals have more than one coordination number, but the most common number in complex ions is 6 with 4 next and 2 less often, being restricted to the positive monovalent ions of the 1B elements.

Chelation is a special form of complex formation where the ligand forms more than one bond with the central atom, or mineral in this case. The complex thus formed is known as a chelate, from the Greek word meaning a crab's claw. The electron-donating atoms in the ligand, often oxygen, sulfur, or nitrogen, hold the mineral in a clawlike manner and form a heterocyclic ring. Rings may contain varying numbers of atoms dependent upon the number of atoms in the ligand which form bonds with the metal. A bidentate ligand, such as an oxalate ion, has two donor atoms, tridentate ligands have three, and so on, with compounds such as EDTA being hexadentate. The central atom of such complexes is likely to be a transition metal, since chelation is involved mostly with the interaction of electrons in the d orbitals, and in some, but not all, cases the metal involved as the central atom attains a noble gas structure.

A further and more complete discussion of chelates and their chemistry may be found in an excellent book by Kratzer and Vohra (1986).

Although chelation and coordinate bonds are extremely important in nutritional chemistry, this section would not be complete without mentioning the other types of bonds which can effectively tie up and precipitate or solubilize a mineral in a complex. Both ionic and covalent bonds are important in mineral complexes and molecules. These bonds are often responsible for the chemical properties of a complex, but in terms of physical properties, including solubility, various types of intermolecular forces are more important than bonding since these must be overcome for solubilization to occur. Such forces include dipoles, hydrogen bonds, and dispersion or London forces. Of the three, dispersion forces are the most common.

In studies of the crystalline structure of polyguluronic acid, it has been shown that the intermolecular hydrogen bond arises from inclusion of water molecules between the chains of macromolecules. It is probable that metal cations might occupy the same sites in polyguluronate salts that water molecules do in the polyacid (Kohn, 1975). This idea is in accordance with the egg-box model shown in Fig. 3.1 (Glicksman, 1982) and

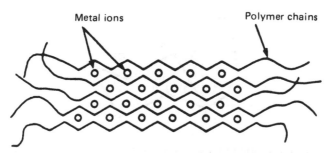

**FIG. 3.1**   Egg-box model for binding polysaccharides and minerals. (*From Glicksman, 1982.*)

could have an effect on metal bioavailability. Intermolecular bonds between calcium ions and the polygalacturonates of pectin are also responsible for complexation.

Angyal (1973) has suggested that metal cations form 1:1 complexes with cyclitols and sugars containing an axial–equatorial–axial sequence of three hydroxyl groups in a six-membered ring, or a cis–cis sequence in a five-membered ring with complex integrity remaining when one of the hydroxy groups is replaced with a methoxy group. These complexes are formed by the replacement of water molecules in the solvation sphere of the cation (e.g., calcium) by at least three hydroxyl groups in the steric arrangements mentioned above. This is shown in Fig. 3.2, where calcium is bound to epi-inositol by three hydroxyl groups in the ax–eq–ax sequence. Also, α-D-allopyranose, which contains the ax–eq–ax sequence, and the α-furanose form of D-allose, which contains a cis–cis sequence on a five-membered ring (shown in Fig. 3.3), were both found to bind calcium strongly, magnesium less so, and zinc, hardly at all.

The complexation of minerals by alginic acid may be a reaction of some nutritional significance, since it is widely used in the food industry. Alginates are linear copolymers of D-mannuronic and L-guluronic acid. Each molecule contains sections of each of these constituent sugars and also regions where the two residues alternate. It had been proposed that the ax–eq–ax sequence in gulopyranosyluronic acid might explain metal bonding (Angyal, 1973), but more recent evidence indicates that an association of either polymannuronic or polyguluronic acid segments could bind minerals in an egg-box fashion as shown in Fig. 3.1 (Glicksman, 1982). Polyguluronic acid, of all the polyuronic acids, is the only one to selectively bind metals, forming stronger complexes with strontium and calcium than with magnesium or sodium (Angyal, 1973). This means that one mineral could be displaced by another and thus potentially affect bioavailability.

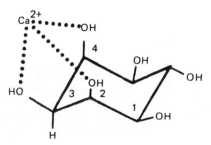

**FIG. 3.2**  Complexation of Ca with epi-inositol by the oxygen atoms in an axial–equatorial–axial configuration. (*From Angyal, 1973.*)

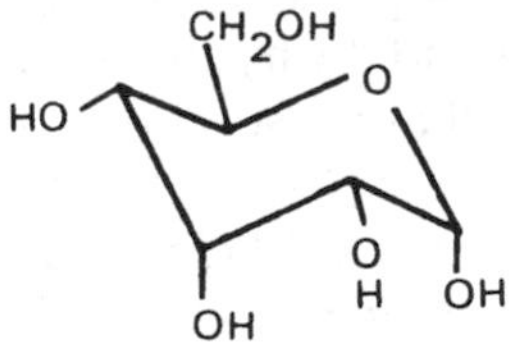

(XI)

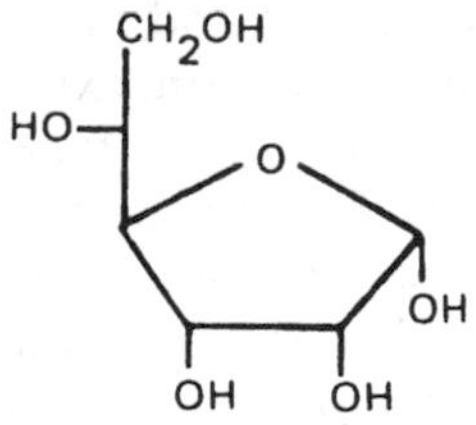

(XII)

FIG. 3.3  α-D-allopyranose (XI), which contains the axial–equatorial–axial sequence of hydroxyl groups, and its anomeric furanose form (XII) containing a cis–cis sequence. (*From Angyal, 1973.*)

It is tempting to postulate that the forces described above and/or those involved in the egg-box model of metal complexation may be common to the gelling carbohydrate polymers and could explain some of their influence on bioavailability. For instance, studies in Rosenberg's laboratory (Kelly et al., 1984) and others (Anon., 1985) have shown that glucose polymers enhance the absorption of calcium. Bei et al., (1986) have shown that they also increase magnesium and zinc absorption. A large part of this enhancing affect may be involved with the ability of glucose polymers to maintain calcium solubility in the presence of inhibitors and/or through the pH extremes of the intestinal tract. We are currently studying the solubilizing effects of glucose and several glucose polymers since a recent study suggests that glucose alone may play an important role in calcium absorption (Wood et al., 1987).

## HYDROLYSIS

If both nutritionally and nonnutritionally important molecular ligands were considered, water and ammonia would be the most frequently encountered.

From a mineral-nutrition perspective, this is important to all metals but particularly to the transition series, since they do not exist in water as ions but as hydrates. These hydrates are a particular case of complex ion formation and include the nutritionally important metals Fe, Co, Ni, Cu, and Zn.

Hydrolysis of the hydrates may occur as the pH is raised, causing them to lose protons and forming less soluble or insoluble hydroxides, which may precipitate and thus become unavailable The most well known example of this is iron, existing in solution at low pH as $Fe(H_2O)_6^{+3}$ and $Fe(H_2O)_6^{+2}$. As the pH is raised, protons are lost and the hydroxides, $Fe(OH)_3$ and $Fe(OH)_2$, are formed in neutral and alkaline solutions. Although $Fe(OH)_2$ is more soluble ($10^{-1}$ M) then $Fe(OH)_3^{-1}$ ($10^{-16}$ M), both are considerably less soluble than their respective hydrates (Clydesdale, 1983). As well as precipitating, the hydrolysis products of ferric iron may form ferric hydroxide sols and then colloidal solutions. These large aggregates may form polynuclear complexes and be unavailable for passage through membranes (Spiro and Saltman, 1969). Berner et al. (1985) isolated ferric hydroxide polymers and found that they were available to rats. However aging, a phenomenon to be discussed in the next section, may be required for the polymers to become unavailable.

## AGING

The aging of minerals is a well-established phenomenon in the geological sciences. Over time, the environment causes changes in mineral-containing compounds, which affect their chemical and physical characteristics.

In the nutritional arena, little work has been done on this phenomenon except in the case of iron. The hydrolysis of iron during storage or aging proceeds from free ions in solution to various iron oxides, and alkalinization, oxidation, and heating further drives the reaction toward dehydrated polymers with increasingly ordered structures. In solution at room temperature, iron hydroxide gels undergo "hardening" due to this increase in order (Gheith, 1952) and become increasingly resistant to acid attact over time (Sommer et al., 1973). Such an occurrence implies that at some time in storage iron sources will be converted to forms that are irreversibly insoluble even when subjected to the acidic conditions of the stomach. This would explain why some forms of iron, such as the ferric orthophosphates and pyrophosphates and some ferric hydroxides, do not fully exchange with the nonheme iron pool (Bothwell et al. 1982) and why $Fe(OH)_3$ was absorbed only half as well in humans as intrinsic iron in maize porridge (Derman et al., 1977).

Eyerman et al. (1987) felt that it would be advantageous to be able to predict at what point in storage such irreversibility might occur. Therefore, they examined the hydrolysis of iron and the effect of aging on the resolubilization of iron hydroxides under simulated gastric pH conditions utilizing reflectance spectroscopy as a predictor of the degree of resistance to resolubilization in acid. When $FeSO_4$ was added to the system under study, both soluble and acid resoluble iron decreased in a linear manner with time over a 10-week period. This was most interesting in light of the fact that Park et al. (1983) found that over a 3-month period the bio-availability of $FeSO_4$ in anemic rats was reduced from 84 to 65%. The cor-relation is evident and indicates that at some time in the future physical techniques, such as reflectance spectroscopy, might be used to predict tially greater final bioavailability.

Such results have interesting implications in mineral fortification. Recommendations have been made that more bioavailable and, there-fore, more reactive forms of iron should be used in food fortification and the subsequent loss in food quality ignored. However, the chemistry in-volved clearly indicates that when the iron source causes a change in quality, it also undergoes a change itself. If this change leads to an irrever-sibly insoluble form, then bioavailability will be not only reduced but un-predictable. Therefore, at the present time, the most predictable and stable form of bioavailable iron is elemental iron, which will not change on storage. Its somewhat decreased initial bioavailability, as compared to $FeSO_4$, will be more than compensated for by its predictable and poten-tially greater final bioavailability.

As concern over acid rain generates more research on mineral aging and acid resolubilization, more knowledge and techniques will become available for the food scientist and nutritionist to apply to minerals in the food system.

## pH AND REDUCTION POTENTIAL

Although they are not necessarily discussed together in a classic chemistry textbook, I believe it is important to consider both pH and reduction potentials simultaneously when referring to their effects on minerals in food systems.

When electrons are lost from an element the process is known as oxida-tion, and when they are gained, as reduction. When the two reactions occur simultaneously, it is called an oxidation-reduction reaction or, more simply, a redox reaction. The ease of occurrence of these reactions is a measure of the addition or removal of electrons and provides a valuable tool in assessing reactivity alone and in the presence of other compounds. Such reactions may be measured in a voltaic cell since in a properly designed cell, at a given temperature, the voltage depends upon the nature of the cell reaction and the concentration of the various species, ions or molecules, participating. Thus we can define the ease of removal or addition of electrons in terms of the standard reduction potential (SRP), in volts, of the reduction half-reactions of the elements. Thus the SRP in water of the reaction $Zn^{+2} + 2e^- \rightarrow Zn^0$ is $-0.76$ v. This means that whenever $Zn^{+2}$ is in a system with a redox potential less than $-0.76$, this reaction will occur; if greater than $-0.76$, the reverse reaction will occur spontaneously. Most food systems have potentials within $\pm 0.10$ V of $+0.40$ V (Nojeim et al., 1981) and, therefore, would have little effect. However, pH has a dramatic effect on the SRP of a given half-reaction and would affect the formation of compounds which may or may not be bioavailable. This is the reason that there is an important linkage between pH and reduction potential in nutritional terms. In order to clarify this further, consider the following reactions:

$$Fe^{+3} \rightarrow Fe^{+2} \text{ (0.77 V in an aqueous system)} \qquad (1)$$

$$Fe(OH)_3 \rightarrow Fe(OH)_2 \text{ (}-0.56\text{ v in a basic solution)} \qquad (2)$$

$$Zn^{+2} \rightarrow Zn \text{ (}-0.76\text{ v in an aqueous system)} \qquad (3)$$

$$Zn(OH)_4^{-2} \rightarrow Zn \text{ (}-1.22\text{ v in a basic solution)} \qquad (4)$$

It is obvious that the end products will be very different in reactions (1) versus (2) and in (3) versus (4), indicating the importance of pH control when fortifying foods as well as in studies on bioavailability.

Physiologically, the importance of pH in mineral metabolism is well known with the various mineral disorders caused by achlorhydria. Its importance in the food system in relation to mineral bioavailability, however, is less well recognized. We have discussed hydrolysis as it relates to mineral hydrates and their solubility, but it is also important to note that the insoluble mineral hydroxides, which form as pH is increased, are in equilibrium with the metal ion at any given pH in accordance with the following equations for the solubility product (Ksp):

$$M^{+2} + 2\,OH^- \rightleftharpoons M(OH)_2$$

$$Ksp = [M^{+2}]\,[OH-]^2$$

Since Ksp is constant, an increase in [OH⁻] or pH causes a decrease in the free metal ion [M] which follows first, second, and third powers of hydroxyl ion concentration with the mono-, di-, and trivalent metal hydroxides, respectively (Chaberek and Martell, 1959).

The mineral salts of the phosphates, carbonates, and oxalates have minimum solubility and potentially minimum bioavailability at high pH. Solubility is affected because a decrease in pH decreases the concentration of the anion and, therefore, increases the concentration of the metal ion in solution in order to maintain a constant Ksp. Of course if the reverse happens, as mentioned earlier, mineral solubility decreases.

Oxalic acid forms soluble salts with alkali metals, but with other metals the salts formed are sparingly soluble in water, which would affect bioavailability. At neutral or alkaline pH oxalate forms a practically insoluble salt with calcium (0.67 mg/100 ml), while $Fe^{+2}$ and $Fe^{+3}$ salts are very soluble (Clydesdale, 1987). Heaney (1986) has noted that calcium balance studies in humans have shown conflicting results when an excess of oxalic acid or spinach has been mixed into food. These results may be due to a lack of control of the pH of the food and/or the presence of enhancers.

pH may also affect the degree and strength of binding of minerals to other enhancers and inhibitors of absorption. Clydesdale (1983) tabulated the results of various studies of the binding of ascorbate, citrate, phytate, EDTA, and cysteine to iron, which show the effects of pH on complex solubility. Further, Lee and Garcia-Lopez (1985) found that iron binding by NDF (neutral detergent fiber) and ADF (acid detergent fiber) increased with pH, and quite dramatically above pH 5. Berner and Hood (1983) illustrated the effect of pH on iron binding by alginate in the presence of various chelating agents, and Thompson and Weber (1979) evaluated its effect on Cu, Zn, and Fe in six fiber sources. Expectedly, the pH effects on solubility were modified by the presence of chelators, which would explain Heaney's (1986) comments on calcium as mentioned above.

Finally, it should be stressed that the addition of many mineral salts to a food or model system will lower the pH, with the decrease being dependent on the mineral salt and/or the buffering capacity of the system to which it is added. However, decreases of 1 to 2 pH units are not uncommon, which would affect solubility and, therefore, bioavailability. Therefore, when assessing bioavailability, great care should be taken in controlling pH or at least recording the pH of every sample just prior to feeding.

## ENHANCERS AND INHIBITORS

### Definition

In the literature, mineral enhancers are often described simply as ligands, and inhibitors are generally not described. As noted previously, however, it is not as simple as that.

Enhancers are molecular species in food that form a compound with the mineral which is soluble and can be absorbed by the mucosal cells as such and/or may undergo cleavage to release the mineral in a soluble form and/or has constants which allow the mineral to be transferred to a mucosal or serosal acceptor.

On the other hand, inhibitors are molecular species in food which form an insoluble compound, and/or cannot be absorbed as such, and/or cannot be cleaved and/or releases the mineral in an insoluble form, and/or has thermodynamic constants which do not allow transfer to a mucosal or serosal acceptor.

Obviously, the description of an enhancer simply as a ligand is inadequate, and therefore the other qualifications described above will be discussed in the sections that follow.

## THERMODYNAMIC CONSIDERATIONS

Since complexation and subsequent release of a mineral are important considerations for both enhancers and inhibitors, it is important to consider the thermodynamics of the complex.

Implicit in the definition of an enhancer is the need to form a complex with a mineral that is not only soluble but that has a stability constant greater than that of an inhibitor. Further, it is necessary that the stability constant be such that a mucosal acceptor molecule will have a greater stability constant with the mineral than that of the enhancer. Also, it is essential that this transfer take place within a reasonable period of time, a characteristic described by the kinetic stability constant of a complex which Forth and Rummel (1973) defined as the time required to achieve half-saturation or exchange of a mineral between complexes.

Thermodynamic stability constants of mineral amino acid complexes established by Albert (1950) followed the order:

$$Cu^{+2} > NI^{+2} > Zn^{+2} > Co^{+2} > Cd^{+2} > Fe^{+2} > Mn^{+2}$$

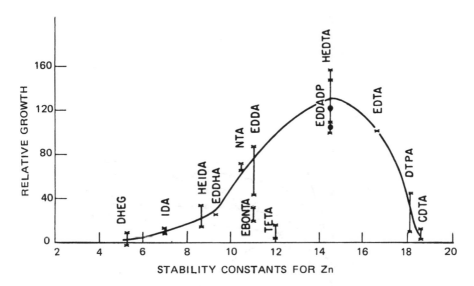

**FIG. 3.4** Relationship of the stability constants of several chelating agents with Zn to their growth-promoting effect in poults when added to a Zn-deficient diet. (*From Kratzer and Vohra, 1986.*)

but for any one metal, the stability constants for the various amino acids varied over a 100-fold range.

Vohra and Kratzer (1964) evaluated the ability of various chelating agents to improve the availability of Zn from isolated soybean protein for turkey poults. They plotted relative growth versus the stability constants for the various chelating agents as shown in Fig. 3.4. The data suggested that chelating agents with stability constants below 13 were ineffective in removing the Zn from the Zn inhibitor in soybean protein. They speculated that the stability constant of the inhibitor was higher than 13 but lower than 14.5 and that another chelating system in the body with a stability constant lower than 16.5 but higher than 14.5 transported Zn to various tissues in the body.

Both Kratzer and Vohra (1986) and Forth and Rummel (1973) caution against direct correlations between thermodynamic stability constants in a defined medium and what might happen in food. Forth and Rummel suggest the use of the kinetic stability constant in place of the thermodynamic stability constant.

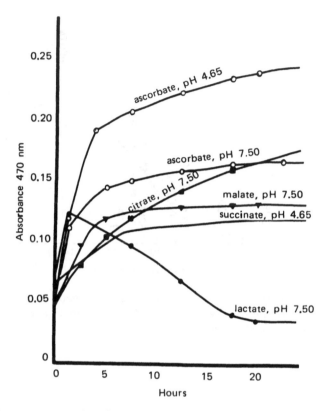

**FIG. 3.5** Curves representing the rate of transfer of iron from ascorbate, citrate, malate, succinate, and lactate iron complexes to apotransferrin. (*From Gorman and Clydesdale, 1984.*)

Gorman and Clydesdale (1984) established both thermodynamic and kinetic stability constants for $Fe^{+3}$ with ascorbic acid and varous carboxylic acids. The kinetic constants for ascorbate, lactate, and malate were 2.7, 0.75, and 2 hours, respectively. Citrate was found to have a constant of 8 hours in this study, which was the same as that reported by Bates et al. (1967).

Upon examination of the iron exchange curves, as shown in Fig. 3.5, Gorman and Clydesdale (1984) suggested that the slope of the initial portion of the curves, prior to saturation, might be the most meaningful measure of the exchange rate, since it more closely approximates the short time intervals involved with duodenal absorption.

Forth and Rummel (1973) were concerned over the presence of competing ligands in food, their effect on mineral bioavailability, and attempts to relate a stability constant to availability. Certainly the fact that a ligand forms a complex with a mineral is not enough to label it either an enhancer or an inhibitor, if such complexation has not been evaluated in the presence of competing ligands. For instance, cellulose has been found to bind iron to varying degrees in model systems, particularly at pH values near neutrality (Thompson and Weber, 1979; Camire and Clydesdale, 1981; Fernandez and Phillips, 1982). However, there seems to be little effect on iron absorption in animals and humans (Monnier et al., 1980; Buth and Mehta, 1983; Gilooly et al., 1984). One of the reasons for this may be that cellulose is displaced easily by other components in a food system. Platt and Clydesdale (1984) found that both cellulose and lignin alone bound a high percentage of iron in an insoluble form near neutral pH and much less at pH 2. However, in the presence of other binding components, cellulose, unlike lignin, had a minimal effect on iron, giving it up to Na phytate and beta-glucan where it became soluble. One must assume that similar results would be found in a mixed system with other minerals and other ligands dependent upon the specificity of the binding site, its stability constant with the mineral in question, and the amount of mineral bound by each site on the inhibitor. Platt and Clydesdale (1987a) have investigated these characteristics with several fiber–mineral combinations, and their results are shown in Table 3.1. Interestingly, the values found correlate with the majority of in vivo and in vitro studies in the literature, which show that lignin inhibits iron absorption, while cellulose may or may not. It also should be noted that lignin has two binding sites for Fe, Cu, and Zn with the high affinity sites binding them in the order Fe > Cu > Zn but with twice as much Cu bound as either Fe or Zn. This has obvious implications in the displacement and inhibition of one mineral over another when both are in the presence of lignin. Therefore, it must be concluded that thermodynamic constants are extremely important indicators when used with other parameters to predict bioavailability, but if used alone might lead to erroneous judgments.

**TABLE 3.1** Binding Characteristics of Various Fiber–Mineral Combinations

| Fiber–mineral system | Characteristics of binding sites | $K_{eff}$ | Bound mineral (mol per g fiber) |
|---|---|---|---|
| Control + Fe(II) | Nonspecific | — | — |
| Cellulose + Fe(II) | Nonspecific | — | — |
| Guar gum + Fe(II) | (a) Specific | $6.27 \times 10^6$ | 8.4 |
| | (b) Nonspecific | — | — |
| NDF + Fe(II) | (a) Specific | $3.43 \times 10^5$ | 5.0 |
| | (b) Nonspecific | — | — |
| High methoxy pectin + Fe(II) | (a) Specific | $5.90 \times 10^5$ | 5.0 |
| | (b) Nonspecified | — | — |
| Low methoxy pectin + Fe(II) | Specific | $1.84 \times 10^4$ | 39.9 |
| Lignin + Fe(II) | (a) Specific | $7.68 \times 10^5$ | 19.5 |
| | (b) Specific | $2.07 \times 10^4$ | 118.9 |
| Lignin + Cu(II) | (a) Specific | $4.67 \times 10^5$ | 32.3 |
| | (b) Specific | $4.58 \times 10^4$ | 81.6 |
| Lignin + Zn(II) | (a) Specific | $1.21 \times 10^5$ | 15.4 |
| | (b) Specific | $7.84 \times 10^3$ | 72.2 |

*Source: Platt and Clydesdale (1987a).*

## EXAMPLES OF ENHANCERS AND INHIBITORS IN FOOD

There is an almost limitless number of enhancers and inhibitors in food, and to select a few for discussion is difficult. The problem is somewhat simplified because many of the more important examples will be covered in other chapters of this book by experts in those particular fields. This provides me with the luxury of choosing a few examples that I consider interesting from an interactive viewpoint.

### Ascorbic Acid

One of the best known mineral enhancers is ascorbic acid, and its effect on iron bioavailability has been widely studied. Its effect may be attributed to one or more of three physicochemical factors, namely, a reduction in pH, complexation, and/or reduction potential (Clydesdale, 1982). In order to evaluate the effect of pH, Rizk and Clydesdale (1984) compared a series of soy concentrates with 10–50 mm added ascorbate versus a series brought to the same pH with HCl. It was clear that the ascorbate produced significantly more soluble iron due to both reductive release of iron from the soy concentrate forming $Fe^{+2}$ and the formation of $Fe^{+3}$-ascorbate complexes. In this case pH was a minor factor. Conrad and Schade (1968) demonstrated the formation of an $Fe^{+3}$-ascorbate complex, which formed at low pH but remained stable as the pH was raised through neutrality. This prompted a series of experiments by Gorman and Clydesdale (1983, 1984) on this and other reactions, which might better explain the effect of ascorbate. Thermodynamic constants established by titrimetric techniques (Martell and Calvin, 1952) indicated that $Fe^{+3}$-ascorbate was much more stable (K = $1.90 \times 10^3$ to $2.61 \times 10^4$) than $Fe^{+2}$-ascorbate (K = $7.69 \times 10^{-3}$ to $6.95 \times 10^{-2}$) over pH ranges and reduction potentials generally found in food and the gastrointestinal tract. Also, as mentioned previously, exchange parameters with human transferrin were established with ascorbic acid and several carboxylic acids, as shown in Table 3.2. It was obvious from these results that the thermodynamic constants did not completely explain the activity of ascorbic acid, since it is a superior enhancer as compared to the carboxylic acids in Table 3.2. Therefore, the redox couples between ascorbate and iron were also studied. At acid pH (pH = 1.5) the redox potential of the $Fe^{+3}$-ascorbate system was +340 mv, and at neutral pH +15 mv. Similarly, the net redox potential values for the $Fe^{+2}$-ascorbate system were +269 mv and +32.0 mv at acid (pH = 2.5) and neutral pH, respectively. The actual values obtained at acid pH for the

**TABLE 3.2** Kinetic Stability Constants, Time to Maximize Absorption, and Slope of the Exchange Curves for Ascorbate, Lactate, Malate, Succinate, and Citrate Ferric Iron Complexes with Human Apotransferrin at pH 4.65 and 7.50 at 470 nm

| Complex | pH | $A_{max}$ | Time to max absorbance (hr) | Time to half-saturation of transferrin (kinetic stability constant) (hr) | Slope |
|---|---|---|---|---|---|
| $Fe^{+3}$–ascorbate | 4.65 | 0.240 | 17 | 2 | 0.044 |
| $Fe^{+3}$–ascorbate | 7.50 | 0.170 | 16 | 2 | 0.040 |
| $Fe^{+3}$–lactate | 7.50 | 0.125 | 2 | 0.75 | 0.040 |
| $Fe^{+3}$–malate | 7.50 | 0.134 | 8 | 2 | 0.108 |
| $Fe^{+3}$–succinate | 4.65 | 0.122 | 8 | 2.3 | 0.011 |
| $Fe^{+3}$–citrate | 7.50 | 0.180 | 20 | 8 | 0.014 |

*Source: Gorman and Clydesdale (1984).*

$Fe^{+3}$ and $Fe^{+2}$–ascorbate solutions were close to the value theoretically expected ($+380$ mv). However, the net emf of $+15$ mv obtained experimentally for $Fe^{+3}$ at neutral pH differed greatly from the theoretical value of $+640$ mv, indicating that iron was much less susceptible to oxidation at neutral pH in the presence of ascorbic ascid. The discrepancy between actual and theoretical values may be due to the fact that the theoretical half-reaction values do not consider the complexation that occurs between ascorbate and iron. This is an interesting observation, because it implies that the enhancing effects of ascorbic acid are not due to complexation and reducing ability as two separate entities but are unique to ascorbic acid as they act synergistically to promote optimal conditions for the maintenance of both $Fe^{+3}$ and $Fe^{+2}$ in solution.

## Effect of Acid Pretreatment

These studies all indicated that the complexes, although stable over a wide pH range, formed at low pH values. This led to a series of studies in our laboratory investigating the effect of an acid pretreatment on the solubility of several iron–ligand complexes. This was accomplished by an initial incubation at pH 2.0 and neutralization to pH 6.0 prior to addition of the complex to a wheat flake cereal and a sequential pH treatment from the endogenous cereal pH (E) to pH 2 to pH 6 in order to simulate gastrointestinal pH changes.

In the case of ascorbic acid the incubation increased solubility throughout the pH treatment (Clydesdale and Nadeau, 1985) as predicted from the model system studies by Gorman and Clydesdale (1983). In an evaluation of citric and malic acids, Nadeau and Clydesdale (1986) demonstrated that citrate was more effective than malate in maintaining solubility with several iron sources hydrogen (HRI) or electrolytically reduced elemental iron (ERI), ferrous sulfate ($FeSO_4$), ferric chloride ($FeCl_3$), or ferric orthophosphate (FOP)) following acid incubation and the sequential pH treatment. However, malate–iron complexes showed superior pH stability following pH reduction to 2. In a similar study utilizing EDTA, cysteine, lactic and succinic acids, acid incubation significantly enhanced the iron-solubilizing potential of EDTA at each sequential pH with HRI and ERI, while lactic and succinic acids affected $FeSO_4$ and $FeCl_3$ in a similar manner at pH 2 (Nadeau and Clydesdale, 1987b). Cysteine, possessing both a ligand and a modest reducing potential, generated substantial amounts of $Fe^{+2}$ at pH 2 at the apparent expense of complexed iron. This duality may explain the superiority of iron enhancement often shown by cysteine over some of the other ligands evaluated. In all cases, FOP exhibited the least solubility, thereby minimizing potential beneficial complexation reactions.

## Citric Acid

Citric acid has been identified as the low molecular weight Zn-binding ligand in human milk by Hurley and Lonnerdal (1982) and is involved with Ca in bone metabolism (Neuman and Neuman, 1958). With calcium citrate, a more optimum calcium bioavailability is observed than with calcium carbonate in both normal humans (Nicar and Pak, 1985) and patients suffering from achlorhydria (Recker, 1985). Citric acid has been found to both inhibit and enhance iron absorption, depending on the meal and the investigator (Hallberg and Rossander, 1984). It is, therefore, an interesting compound to discuss from a chemical point of view.

Lyon (1984) evaluated the ability of citric acid to promote mineral solubility in $HCO_3^-$-neutralized extracts from cereals, and his results are shown in Table 3.3. The addition of citric acid had a dramatic effect in decreasing the level of precipitation and rendered by far the greater bulk of the trace minerals soluble. The solubilizing effect of other complexing agents and lactose was also studied, and the results are also shown in Table 3.3. EDTA, as might be expected, formed soluble chelates and solubilized all of the Zn. Citric acid was also very efficient at solubilizing Zn and Fe but was somewhat poorer at solubilizing Ca and Mg. Histidine partially solubilized Zn (51%) but did not affect Fe, Ca, or Mg. The effect of lactose was part of a different study but was included for information. Various other studies, including those of Kojima et al. (1981), Leigh and Miller (1983), Nadeau and Clydesdale (1986), and Rizk and Clydesdale (1985a,b), have shown citrate to be an effective solubilizing agent with iron but dependent upon pH and the ratio of iron:citrate studied. Nadeau and Clydesdale (1986) found an increasing iron-solubilizing effect in cereal samples subjected to a sequential pH treatment from the cereal pH (E) to 2 to 6 with iron:citrate ratios from 0:1 to 150:1. Leigh and Miller (1983) found that the ratio of iron:citrate, in three ratios (1:0.4, 1:1, 1:25) studied, had little effect on the binding of iron to fiber at pH values from 2 to 7. However, significantly higher concentrations of low molecular weight iron species (MW < 25,000) were produced at pH 6 and 7 by the higher ratios of citrate:iron. These differences indicate that there are chemical reasons for the variability of the effect of citrate on iron bioavailability noted by Hallberg and Rossander (1984), Berner et al. (1985), and a long list of other investigators.

**TABLE 3.3**  Solubilizing Effect of Various Reagents

| | % Soluble mineral in the acid extract precipitated when pH was raised to 7 with $HCO_3^-$ | | | | |
|---|---|---|---|---|---|
| | Zn | Fe | Ca | Mg | Cu |
| EDTA | 0 | 0 | 0 | 19 | 0 |
| Citric acid | 6.9 | 0 | 37 | 16 | 0 |
| Histidine | 51 | 62 | 87 | 49 | 0 |
| Ascorbic acid | 98 | 71 | 87 | 45 | 0 |
| Lactose | 99 | 78 | 83 | 44 | 0 |
| $HCO_3^-$ (control) | 99.5 | 75 | 83 | 62 | 0 |

*Source: Lyon (1984).*

Warner and Weber (1953) examined the complexes of both ferric and cupric ions with citric acid as a function of pH. In each case they found a chelate with a metal ion:citrate ratio of 1:1 in which four protons were displaced from citric acid at pH 3 for iron and at pH 6–7 for copper. At pH values below these, other 1:1 complexes were present but with less than four protons displaced. Similar ratios were found by Spiro et al. (1967) with iron:citrate complexes. They were prompted to investigate this complex, since in another study they had demonstrated that the addition of base to a solution containing equimolar iron III and citrate at low pH led to the formation of an anionic monocitrate–iron chelate, postulated to be FeCit⁻. This compound then polymerized at higher pH values as more base was added. The product had a MW of 200,000 and was spherical with a diameter of 72A. They then found (Bates et al., 1967) that excess citrate accelerated the rate of incorporation of iron into transferrin. They concluded that the excess citrate must cause the formation of a low molecular weight species rather than the polymer and set out to show its existence in a series of experiments (Spiro et al., 1967). From this study, they postulated two competing reaction paths when hydroxide is added to solutions containing iron and citrate, beyond the point where the 1:1 chelate, FeCit⁻, is formed.

$$FeCit^- + 1.80H \rightarrow [(FeO)_{0.95}(OH)_{0.75}(Cit)_{0.15}]_{(polymeric)}^{-0.24} + 0.85(HCit)^{-3} + 0.1H_2O \qquad (5)$$

and

$$FeCit^- + HCit^{-3} + OH^- \rightarrow [FeCit_2]^{-5} + H_2O \qquad (6)$$

However, when the concentration of citrate ion, $HCit^{-3}$, is increased, the rate of reaction (6) increases at the expense of reaction (5) until, at about 0.02 M excess citrate, reaction (1) becomes negligible.

This means that in a meal where the products of reaction (5) are predominant, citrate might act as an inhibitor. This could be due to incomplete association of citrate and iron, low concentrations of citrate, high concentrations of iron, or the meal providing too great a buffering capacity so that neither the mono- nor the dicitrate complexes are allowed to form. However, under other circumstances it would act as an enhancer.

Chemically, it would also explain why Leigh and Miller (1983) found iron binding to be independent of citrate concentration, since it would include both $FeCit^{-1}$ and $FeCit_2^{-5}$, but not the formation of a low molecular weight species (MW < 25,000), which would be the $FeCit_2^{-5}$ only.

## The Meat Factor

Although a subsequent chapter will address iron protein interactions, I would like to mention briefly some work that casts some more light on our understanding of the meat factor. This factor enhances both heme and nonheme iron absorption and was first demonstrated in humans by Layrisse et al. (1968). Since that time, many studies have attempted to link protein concentration, quality, and specific amino acids without much success. Kane and Miller (1984) concluded that protein digestability by pepsin-pancreatin was not directly related to iron bioavailability in eight protein sources studied. When these sources were dialyzed after pepsin-pancreatin digestion, it was found that the small molecular weight (SMW) fraction (<6000–8000 MW) produced substantially more dialysate (soluble) iron with BSA and beef and slightly more with egg albumin than with a blank. SMW fractions of casein and soy did not enhance iron dialysis. Also mixing soy protein isolate with the SMW fractions in another study inhibited iron dialysis. Dialysate iron from the large molecular weight (LMW) fraction (>6000–8000 MW) was very low in all cases, and the complete fraction (SMW + LMW) resulted in increased iron dialysis in the presence of some proteins but not others.

In order to further investigate the fractions involved, several studies were undertaken in our laboratories. Bonner and Clydesdale (1988) examined the effects of fractionation, pepsin digestion products, pepsin-pancreatin digestion products, and the effects of pH when iron was added on iron solubility in chicken breast muscle. The solubility of added iron was significantly affected only by an acid-insoluble fraction and not by a water-soluble or acid-soluble extract. Samples were assayed for soluble iron during pepsin digestion, and solubility was found to increase linearly with time over a 4-hour period. However, when a 2-hour pepsin treatment was followed by a 4-hour pancreatin treatment, maximum iron solubilization was found after 1 hour. As can be seen in Fig. 3.6, iron solubilization increased pancreatin digestion up to 1 hour and then decreased with continuing time indicating that iron solubilization is due to a particular peptide fragment(s) and/or profile of amino acids which are not as effective if smaller or larger.

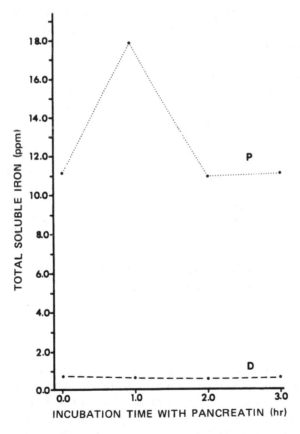

**FIG. 3.6** Effect of pancreatin digestion time following a 2-hr pepsin digestion on total soluble iron at pH 5 in an acid-insoluble protein extract of chicken muscle. (*From Bonner and Clydesdale, 1988.*)

Pepsin digestion products with MW < 10,000 solubilized significantly more iron than those with MW > 10,000. However, pepsin-pancreatin digestion products of MW < 10,000 were not as effective as those >10,000. This is in contrast to the results of Kane and Miller (1984) and is probably due to the fact that they added their iron and pancreatin together at pH 2, whereas we brought them together at pH 7.5 assuming that pepsin digestion products plus iron would only come into contact with pancreatin at pH values approximating those in the intestine. The effect of pH should be reiterated as in this case we found dramatic differences due to the pH at the time of addition of iron regardless of the later incubation pH (Table 3.4).

Ghia and Clydesdale (1988) continued these studies in an attempt to identify the molecular weight of the pepsin fractions that created the most effective iron-solubilizing pepsin-pancreatin end products. Gel chromotography on Sephadex G-25 indicated that the pepsin-pancreatin digestion products that solubilized the most iron were formed from pepsin digestion products in the molecular weight range 2550–6250. Since free sulfhydryl groups have been suggested as possible contributors to the enhancing effect of meat, we decided to evaluate the presence of free sulfhydryl groups (DTNB reagent) in the various fractions after pepsin digestion but both before and after pancreatin digestion and with and without added iron in the latter case. This provided information on the effects of both pancreatin digestion and added iron on free sulfhydryl groups. However, there were no significant differences in the content of free sulfhydryl groups between the three treatments in any of the protein fractions of interest, indicating that iron solubilization was not related to binding by free sulfhydryl groups.

Obviously, we have a way to go before the mechanism by which the meat factor enhances iron bioavailability is fully understood. However, the pieces are beginning to fit together and the puzzle is well on the way to being solved.

## Effects of Processing

It is obvious from the preceding discussion that there are many reactions that may effect the bioavailability of minerals both positively and negatively. These are just as likely, or more likely, to occur in a food as in a model system. Changes in pH, reduction potential, charge density, complexation, cleavage—all of these are likely to be affected in a food, and particularly so under the rigors of processing.

Many of the changes in bioavailability that iron undergoes during processing have been reviewed by Lee (1982). However, in most cases, the changes that occur can be ascribed to those factors discussed in this paper. For instance, Moeljopawiro et al. (1987) reported that fermenting soybeans increased the bioavailability of iron in rats. It is almost certain that the drop in pH reported played a large part in this increase as it would cause greater solubilization of iron and tend to create a reduction potential that favored the $Fe^{+2}$ form as well as ionization of both $Fe^{+2}$ and $Fe^{+3}$. Erdman has been involved in several studies, including that of Ketelsen et al. (1984), which found that neutralization of soy protein concentrates reduced zinc bioavailability to the rat. It has been suggested that this reduced bioavailability may be the result of the formation of stable protein–phytic acid–zinc complexes in the dried neutral product (Erdman et al., 1980).

Camire and Clydesdale (1981) found that, in general, increases in pH from 5 to 7 with wheat bran, cellulose, and lignin caused an increase in binding and decrease in solubility of Ca, Mg, Zn, and Fe. However, in a similar study with soy flour, the same effect was noted for Ca, Fe, and Zn, but the reverse occurred with Mg (Clydesdale and Camire, 1983). Similar insolubilization of minerals with rice bran as pH increased was found by Champagne et al. (1985a,b) along with other studies with different substrates too numerous to mention. Clearly, pH control is an important processing parameter to consider in terms of mineral chemistry and ultimate bioavailability.

When the modification of pH, reduction potential, and/or other environmental factors is not feasible, foods that are known to contain inhibitors of mineral absorption may also be modified by formulation or processing to increase bioavailability. We have already discusssed enhancers, such as ascorbate, meat, citric acid, and other ligands, which may be used in formulations to overcome the inhibition of mineral absorption by food, but many other combinations are available. Based on studies we conducted with milk with both iron in cereals (Clydesdale and Nadeau, 1984) and Fe, Zn, and Ca in model systems (Platt et al., 1987), we have proposed that there is a "milk factor" which facilitates the solubilization of minerals even in the presence of inhibitors such as fiber and sodium phytate. Clydesdale and Nadeau (1984) evaluated the ability of milk and several of its fractions to solubilize iron in three cereals. They found that the milk factor seemed to be in the protein fraction except in the case of the corn cereal. Platt et al. (1987) evaluated the effect of Na phytate on the solubility of calcium in milk versus $CaCl_2$ in the presence of Zn and Fe.

**TABLE 3.4** Effect on Total Soluble Iron of the Initial pH at Which Exogenous Iron Is Added to the Digestion Products Prior to Their Incubation at Intestinal pH (7.5)

| Initial pH | Enzymatic digestion total soluble iron (ppm)[a,b,c] after incubation at pH 7.5 | |
| --- | --- | --- |
|  | Pepsin | Pepsin-Pancreatin |
| 2.0 | 3.31 ± 0.10* | 17.81 ± 0.31* |
| 5.0 | 6.53 ± 0.06** | 10.61 ± 0.28** |
| 7.5 | 4.93 ± 0.12*** | 18.73 ± 0.22* |

[a]25 ppm iron were added as $FeCl_3$ at the pH shown after either a 2-hr pepsin or 2-hr pepsin–1-hr pancreatin digestion, then incubated at pH 7.5.
[b]Values represent x ± SEM.
[c]Means within a column followed by different superscripts are significantly different ($p <$ 0.05).
*Source: Bonner and Clydesdale (1988).*

The results are shown in Table 3.5, where it can be seen that a milk factor does exist, which contributes to mineral solubility even at the unfavorable pH values shown, which are all above 6.

Even added protein can be significant as Ranhotra et al. (1983) have shown in iron-enriched bread, where both milk and soy supplementation increased the bioavailability of iron in rats over gluten in supplemented bread. Albright et al. (1984) found that the Fe–phosvitin complex in egg yolk did not release iron when heated, but EDTA was able to solubilize it. Therefore, formulation may be more important than the physical process in some cases.

At times, changes in formulation seem to make little difference, even when theoretical premises indicate they should. One such example involves fructose, thought to increase iron absorption by virtue of its iron-chelating properties and the apparent ability of the ferric–fructose complex to cross the intestinal mucosa. However, Nadeau and Clydesdale (1987a) have compared the iron-solubilizing effect of sucrose, fructose, and aspartame under a simulated digestion pH procedure in a wheat flake cereal and found little effect, regardless of pH, iron, or sweetener source. It was speculated that theoretical effects may have been masked by other cereal components.

Physiological conditions must also be considered in terms of bond breakage and the release of minerals from inhibitor complexes. Mod et al. (1981) showed that digestive enzymes (hemicellulase, pepsin, and trypsin) released copper, iron, and zinc from rice hemicellulose to varying degrees during in vitro incubation, indicating that it may not be a serious inhibitor of absorption.

Physical processes often release minerals from bound complexes and, therefore, should be considered when attempting to maximize bioavailability. Clydesdale and Camire (1983) showed that boiling caused a significant increase in the binding of Zn and Mg to soy flour at pH 5.0 and 6.8 but was pH-dependent for Fe and Ca. Toasting increased the binding of both Zn and Ca at both pH values but was pH-variable with Fe and Mg. With cellulose, toasting had no effect on metal binding but did with lignin and wheat bran, while boiling had an effect with all of these potentially inhibitory compounds (Camire and Clydesdale, 1981). The effects of boiling can be corroborated by the results of an entirely different study by Muldoon and Liska (1970), who found that a decrease in ionized Ca occurred when milk was heated, with the effect being proportional to the intensity of heat. There are many other examples of processing effects on potential bioavailability too numerous to discuss here. However, it should be stressed that the optimization capabilities of technology should be utilized more often to increase mineral bioavailability. After all, the aim of technology is to maximize factors like availability and nutritional value while minimizing risk and cost.

## Mineral–Mineral Reactions

**Potential effects.**  The interaction of minerals in the intestine and the physiological consequences of competition at mucosal absorption sites are well known and are discussed in some detail in Chapter 2. However, the interaction of minerals with a food system also has potential consequences that are less studied. The complexity of both natural and formulated foods creates an environment where minerals may compete for binding sites on the ligands present, thus influencing their bioavailability when eaten. Unfortunately, this same complexity has required that most studies be carried out in model systems that often use only one mineral. However, more recent investigations are taking such mineral competiton into consideration, and the results indicate that the effects may be potentially dramatic.

**Mechanisms.**  The mechanisms involved in mineral–mineral interactions that have the potential to affect bioavailability may be broadly defined in four categories:

1. displacement of a mineral from a complex with another mineral to form a soluble (available) or insoluble (unavailable) complex

2. the addition of a second or third mineral to a soluble mineral-ligand complex causing precipitation

3. the addition of a mineral causing a mineral–ligand complex to bind to another substrate (ligand) and form a poly mineral–poly ligand complex

4. the formation of a poly mineral–ligand complex, which changes the susceptibility of the mineral–ligand bonds to cleavage by digestive enzymes

The basic thermodynamic characteristics involved in mineral displacement have been mentioned before. They are the specificity of binding sites, the stability constants, and the amount of mineral bound by each site as was shown for some selected combinations in Table 3.1 (Platt and Clydesdale, 1987a).

The second and third mechanisms involving the formation of an insoluble poly mineral–ligand or poly ligand complex are exemplified by various mineral reactions with ligands, including phytic acid. I would like to mention a few examples involving this compound. Mills (1985) has written an excellent review on dietary interactions in which he discusses at some length the need for the presence of both Ca and Zn in order for phytate to act as an antagonist of Zn absorption by forming a poly mineral–ligand complex. Also, he points out that while the (Ca) (phytate):(Zn) ratio might be better than the phytate:Zn ratio as a predictor of bioavailability, both have limitations in a mixed diet that includes amino acids. There is evidence that dietary protein increases Zn bioavailability in the presence of phytate, and this may be due to the desorbtion of Zn from phytate complexes by histidine, cysteine, and methionine, which also desorb other minerals in the order: Cu > Cd > Mn > Zn > Pb.

Fordyce et al. (1987) have also shown the efficacy of the (Ca) (phytate):(Zn) ratio but found in soy products that its predictive value was reduced due to processing, since the bioavailability of Zn in neutralized soy products is lower than that of acid-neutralized products, as discussed previously. Therefore, both dietary components and processing affect mineral interactions as would be expected. The complexity of the phytic acid systems has been further demonstrated by Champagne et al. (1985b), who found that the solubility of potassium, magnesium, and calcium phytates was affected by pH but not by the additon of $Fe^{+2}$, $Fe^{+3}$, $Zn^{+2}$, or $Cu^{+2}$.

The effect of a poly mineral, poly ligand, or mineral–poly ligand complex can be seen clearly in Table 3.5 (Platt et al., 1987). In the cases where a mineral–ligand (phytate and Ca) complex existed, Ca was insoluble, but when milk was added to provide the possibility of formation of a mineral–poly ligand or poly mineral–poly ligand complex, it became soluble. Similarly when poly mineral–ligand complexes (Ca and Fe or Ca and Zn, with Na phytate, respectively) were allowed the opportunity to form mineral or poly mineral–poly ligand complexes with milk, the solubility increased.

Another study by Platt and Clydesdale (1987b) compared the effects of Na phytate on a single mineral or combination of minerals (mineral or poly mineral–ligand complex) versus the effect of a phytate-rich insoluble fiber-rich (PRFR) isolate on the same mineral(s). Fe formed a soluble complex with Na phytate, but it became 70% insoluble in the presence of the PRFR isolate, indicating either an insoluble fiber complex or a mineral–poly ligand (phytate–fiber) complex. A similar solubility pattern was followed when both Fe and Cu were used, but when Fe and Zn were added to the PRFR isolate there was a decrease in soluble phytic acid, phosphorus, endogenous Ca, and Fe but little change when they were added to Na phytate. This suggests that Zn has a greater tendency than Cu to form insoluble poly mineral– poly ligand complexes and explains some of the results from clinical studies that have used diets that would allow such a complex to form.

When Ca and Fe were added, there was a significant decrease in solubilities with both Na phytate and PRFR isolate. The synergy noted here is in agreement with other studies involving Ca and other minerals in the presence of phytates and the complexes discussed by Mills (1985). Of course the protein–phytic acid–zinc complexes postulated by Erdman et al. (1980) are also examples of (poly)mineral–(poly)ligand complexes.

The last mechanistic category involves mineral binding, which affects enzyme resistance. Phytase is perhaps the most studied enzyme in terms of mineral bioavailability, but it does not seem to attack bound phytic acid, although I am unaware of any studies which specifically examine this.

Mod et al. (1981), however, have found that a Cu–Zn–water-soluble rice hemicellulose, when treated with hemicellulase, did not release Cu but did release 69% of the zinc, which they postulated would then be available for absorption. On the other hand, they found that treatment with trypsin released 20% of the Cu and 72% of the Zn. Interestingly, different results were found with alkali-soluble hemicellulose, which shows that processing would have an effect on the substrate as well as the enzyme in this mechanism.

**TABLE 3.5** Effects of Na Phytate on Mineral Solubility in (I) milk and (II) CaCl$_2$ Solutions, Alone and in the Presence of Fe or Zn

| System | Final pH | Total (mmol)[b] | | | Phytate:Ca:Fe:Zn molar ratios | Percent soluble[b] | | |
|---|---|---|---|---|---|---|---|---|
| | | Ca | Fe | Zn | | Ca | Fe | Zn |
| 1. Milk (control) | 6.65 | 2.88 | — | — | | 97.3 | — | — |
| 2. (I) Milk + Na phytate[a] | 6.35 | 2.83 | — | — | 3:6:0:0 | 99.0 | — | — |
| (II) CaCl$_2$ + Na phytate | 6.40 | 2.77 | — | — | 3:6:0:0 | 33.9* | — | — |
| 3. (I) Milk + Fe + Na phytate | 6.16 | 2.91 | 0.51 | — | 3:6:1:0 | 102.2 | 100.7 | — |
| (II) CaCl$_2$ + Fe + Na phytate | 6.17 | 2.85 | 0.50 | — | 3:6:1:0 | 29.8* | 55.4* | — |
| 4. (I) Milk + Zn + Na phytate | 6.21 | 2.96 | — | 0.51 | 3:6:0:1 | 97.7 | — | 93.3 |
| (II) CaCl$_2$ + Zn + Na phytate | 6.23 | 2.84 | — | 0.51 | 3:6:0:1 | 19.0* | — | 50.2* |

[a]1.44 mmol of Na phytate was added to each of the systems.
[b]The symbol (—) means that the levels were below the detection level using AAS.
*Significantly different $p < 0.01$ between groups I and II within each system.
Source: Platt et al. (1987).

**Competition for binding sites.**   The discussion in this section could have been included in the last section under the first-proposed mechanism involving displacement of a mineral by another mineral from a binding site on a component of food. However, such a displacement might occur only when there was a common binding site. For this reason and because this is a relatively recent consideration in mineral bioavailability, it was decided to discuss this phenomenon in a separate section.

The competition of minerals for binding sites on a particular ligand is dependent on several factors, including their effective stability constants ($K_{eff}$). Platt and Clydesdale (1987a) calculated the specifity, $K_{eff}$, and number of binding sites via Scatchard analysis for several combinations of ligands and Fe as well as for the combination of lignin with Fe, Cu, and Zn, as shown in Table 3.1. In this study iron was found to have only nonspecific (i.e., very weak) binding with cellulose, which indicates that cellulose would not be of great concern as an inhibitor of iron in a mixed diet since many other compounds, including those in Table 3.1, have at least one specific (i.e., stronger) binding site for cellulose. Guar gum, NDF, and high methoxy pectin all had both a specific and nonspecific binding site for iron. Lee and Garcia-Lopez (1985) found two sites for Zn and Cu on pinto bean NDF but only one for Fe. However, they used only four points on the Scatchard plot, and the very weak second nonspecific binding site we found would not be evident since it took ten points in our work to show a clear break in the binding line, thus indicating two sites. We also found (Platt and Clydesdale, 1987a) that iron had one specific site with low methoxy pectin and two with lignin, the latter finding confirming previous work by Fernandez and Phillips (1982) and Platt and Clydesdale (1984). Cu and Zn were also found to have two specific binding sites with lignin.

If we were dealing with chemically defined systems in controlled standard environments, it would be possible to predict the outcome of adding two or more minerals to a particular compound or food system simply by knowing $K_{eff}$. However, this is not the case, and the mineral–mineral competition that exists with food components is not that predictable. For instance, when Fe and Cu were added to NDF, the amount of iron bound was decreased, thus suggesting a common binding site (GArcia-Lopez and Lee, 1985). However, when Fe and Cu were added to wheat bran systems (Platt and Clydesdale, 1986) or to phytate-rich fiber-rich fractions of wheat bran (Platt and Clydesdale, 1987b), there was no effect on solubility. This was likely due to the formation of soluble phytic acid poly mineral complexes in the latter study. Garcia-Lopez and Lee (1985) also found that zinc decreased the amount of Fe bound to NDF but was not effected by cobalt or magnesium. With ADF they found that Cu increased the Fe bound, while magnesium and zinc had no effect. Again Platt and Clydesdale (1986) found conflicting results with wheat bran and Zn and also found that Ca decreased iron solubility, which might be expected with the increased complexity of wheat bran.

Platt and Clydesdale (1987b) found some very interesting results in mineral–lignin systems when the order of addition of the minerals was changed. When Zn, Mg, or Ca was added at a concentration equivalent to the RDA to a lignin sample that contained 9 mg iron, there was an increase in soluble iron of 12%, 21%, and 22%, respectively. However, when these same levels were used but the order of addition changed (i.e., the iron added last) there was a greater increase in soluble iron of 21%, 29%, and 33%, respectively. This means that each of Ca, Mg, and Zn was able to dislodge the iron from a site on the lignin molecule, thus increasing its solubility. Although we do not have data indicating the $K_{eff}$ with lignin for Ca and Mg, it is interesting that Zn dislodged the iron since one of its binding sites with lignin is weaker than either of iron's and its other site is stronger than only one of iron's (Table 3.1). It is difficult to postulate whether one or both sites are common to both Zn and Fe, but it would seem awfully coincidental if only the site where Zn had a greater $K_{eff}$ was a common one. Another explanation may be that certain binding sites of lignin are irreversibly bound by minerals, while other sites may be reversibly bound and thus susceptible to displacement by other minerals.

These results also indicate that the order of addition of minerals to a food may have some significance in its later bioavailabiity, as the amount of soluble Fe was dependent upon this parameter.

This type of cationic displacement and consequent increased solubility may help to explain controversies that arise in human and animal studies where the levels of all minerals, other than the one under study, are not constant between studies or within different diets of the same study. Such an occurrence could produce differential solubility and bioavailability of the mineral under study.

Although not involving binding sites, another interesting example of mineral–mineral interactions was the report of a conditioned Cu deficiency due to the ingestion of large amounts of the oxides of bismuth, aluminum, silica, magnesium, and sodium in the form of antacid tablets. Cupric salts are precipitated by alkali oxides at alkaline pH, and apparently the large amount of alkali oxides taken by this patient along with a decreased gastric emptying time reduced the availability of dietary Cu to essentially nil (Anon., 1984).

## CONCLUSIONS

There is no need to reiterate the various chemical interactions that affect bioavailability. It should be obvious that there are many options open to the food processor where steps might be taken to increase the bioavailability not only of endogenous minerals but of those present due to fortification. Enhancers, inhibitors, pH, processing procedures, and even the concentration and order of addition of minerals are all considerations in formulating a food product with optimal mineral nutrition. The day should be over when a mineral or its salt is added to a food simply because it doesn't affect quality and is somewhat bioavailable. Our knowledge of mineral interactions is becoming such that a nutrient package can be created for a specific food and process. Thus calcium addition might consist of the addition of not only calcium but enhancers in the appropriate molar ratio to optimize its bioavailability in that system.

As our knowledge base increases, we will learn the chemistry of the meat factor, the milk factor, and why some carbohydrates increase Ca absorption. Such knowledge will allow the creation of more and better mineral nutrient packages to offer to consumers so that real nutritional choices are available in the marketplace.

## REFERENCES

Albert, A. 1950. Quantitative studies of the avidity of naturally occurring substances for trace metals. *Biochem. J.* 47: 531.

Alberts, B., Bray, D., Lewis, J., Raff, M., Roberts, K., and Watson, J. D. 1983. *Molecular Biology of the Cell.* Garland Publ. Inc., New York.

Albright, K. L., Gordon, D. T., and Cotterill, O. J. 1984. Release of iron from phosvitin by heat and food additives. *J. Food Sci.* 49: 78.

Angyal, S. J. 1973. Complex formation between sugars and metal ions. *Pure & Applied Chem.* 35: 131.

Anon. 1984. Conditioned copper deficiency due to antacids. *Nutr. Rev.* 42: 319.

Anon. 1985. Glucose polymers enhance the intestinal absorption of calcium. *Nutr. Rev.* 43: 201.

Bates, G. W., Billups, C., and Saltman, P. 1967. The kinetics and mechanism of iron (III) exchange between chelates and transferrin. I. The complexes of citric and nitrilofriacetic acid. *J. Biol. Chem.* 242: 2810.

Bates, G. W., Cuccaro, P., Murphy, C., Conception-Calvo, M., Chanona, B., Bourges, H., and Moisterrena, J. 1983. In vitro studies of iron bioavailability with emphasis on the Mexican and Mexican-American diet. *Fed. Proc.* 42: 1068.

Beal, L. and Mehta, T. 1985. Zinc and phytate distribution in peas. Influence of heat treatment, germination, pH, substrate, and phosphorus on pea phytate and phosphorus. *J. Food Sci.* 50: 96.

Bei, L., Wood, R., and Rosenberg, I. H. 1986. Glucose polymer increases jejunol calcium, magnesium, and zinc absorption in humans. *Am. J. Clin. Nutr.* 44: 244.

Berner, L.A. and Hood, L. F. 1983. Iron binding by sodium alginate. *J. Food Sci.* 48: 755.

Berner, L. A., Miller, D. D., and VanCampen, D. 1985. Availability to rats of iron in ferric hydroxide polymers. *J. Nutr.* 115: 1042.

Bonner, C. A. and Clydesdale, F. M. 1987. The effects of proteolytic digestion products of chicken breast muscle on iron solubility. *Am. J. Clin. Nutr.* (in press).

Bothwell, T. H., Clydesdale, F. M., Cook, J. D., Dallman, P. R., Hallberg, L., Van Campen, D., and Wolf, W. J. 1982. *The Effects of Cereals and Legumes on Iron Availability.* Intntl. Nutr. Anemia Cons. Group. The Nutrition Foundation, Washington, DC.

Buth, M. and Mehta, T. 1983. Effect of psyllium husk on iron bioavailability in monkeys. *Nutr. Rep. Intl.* 28: 743.

Camire, A. L. and Clydesdale, F. M. 1981. Effect of pH and heat treatment on the binding of calcium, magnesium, zinc, and iron to wheat bran and fractions of dietary fiber. *J. Food Sci.* 46: 548.

Chaberek, S. and Martell, A. E. 1959. *Organic Sequestering Agents.* John Wiley and Sons, Inc., New York.

Champagne, E. T., Rao, R. M., Liuzzo, J. A., Robinson, J. W., Gale, R. J., and Miller, F. 1985a. Solubility behaviours of the minerals, proteins, and phytic acid in rice bran with time, temperature, and pH. *Cereal Chem.* 62: 218.

Champagne, E. T., Rao, R. M., Liuzzo, J. A., Robinson, J. W., Gale, R. J., and Miller, F. 1985b. The interactions of minerals, proteins, and phytic acid in rice bran. *Cereal Chem.* 62: 231.

Clydesdale, F. M. 1982. The effects of physicochemical properties of food on the chemical status of iron. In *Nutritional Bioavailability of Iron.* Kies, C. (Ed.), p. 55. ACS Symposium Series 203, Washington, DC.

Clydesdale, F. M. 1983. Physicochemical determinants of iron bioavailability. *Food Technol.* 37(10): 133.

Clydesdale, F. M. 1988. Minerals: their chemistry and fate in food. In *Handbook of Trace Minerals in Foods: Their Relationship to Health and Nutrition.* Smith, K. T. (Ed.). Marcel Dekker, Inc., New York.

Clydesdale, F. M. and Camire, A. L. 1983. Effect of pH and heat on the binding of iron, calcium, magnesium, and zinc and the loss of phytic acid in soy flour. *J. Food Sci.* 48: 1272.

Clydesdale, F. M. and Nadeau D. B. 1984. Solubilization of iron in cereals by milk and milk fractions. *Cereal Chem.* 61: 330.

Clydesdale, F. M. and Nadeau, D. B. 1985. The effect of acid pretreatment on the stability of ascorbic acid iron complexes with various iron sources in a wheat flake cereal. *J. Food Sci.* 50: 1342.

Conrad, M. E. and Schade, S. G. 1968. Ascorbic acid chelates in iron absorption: A role for hydrochloric acid and bile. *Gastroenterology* 55: 35.

Derman, D., Sayers, M., Lynch, S. R., Charlton, R. W., Bothwell, T. H., and Mayet, F. 1977. Iron absorption from a cereal-based meal containing cane sugar fortified with ascorbic acid. *Brit. J. Nutr.* 38:261.

Emery, T. 1982. Iron metabolism in humans and plants. *Am. Scientist.* 70: 626.

Erdman, J. W. Jr., Weingartner, K. E., Mustakas, G. C., Schnutz, R. D., Parker, H. M., and Forbes, R. M. 1980. Zinc and magnesium bioavailability from acid-precipitated and neutralized soybean protein products. *J. Food Sci.* 45: 1193.

Eyerman, L. S., Clydesdale, F. M., Huguenin, R., and Zajicek, O. T. 1987. Characterization of solution properties of four iron sources in model systems by solubility studies and IR/VIS reflectance spectrophotometry. *J. Food Sci.* 52: 197.

Fernandez, R. and Phillips, S. F. 1982. Components of fiber bind iron in vitro. *Am. J. Clin. Nutr.* 35: 100.

Flynn, S. M., Clydesdale, F. M., and Zajicek, O. T. 1984. Complexation, stability and behaviour of 1-cysteine and 1-lysine with different iron sources. *J. Food Prot.* 47: 36.

Fordyce, E. J., Forbes, R. M., Robbins, K. R., and Erdman, J. W. Jr. 1987. Phytate × calcium/zinc molar ratios: are they predictive of zinc bioavailability. *J. Food Sci.* 52: 440.

Forth, W. and Rummel, W. 1973. Iron absorption. *Phys. Rev.* 53: 724.

Garcia-Lopez, J. S. and Lee, K. 1985. Iron binding by fiber is influenced by competing minerals. *J. Food Sci.* 50: 424.

Gheith, M. A. 1952. Differential thermal analysis of certain iron oxides and oxide hydrates. *Am. J. Sci.* 250: 677.

Ghia, M. L. and Clydesdale, F. M. 1987. Effect of enzymatic digestion, pH, and molecular weight on the iron solubilizing properties of chicken muscle. *J. Food Sci.* (In press).

Gilooly, M., Bothwell, T. H., Charlton, R. W., Torrance, J. O., Bezwoda, W. R., MacPhail, A. P., Derman, D. P., Novelli, L., Morral, P., and Mayet, F. 1984. Factors affecting the absorption of iron from cereals. *Br. J. Nutr.* 51: 37.

Glicksman, M. 1982. Food application of gums. In *Food Carbohydrates,* Lineback, D. R. and Inglett, G. E. (Ed.), p. 270. AVI Publ. Co., Westport, CT.

Gorman, J. E. and Clydesdale, F. M. 1983. The behavior and stability of iron-ascorbate complexes in solution. *J. Food Sci.* 48: 1217.

Gorman, J. E. and Clydesdale, F. M. 1984. Thermodynamic and kinetic stability constants of selected carboxylic acids and iron. *J. Food Sci.* 49: 500.

Grant, G. T., Morris, E. R., Rees, D. A., Smith, P. J. C., and Thom, D. 1973. Biological interactions between polysaccharides and divalent cations. The egg-box model. *FEBS Lett.* 32: 195.

Hallberg, L. and Rossander, L. 1984. Improvement of iron nutrition in developing countries: comparison of adding meat, soy protein, ascorbic acid, citric acid, and ferrous sulphate on iron absorption from a simple Latin American-type of meal. *Am. J. Clin. Nutr.* 39: 577.

Heaney, R. P. 1986. Calcium bioavailability. *Contemporary Nutr.* 11: 8. (General Mills Inc., Minneapolis, MN.)

Hurley, L. S. and Lonnerdal, B. 1982. Zinc binding in human milk: citrate versus picolinate. *Nutr. Rev.* 40: 65.

Kane, A. P. and Miller, D. D. 1984. In vitro estimation of the effects of selected proteins on iron bioavailability. *Am. J. Clin. Nutr.* 39: 393.

Ketelsen, S. M., Stuart, M. A., Weaver, C. M., Forbes, R. M., and Erdman, J. W., Jr. 1984. Bioavailability of zinc to rats from defatted soy flour, acid-precipitated soy concentrate and neutralized soy concentrate as determined by intrinsic and extrinsic labelling techniques. *J. Nutr.* 114: 536.

Kelly, S. E., Chawla-Singh, K., Sellin, J. H., Yasillo, N. J., and Rosenber, I. H. 1984. *Gastroenterology* 87: 596.

Kohn, R. 1975. Ion binding on polyuranates-alginate and pectin. *Pure Applied Chem.* 42: 371.

Kojima, N., Wallace, D., and Bates, G. W. 1981. The effect of chemical agents, beverages and spinach on the in vitro solubilization of iron from cooked pinto beans. *Am. J. Clin. Nutr.* 34: 1392.

Kratzer, F. H. and Vohra, P. 1986. *Chelates in Nutrition.* CRC Press Inc., Boca Raton, FL.

Layrisse, M., Martinez-Torres, C., and Roche, M. 1968. Effect of various foods on iron absorption. *Am. J. Clin. Nutr.* 21: 1175.

Lee, K. 1982. Iron chemistry and bioavailability in food processing. In *Nutritional Bioavailability of Iron,* Kies, C. (Ed.), p. 27. Am. Chem. Soc. Symp. Series No. 203. Washington, DC.

Lee, K. and Garcia-Lopez, J. S. 1985. Iron, zinc, copper and magnesium binding by cooked pinto bean (phaseolus vulgaris) neutral and acid detergent fiber. *J. Food Sci.* 50: 651.

Leigh, M. J. and Miller, D. D. 1983. Effects of pH and chelating agents on iron binding by dietary fiber; implications for iron bioavailability. *Am. J. Clin. Nutr.* 38: 202.

Lyon, D. B. 1984. Studies on the solubility of Ca, Mg, Zn, and Cu in cereal products. *Am. J. Clin. Nutr.* 39: 190.

Martell, A. E. and Calvin, M. 1952. *Chemistry of the Metal Chelate Compounds.* Prentice Hall, Inc., New York.

Miller, D. D., Schriker, B. R., Rassmussen, R. R., and Van Campen, D. R. 1981. An in vitro method for estimation of iron availability from meals. *Am. J. Clin. Nutr.* 34: 2248.

Mills, C. F. 1985. Dietary interactions involving the trace elements. In *Annual Review of Nutrition,* Vol 5, Olson, R. E., Beutler, E., and Broquist, H. P. (Ed.), p. 173. Annual Reviews Inc., Palo Alto, CA.

Mod, R. R., Ory, R. L., Morris, N. M., and Normand, F. L. 1981. Chemical properties and interactions of rice hemicellulose with trace minerals in vitro. *J. Agric. Food Chem.* 29: 449.

Moeljopawiro, S., Gordon, D. T., and Fields, M. L. 1987. Bioavailability of iron in fermented soybeans. *J. Food Sci.* 52: 102.

Monnier, L., Colette, C., Aquirre, L., and Mirouze, J. 1980. Evidence and mechanism for pectin-reduced intestinal inorganic iron absorption in idiopathic hemachromatosis. *Am. J. Clin. Nutr.* 33: 1225.

Morck, T. A., Lynch, S. R., and Cook, J. D. 1982. Reduction of the soy induced inhibition of nonheme iron absorption. *Am. J. Clin. Nutr.* 36: 219.

Motzok, I., Ramesh, S. V., Chen, S. S., Rasper, J., Hancock, R. G. V., and Ross, H. U. 1978. Bioavailability, in vitro solubility, and physical and chemical properties of elemental iron powders. *J. Assoc. Off. Anal. Chem.* 61: 887.

Muldoon, P. J. and Liska, B. J. 1970. Effects of heat treatment and subsequent storage on the concentration of ionized calcium in skim milk. *J. Dietary Sci.* 5: 35.

Nadeau, D. B. and Clydesdale, F. M. 1986. Effect of acid pretreatment on the stability of citric and malic acid complexes with various iron sources in a wheat flake cereal. *J. Food Biochem.* 10: 241.

Nadeau, D. B. and Clydesdale, F. M. 1987a. Effect of sucrose, fructose and aspartame on fortificant iron solubility in a wheat flake cereal. *J. Food Prot.* 50: 21.

Nadeau, D. B. and Clydesdale, F. M. 1987b. Effect of acid pretreatment on the stability of EDTA, cysteine, lactic and succinic acid complexes of various iron sources in a wheat flake cereal. *J. Food Prot.* 50: 587.

Neuman, W. F. and Neuman, H. W. 1958. *The Chemical Dynamics of Bone Mineral.* Univ. Press., Chicago, IL.

Nicar, M. J. and Pak, C. Y. 1985. Calcium bioavailability from calcium carbonate and calcium citrate. *J. Clin. End. Met.* 61: 391.

Nojeim, S. J. , Clydesdale, F. M., and Zajicek, O.T. 1981. Effect of redox potential on iron valence in model systems and foods. *J. Food Sci.* 46: 1265.

Park, Y. W. Mahoney, A. W., and Hendricks, D. G. 1983. Bioavailability of different sources of ferrous sulfate iron fed to anemic rats. *J. Nutr.* 113: 2223.

Perrin, D. D. 1958. The stability of complexes of ferric ion and glycine. *J. Chem. Soc.* 290: 3125.

Platt, S. R. and Clydesdale, F. M. 1984. Binding of iron by cellulose, lignin, sodium phytate and beta-glucan, alone and in combination, under simulated gastrointestinal pH conditions. *J. Food Sci.* 49: 531.

Platt, S. R. and Clydesdale, F. M. 1986. Effects of iron alone and in com-with calcium, zinc, and copper on the mineral-binding capacity of wheat bran. *J. Food Prot.* 49: 37.

Platt, S. R. and Clydesdale, F. M. 1987a. Mineral binding characteristics of lignin, guar gum, cellulose, pectin, and NDF under simulated duodenal pH conditions. *J. Food Sci.* 52: 1414.

Platt, S. R. and Clydesdale, F. M. 1987b. Interactions of Fe alone and in combination with Ca, Zn, and Cu with a phytate-rich, fiber-rich fraction of wheat bran under gastrointestinal pH conditions. *Cereal Chem.* 64(2): 102.

Platt, S. R., Nadeau, D. B., Gifford, S. R., and Clydesdale, F. M. 1987. Protective effect of milk on mineral precipitation by Na phytate. *J. Food Sci.* 51: 240.

Ranhotra, G., Gelroth, J., Novak, F., Bock, A., and Bohannon, F. 1983. Iron-enriched bread: Interaction effect of protein quality and copper on iron bioavailability. *J. Food Sci.* 48: 1426.

Rao, N. and Prabhavthi, T. 1978. An in vitro method for predicting bioavailability of iron from foods. *Am. J. Clin. Nutr.* 31: 169.

Recker, R. R. 1985. Calcium absorption and achlorhydria. *New England J. Med.* 313(2): 70.

Rizk, S. W. and Clydesdale, F. M. 1984. The effects of ascorbic acid, pH, and exogenous iron on the chemical if a soy protein concentrate. *J. Food Biochem.* 8: 91.

Rizk, S. W. and Clydesdale, F. M. 1985a. The effects of baking and boiling on the ability of selected organic acids to solubilize iron from a corn-soy-milk food blend fortified with exogenous iron sources. *J. Food Sci.* 50: 1088.

Rizk, S. W. and Clydesdale, F. M. 1985b. Effects of organic acids in the in vitro solubilization of iron from a soy-extended meat patty. *J. Food Sci.* 50: 577.

Somer, B. A., Margerum, D.W., Renner, J., Saltman, P., and Spiro, T. G. 1973. Reactivity and aging in hydroxy iron (III) polymers, analogs of ferritin cores. *Bioinorg. Chem.* 2: 295.

Spiro, T. G., Bates, G., and Saltman, P. 1967. The hydrolytic polymerization of ferric citrate. II. The influence of excess citrate. *J. Am. Chem. Soc.* 89: 5555.

Spiro, G. and Saltman, P. 1969. Polynuclear complexes of iron and their biological considerations. In *Structure and Bonding,* Hemmerich, P., Jorgensen, C. K., Nielands, J. B., Nyholm, Sir Ronald S., Reinen, D., and Williams, R. J. P. (Ed.), 6: 116. Springer-Verlag, New York.

Thompson, S. A. and Weber, C. W. 1979. Influence of pH on the binding of copper, zinc and iron in six fiber sources. *J. Food Sci.* 44: 753.

Vohra, P. and Kratzer, F. H. 1964. Influence of various chelating agents on the availability of zinc. *J. Nutr.* 82: 249.

Warner, R. C. and Weber, I. 1953. The cupric and ferric citrate complexes. *J. Am. Chem. Soc.* 75: 5086.

Wood, R. J., Gerhardt, A., and Rosenberg, I. 1987. *Am. J. Clin. Nutr.* 46: 699.

# 4

# Physiological Interaction of Minerals

**Noel W. Solomons**

Center for Studies of Sensory Impairment, Aging and
    Metabolism (CeSSIAM)
Guatemalan National Committee for the Blind and Deaf
Hospital de Ojos y Oidos
Guatemala City, Guatemala

## INTRODUCTION

There is no doubt that mineral–mineral* interactions in living organisms
have been important in physiology since the dawn of creation, as interactions among elements are noted in primitive bacterial systems. It has only
been in the last two decades, however, that analytical capabilities in the
laboratory and conceptual notions in biology have combined to allow for
scientific exploration of mineral–mineral interaction in nature.

---

*The term "mineral" is used somewhat imprecisely as a comprehensive term for
biologically relevant *elements.* In the strictest (metallurgical) sense, a mineral is
any inorganic substance found in nature, such as metallic ore. Some of the inorganic species discussed here are indeed metals but others, such as selenium, are
amphoteric. We should bear in mind the license being exercised in our use of the
term "mineral."

Only a limited amount of data are available related to human medicine and biology. The majority of observations have been made in in vitro systems or in laboratory animals or domestic livestock. It is the intention of this review to focus our attention on the health issues of physiological interactions of minerals in men and women. In a few recent publications some authors have begun to synthesize the concepts in this field (Levander and Cheng, 1980; Mills, 1985; Solomons, 1986a; Spencer, 1986). Herein, we shall move through an update of our understanding today, using human data where available and extrapolating from in vitro and animal experimentation when necessary. As appropriate, we shall project toward the critical questions that remain to be addressed in the immediate future.

## The Hill and Matrone Thesis

The recognized authors of the conceptual thesis that crystallized and galvanized thinking regarding the interaction of chemically similar inorganic elements in biological systems were Charles H. Hill and Gennard Matrone (Hill and Matrone, 1970). To quote from that treatise, "The thesis may be stated simply as, 'those elements whose physical and chemical properties are similar will act antagonistically to each other biologically.'" Given that the physical and chemical properties of elements depended on their electronic structure, Hill and Matrone relied on the electronic configuration as the predictive paradigm.

For instance, Hill and Matrone (1970) noted that certain valence states of selenium ($Se^{4+}$), arsenic ($As^{3+}$), tin ($Sn^{2+}$), and tellurium ($Te^{4+}$) had a similar outer-shell electron configuration in terms of filling of orbitals and electron-spin complementarity. The former two elements have two electrons in the 4s orbital with the 4p and 4d orbitals empty, whereas the latter two have the 5s orbital filled and the 5p and 5d orbitals vacant. Thus, they *predicted* that mutual interactions would exist; then, they proceeded to *demonstrate* this in a chick model, showing that arsenic, tin, and tellurium would inhibit the intestinal absorption of $^{75}Se$. Similarly, noting that both vanadate and chromate had the 3p orbitals filled with electrons, they predicted and demonstrated competition, this time at the level of a physiological suspension of liver mitochondria. They showed that the toxic uncoupling of oxidative phosphorylation by $V^{5+}$ was blocked by progressive addition of $Cr^{6+}$.

## Types of Mineral–Mineral Interactions

Time has shown that the Hill and Matrone theorem need not be adhered to in strict orthodoxy. These original proposers, in fact, found an exception to that rule in *non*interaction of mercury with either zinc or copper, species with comparable electron configuration (Hill and Matrone, 1970). Moreover, biological interactions among similarly sized or similarly charged ions—without identical electron configuration—have been widely confirmed (Pollack et al., 1965). Finally, not only *antagonist,* but also *synergistic* and *combined, multiple* interactions, can be described between or among minerals in mammalian systems.

**Antagonistic interactions.**   As noted by Hill and Matrone (1970), the majority of interactions between chemically similar ions tend to be antagonistic, with the presence of one reducing the movement or biological efficacy of the other. They point out: "In the studies reported here on the effect of arsenic, tellurium, and tin on the absorption of selenium, as well as those studies on the mitochondrial uptake of various anions, the site of biological interaction appears to be at the level of ion penetration of membranes." Most, but not all, antagonistic mineral–mineral interactions involve competition at a membrane.

When true *competition* is the mechanism, the interaction should respond to the relative proportions of the species. That is, the degree of inhibition of Species A by Species B can be modified either by increasing the concentration of A or by decreasing that of B. Under some circumstances, however, a certain dose of Species B will have a *toxic* effect on the response, permanently impeding Species A from participation.

All of the examples cited by Hill and Matrone (1970) are antagonistic in nature. Recent experiments in rats at the University of Wisconsin have explored the interaction of aluminum or tin and the nutritionally important metals—copper, iron, and calcium (Greger et al., 1985; Johnson and Greger, 1985). The iron–zinc and zinc–copper interactions, major examples of health-relevant issues cited in this chapter, are expressly antagonistic.

**Synergistic interactions.**   In synergistic interactions, the two minerals act in a complementary fashion. We use the term here to describe interactions in which one mineral spares, or substitutes for, the action of the other, or in which the two minerals together mutually enhance a biological function. Riordan and Vallee (1976), reviewing basic studies in enzymology,

noted that certain divalent ions, e.g., $Ni^{2+}$, $Co^{2+}$, and $Mn^{2+}$, could sub-
stitute for the native cations in some metalloenzymes. Dietary nickel
seems to have an iron-sparing effect, retarding the development of anemia
in laboratory animals (Nielsen et al., 1984; Nielsen, 1985). The cooperation of calcium and strontium in preserving skeletal mineralization has
been demonstrated by some investigators (Shorr and Carter, 1950).

**Combined interactions.** The mathematically simplest model for mineral
interactions is one on one, with one element either opposing or supporting
the action of the other. When three or more elements are in an interaction
mode, the possibilities are much more complex. For instance, in a three-
way interaction, they could all be mutually antagonistic. Thus, viewed
from the perspective of either of the two "original" elements in competition, the addition of a third merely sets up more competition for the
transport or functional pathway. In the human research on the iron–tin–
zinc interaction, a universal antagonism was the underlying hypothesis
(Solomons et al., 1983b; Valberg et al., 1984). Conversely, one could envi-
sion general mutual synergism, a state in which all elements would be
working collaboratively toward the same goal. A theoretical scenario for a
three-way synergism would be calcium, flouride, and strontium all acting
in concert to provide increased bone mineral strength.

The possibilities become even more complex when some mixture of antagonistic and synergistic reactions is occurring when three or more
minerals exhibit mutual interactions. Such was the basis of the hypothesis
that guided a recent clinical study performed in Guatemala by Dr. Carlos
Castillo-Duran of Chile and myself (Castillo-Duran and Solomons: unpublished observations). In these experiments we examined the effect of
inorganic, soluble calcium on the interaction of inorganic iron and the
zinc in cooked beef. We suspected that the calcium might, in fact, release
the zinc from some of its inhibition by iron. However, in the dosage system we used, we encountered an independent inhibitory effect of our dose
of calcium on zinc, making it difficult to conclude that calcium had acted
primarily on iron in our three-way interaction model in human volunteers.

## LEVELS OF MINERAL–MINERAL INTERACTIONS

Along the chain of events in which minerals are obtained from the environment, processed by the body, and returned to the environment,
mineral–mineral interactions can occur at many levels. These levels are
enumerated in Table 4.1. First is an interaction in beverages or foods

**TABLE 4.1** Loci of Mineral-Mineral
Interactions of Physiological Importance

In foods and beverages
In the alimentary canal
   competition for uptake sites
   "coadaptation" of the uptake mechanism
At the tissue level
   at storage sites
   in functional proteins or enzymes
In transport within the organism
In pathways of excretion

themselves, during cooking, preparation, or combination into meals. This topic is treated extensively in Chapter 3 (Clydesdale, 1988). The basic physiology at the other levels is surveyed below.

## In the Alimentary Canal

Present evidence suggests that the majority of important mineral-mineral interactions occur in the alimentary tract. As reviewed by Rosenberg and Solomons (1982, 1984), the interactions are either through competition or "coadaptation." With regard to the former, they state: "Apparently, chemically similar minerals share 'channels' for absorption, and the simultaneous ingestion of two or more such minerals will result in competition for absorption." This mechanism is probably represented in the inhibited intestinal uptake of copper in the presence of cadmium and silver (Van Campen, 1966). This is also the mechanism proposed for the zinc-iron interaction in humans (Aggett et al., 1983; Solomons, 1983)

The mechanism—"coadaptation"—is more complex. As defined by Rosenberg and Solomons (1984):

> The locus of certain of the mineral-mineral interactions may be intracellular, perhaps in relation to a common binding protein, rather than intraluminal. Moreover, dietary intake of one mineral may *induce* an intracellular transfer mechanism or mucosal trapping mechanism that—not entirely specific—may influence the absorption of the other minerals. Such "coadaptation" may presumably derive from a direct intracellular action by a mineral or by some hormonal signal reflecting its tissue stores. In any case, deficiency or sufficiency of one mineral can determine the absorptive efficiency of another.

In other words, when the dietary supply of a nutrient and/or the body reserves of a mineral are low, the intestine adapts to improve the efficiency of uptake and transfer. When nonspecific, other similar dietary species are also enhanced in their absorption. A classic example is in iron deficiency, where the upregulation of iron absorption also produced increased uptake of lead (Watson et al., 1980: Anonymous, 1981). On the other hand, when body stores are overloaded with a mineral, the intestine often adapts to *block* its absorption from the diet. When this barrier mechanism is less specific, the uptake of the other chemically similar ions is diminished. This occurs with lead in iron overload states and is prominent in the instance of copper-absorption inhibition by excess zinc (Smith and Cousins, 1980; Fischer et al., 1981).

Finally, since the alimentary tract is temporarily susceptible to stimulation/inhibition influences within a meal, more exotic types of mineral-mineral interactions could be envisioned. For instance, a high oral dose of calcium carbonate might produce rebound gastric hyperacidity. Such hyperacidity might actually enhance the absorption of inorganic iron consumed in the peak acid secretion period, as nonheme iron absorption is favored by a low intraluminal pH.

## At the Level of the Tissues

Once inside the body, minerals have further opportunities for interaction within the tissues of the body. This can occur at the level of tissue stores or in functional roles.

**At storage sites.** The effect of high dietary iron on copper storage was evaluated by Astrup and Lyso (1986). Ratios of 20:1 and 40:1 of iron:copper reduced hepatic copper levels to less than 50% of control values.

Zinc accumulates in both muscle and bone of mammalian organisms. Their respective roles as a reservoir for metabolically available zinc is still disputed. Zinc for normal rat fetal development apparently comes from the dam's diet (Masters et al., 1986). Teratogenesis in rat fetuses of zinc-restricted dams is overcome with simultaneous dietary calcium restrictions. Presumably, the comobilization of both minerals from maternal skeleton occurs, and the zinc released from bone becomes available for fetal nutrition (Masters et al., 1986).

**In functional proteins and enzymes.** An example of an interaction of minerals at the subcellular level is the reversal by zinc of lead's inhibition of delta-amino-levulinic acid dehydratase (ALA-D) (Haeger-Aronsen et al., 1971). Tin has also been shown to inhibit ALA-D activity, and zinc can modify this effect in vivo in the rat (Zareba et al., 1986).

## In Transport Within the Organism

Transferrin (TF) is the physiological serum transport protein for iron. The circulating complement of TF is generally less than 50% saturated with iron in its transit from site to site. In its functional form, chromium is a soluble entity (Anderson, 1986), but its normal transport protein is also TF (Sargent et al., 1979).

## In Pathways of Excretion

It is now generally accepted that the level of circulating ionized calcium governs release of parathyroid hormone (PTH) from the parathyroid gland. PTH status, in turn, determines the renal tubular handling of filtered phosphate. Recent evidence also points to an interaction of calcium and magnesium at the level of renal excretion (Al-Jurf and Chapman-Furr, 1985; Bataille et al., 1985).

## METHODS FOR ASSESSING MINERAL-MINERAL INTERACTIONS

Given the maturation of conceptual notions (reviewed above) and the availability of increasingly precise and accurate analytical methodology, the experimental and observational pursuit of mineral interactions in man would seem to be advancing. With specific reference to humans, four basic methods for detecting and assessing mineral–mineral interactions are available: (1) metabolic balance studies; (2) single-test absorption studies; (3) analysis of body fluids and tissues; and (4) functional outcomes. Each has its logistical advantages and disadvantages as well as interpretative caveats and limitations.

## Metabolic Balance Studies

A major approach to exploring mineral–mineral interaction in humans has been the *metabolic balance study*. In this technique, subjects are fed a specified diet and complete collection of excreta are made. Two variables—*apparent* retention and *apparent* absorption—are calculated. The former is computed as the difference between oral intake of the element and its excretion in urine and feces. This becomes "true" retention when all additional exchanges—respiratory, sweat, dermal, and hair losses, etc.—are accounted for. Specifically, regarding gastrointestinal handling, net apparent absorption is the difference between oral intake (I) and fecal excretion (F) of the nutrient. Apparent *fractional* absorption (as a percent) is determined by the equation $I - F/I \times 100$. Only when all endogenous contributions of the element to the fecal stream is accounted for can "true" absorption be discussed.

Metabolic balance has been used to explore mineral–mineral interactions for several decades. An early example was the study by White and Gynne (1971) on the effects of iron intake on zinc retention and absorption. The technique has been used extensively to explore calcium's interactions with other elements by Spencer's laboratory (Spencer, 1986). The zinc–copper interaction has been evaluated in both health (Greger et al., 1978; Burke et al., 1980; Taper et al., 1980) and disease (Brewer et al., 1983; Cossack and Bouquet, 1986) using the metabolic balance approach.

Clearly, then, metabolic balance can identify interactions that have a net impact on the whole body economy of the nutrient. When *internal* displacements are the effect of interaction, however, the balance sheet per se is not a reliable window. Moreover, the mechanism of intestinal interaction cannot be dissected readily when both absorptive blockade *and* stimulated re-excretion can participate simultaneously in the interactive process. In such instances, other approaches come to the forefront.

## Single-Meal Absorption Tests

Much simpler and less tedious than the metabolic balance approach are absorption tests with a single-dose administration of the mineral(s) of interest. Whether involving foods or test solutions, we have termed this approach generically single-meal tests. The amounts of minerals administered are precisely known. Detection of the magnitude of the absorptive response is critical to the testing of interactions.

The most basic single-meal absorption test is the so-called tolerance test approach; we have used this extensively in our laboratory to test the iron–zinc interaction (Solomons and Jacob, 1981; Solomons et al., 1983a; Solomons et al., 1983b). It has also been applied to this question in the United Kingdom (Aggett et al., 1983; Meadows et al., 1983; Simmer et al., 1987). It involves the serial measurement of changes in circulating levels of zinc after an oral dose of zinc in the pharmacological range.

Single-meal studies can also be done with radioisotopes by measuring systemic uptake (Sandstrom et al., 1985; Spencer et al., 1985) or their fecal output, controlled by a nonabsorbable marker (Payton et al., 1982; Valberg et al., 1984).

## Mineral Concentrations in Tissue and Biological Fluids

The effect of mineral–mineral interactions can be gauged, at times, by mineral-provoked changes in the distribution of circulating trace metals and/or in tissue levels. Investigators at the University of Colorado have used circulating zinc as an index of the effect of supplemental oral iron on zinc interaction in both cross-sectional (Hambidge et al., 1983) and longitudinal (Krebs et al., 1987) studies in pregnant women. This approach was also followed by other investigators in this field (Breskin et al., 1983; Campbell-Brown et al., 1985; Sheldon et al., 1985).

The tissue levels of copper in the liver parenchyma of patients with Wilson's disease have been used to assess the effect of oral zinc therapy (Van Callie et al., 1985). Magnesium excretion in urine has been used in clinical studies to examine the magnesium–calcium interaction.

### Functional Responses

Specific physiological or pathophysiological effects of a given mineral can be used as a functional measure of intermineral interactions. The reversal of abnormal uric acid metabolism in Wilson's disease has been used as an index of zinc's interaction with copper (Umeki et al., 1986). Also in the zinc–copper interaction, the response of hemoglobin levels, bone marrow cells, and neutrophil response have been used to measure the copper-suppressive response of oral zinc (Patterson et al., 1985.)

## CAVEATS AND LIMITATIONS IN ASSESSING MINERAL–MINERAL INTERACTIONS

The investigative tools for research in mineral interactions, especially in humans, are limited but growing. Limitations also exist in the design of experiments, and there are important caveats in the interpretation of data. These issues are listed in Table 4.2.

When interaction at the level of *absorption* is in question, the dosage of the target minerals in the challenge dose—be it an absorption test or a metabolic balance study—is of crucial importance. The transport system may operate at distinct intraluminal concentrations. A trace amount may move across membranes by different mechanisms than those for usual dietary levels. The handling of pharmacological doses may be distinct from that for either of the lower levels. For biological and nutritional interpretation, it is important to validate the extrapolation from one setting to the other.

**TABLE 4.2** Variables Important
for the Design and Interpretation of
Mineral–Mineral Interaction
Research

Dosage of minerals
  subdietary levels (trace)
  usual dietary levels (physiological)
  supradietary levels (pharmacological)
Dietary context
  soluble
  enbedded in food matrices
Duration of feeding
  single-meal
  long-term feeding
Timing of administration
  minerals fed simultaneously
  minerals fed separately
Health status of subject
  normal
  diseased
  nutritionally deficient
  nutritionally overloaded
Gender
  male hormonal milieu
  female hormonal milieu

The dosage form is similarly important. Whether the minerals presented for intestinal absorption are in a soluble form or embedded in their customary food matrices is of interpretative concern. The duration of feeding through the experimental observation period is also crucial if "adaptation" phenomena are potentially involved. One must intrepret data as to whether a single, discrete dosing or a long-term feeding paradigm was employed. Moreover, in whatever setting, it is important to identify whether the minerals in question were fed simultaneously or separately.

The health status of the subjects being tested or observed is another critical variable. The level of total body nutritional reserves of one or another mineral—depleted, replete, or excessive—will determine any state of "preadaptation." Disease states may influence interactions, especially when abnormal handling of a mineral is a hallmark of the pathophysiology, as occurs in Wilson's disease or idiopathic hypercalciuria. Homeostatic regulation of some minerals may be subject to hormonal control that varies with the gender of subjects. Thus, differences in responses between males and females may be common, specifically with respect to calcium interactions.

## PHYSIOLOGICALLY IMPORTANT MINERAL–MINERAL INTERACTIONS IN HUMANS

Given the experimental approaches available and the caveats required to interpret data, both caution and speculation are needed to synthesize the extant findings. We have chosen seven dyadic mineral-mineral interactions to discuss in the context of human health and nutrition. These include: (1) calcium-magnesium; (2) iron-zinc; (3) iron-manganese; (4) tin-zinc; (5) zinc-copper; (6) chromium-iron; and (7) selenium-iron.

### Calcium–Magnesium Interaction

Calcium and magnesium are major constituents of the body; there is over a kilogram of the former mineral in the adult man and over 25 g of the latter. Both exist as divalent cations and circulate in relatively abundant—and tightly, homeostatically controlled—levels in the peripheral circulation. More than any other pair of minerals, these two provide interactions all the way from the diet in the gut to the subcellular, molecular level.

Evidence regarding their interaction in the intestine is controversial and inconclusive (Bengoa and Wood, 1984). It was noted clinically that high calcium intakes aggravated the symptoms of dietary magnesium depletion (Walser, 1967). The nature of the interaction has been pursued with metabolic balance studies. In young, healthy volunteers, Kim and Linkswiler (1980) observed a significant reduction in the apparent fractional absorption of magnesium by increasing dietary calcium from 800 to 2400 mg/day. Daily magnesium consumption was not specified. By contrast, in older men, Spencer (1979) and Spencer et al. (1985) failed to encounter a difference in magnesium absorption via $^{28}$Mg uptake or apparent magnesium retention with calcium intakes ranging from 200 to 2000 mg/day. Magnesium intake was 250 mg. Whether the diminished absorption of calcium with age helps to harmonize these divergent results is a matter of speculation. Since customary calcium intakes from Western dietaries are in the 800 mg or less range, usual meals would be unlikely to result in important degrees of calcium inhibition of magnesium uptake.

Interesting data on Ca–Mg interaction is also to be found at the level of renal excretion. Al-Jurf and Chapmann-Furr (1985), in a parenteral nutrition model in the rat, showed that addition of graded doses of calcium increased the urinary excretion of magnesium and reduced magnesium balance. From France, Bataille et al. (1985) reported excessive urinary magnesium wasting among subjects with idiopathic hypercalciuria and renal calculus formation, as compared to controls. The *relative* contribution of magnesium to urinary electrolytes, however, was proportionately *diminished* as calcium output increased in the stone-formers.

Finnish investigators compared the magnesium and calcium in autopsy specimens of cortical bone (anterior ileac crest) in patients dying either unexpectedly and suddenly ($n$ = 72) or after a chronic, immobilizing disease ($n$ = 51); 68 were men and 55 were women (Lappalainen and Knuattila, 1985). They found a very robust positive correlation between magnesium and calcium content of bone ($r$ = 0.402). Thus, some 16.4% of the variance in cancellous bone magnesium content was accounted for by the calcium content. In males, the correlation was even greater and was independent of health and mobility status.

The metabolic regulation of bone turnover is governed, in part, by the secretion of the parathyroid gland. The parathyroid cell is also the site for a magnesium–calcium interaction. Shoback et al. (1984) grew bovine parathyroid cells in culture and manipulated the extracellular concentrations of $Ca^{2+}$ and $Mg^{2+}$. Increasing the medium's concentration of $Mg^{2+}$ increased the release of PTH from the cells. This process was accompanied by an increase in intracellular, *cytosolic* calcium content. In fact, when extracellular $Ca^{2+}$ was in lowest concentration, increasing $Mg^{2+}$ could not influence intracellular calcium, and no PTH response was seen. This is interpreted as a *mediation* of hormone secretion by cytosolic calcium levels and the *regulation* of the latter by the extracellular $Mg^{2+}$ concentrations.

Several aspects of calcium and magnesium interrelationships are fundamental to skeletal muscle function (contraction-relaxation) and are well characterized. Salama and Scarpa (1985) state:

> $Mg^{2+}$ plays at least two important roles in the $Ca^{2+}$ uptake process. First $Mg^{2+}$ binds to ATP and the Mg–ATP complex is the substrate for the $(Ca^{2+})$-ATPase; Second, $Mg^{2+}$ is required for the hydrolysis of phosphoenzyme formed from ATP and $(Ca^{2+})$-ATPase which is a necessary step for enzyme turnover and further translocation of $Ca^{2+}$.

The debate had centered on a possible third role proposed for $Mg^{2+}$—that of maintaining electrochemical neutrality during $Ca^{2+}$ fluxed by a countertransposition. Using elegant biophysical techniques, Salama and Scarpa (1985) demonstrated nonsynchronous and nonstoichiometric movements of the two ions in the activation of the skeletal muscle. Still, on the other side of the issue, comes the report of Morsy and Shamoo (1985) in favor of the countertransport hypothesis for calcium uptake into sarcoplasmic reticulum.

The smooth cell is another focus of magnesium–calcium interaction, specifically the *vascular* smooth muscle. The role of both calcium and magnesium in vascular hypertension has become a topic of wide contemporary discussion (Altura and Altura, 1985). Altura et al. (1984) demonstrated in situ in intact rats fed a magnesium-restricted diet the decrease in luminal sizes of terminal arterioles and precapillary sphincters in association with increased blood pressure. These constrictive effects on the *micro*circulation are proposed as mediators of hypertension. As with skeletal muscle the permissive action of magnesium on calcium is proposed as a mediator (Altura et al., 1984). However, in this instance, the lowered extracellular $Mg^{2+}$ *enhances* movement of $Ca^{2+}$ into the vessel walls. Increased intracellular calcium concentrations are the putative source of vasoconstriction in the microcirculation.

## Iron–Zinc Interaction

The nutritional consequences for humans of the interaction between iron and zinc has achieved increasing interest (Solomons, 1986a). It is not unusual, in fact, that doses of oral iron of ten or more times the Recommended Dietary Allowance are administered either therapeutically or prophylactically. Iron is of interest because there are two clear possibilities for an interaction: intestinal coadaptation and direct competition for uptake pathways. The phenomenon of coadaptation was demonstrated for iron in the pioneering animal work of Pollack et al. (1965), who showed enhanced absorption of nonferrous metals, such as $Mn^{2+}$ and $Co^{2+}$, by intestines of iron-deficient rats. The classical reflection of this phenomenon in man is the increased absorption of lead by anemic persons (Watson et al., 1980; Anonymous, 1981). That direct competition for uptake sites occurs between iron and other ions is also demonstrable with iron and lead (Flanagan et al., 1980).

**Epidemiological associations.** Various epidemiological associations and casual observations first drew scientific attention to the operation of an iron–zinc interaction in humans. These are summarized in Table 4.3. These have, in turn, given rise to a series of prospective population studies and experimental investigations in human subjects to define further the strength and nature of the iron–zinc interaction.

**Zinc–iron interaction as coadaptation.** The phenomenon of coadaptation of the intestinal absorption capacity and its regulation in response to the iron status of the individual is best seen when persons have taken high doses of oral iron for a predetermined period of time and presumably away from the meal(s) containing dietary zinc. This is the presumption in the pregnancy studies of Hambidge et al. (1983) and Breskin et al. (1983). Three additional studies of the effects of prenatal iron supplementation on circulating zinc status have been published. Krebs et al. (1987) studied 20 women who received from 164 to 395 mg of iron per day (mean = 261 ± 56 mg) during pregnancy. They found a statistically significant difference in mean plasma zinc levels, with the iron-supplemented women having a concentration on average 4 μg/dl lower than expected for gestational age. Campbell-Brown et al. (1985) found a decidedly lower circulating zinc

**TABLE 4.3** Preliminary Phenomenology and Epidemiology on Iron–Zinc Interactions in Humans

---

Zinc suppression of iron efficacy
  In Iranian schoolboys, the linear growth during a mineral supplementation study was greater with a regimen of iron alone than with the combination of zinc and iron. (*Mahloudji et al., 1975*)
Excessive oral iron aggravation of experimental zinc deficiency
  Volunteers were fed for prolonged periods with dietary zinc intakes of less than 3.5 mg. A more rapid and dramatic decline in circulating zinc levels was seen in the pair of subjects fed 130 mg of iron daily as compared to the pair fed 20.3 mg of iron. (*Prasad et al., 1978*)
High iron contents of infant formulas and depressed plasma zinc levels
  Infants fed cow milk–based infant formulas with lower Fe:Zn ratios had higher circulating zinc levels than infants fed formulas with higher Fe:Zn ratios. (*Walravens and Hambidge, 1976; Craig et al., 1984a, b*)
High levels of prenatal iron supplementation were associated with lower maternal zinc levels in pregnancy
  Women with higher amounts of iron in their prenatal mineral supplementation regimens had lower circulating zinc levels. (*Hambidge et al., 1983; Breskin et al., 1983*)

---

concentration in the three women in a British sample who took more than 100 mg of iron daily as a supplement, as compared to the remaining 137 pregnant women in the study. Shelton et al. (1985), however, found no differences in zinc levels at any time during pregnancy between 15 women with no prenatal iron supplementation and 10 women prescribed 160 mg of elemental iron daily.

The issue of prolonged oral supplementation of iron to infants is another focus of interest. Italian investigators (Salvioli et al., 1986) determined plasma zinc levels of 30 infants of low birth weight (820–2000 g) at 6 or 12 months of age. Sixteen were receiving an iron-fortified formula, with 1.2 mg of iron and 0.5 mg of zinc per kg combined, and a separate *medicinal* supplement of iron (ferrous sulfate) of 2 mg/kg of body weight, up to a total of 15 mg daily; the remaining 14 were receiving only the iron in the formula. No subject had a plasma zinc level below 75 µg/dl, and the distribution of values was identical for the two cohorts. Similarly, Yip et al. (1985), in an article entitled "Does iron supplementation compromise zinc nutrition in healthy infants?," report the results of a randomized assignment of healthy, one-year-old infants to a treatment with 30 mg of iron *before a meal* or a placebo. The amount of zinc in the usual diets of these toddlers was not specified. Neither at baseline nor after 3 months of treatment was there any difference in plasma zinc concentration.

Direct studies of zinc absorption—in the absence of simultaneous iron, but after various periods and levels of iron pretreatment—have been reported. Solomons et al. (1983a) found no difference in the plasma uptake curves for zinc from from a 25-mg oral dose either before or after 4 days of supplementation with 130 mg of oral iron. Meadows et al. (1983), in London, supplemented five young women and five young men for 14 days with 100 mg of iron daily (along with 350 µg of folic acid). Plasma appearance studies with 50 mg of oral zinc before and after the intervention showed a 50% reduction in the area under the curve. In the same laboratory, Simmer et al. (1987) confirmed the findings, this time using a 25-mg oral challenge dose of zinc in the absorption test, in *pregnant* women in their second trimester. The same iron-folate supplement was again taken for a 2-week period, and again about a 50% reduction in the plasma zinc response was seen after the intervention. (Interestingly, in this same report, an *independent* effect of 14 days of supplementation with the folate alone on zinc absorption was found; this confounds the interpretation of the *iron* effect in these two London studies.) Finally, Sandstrom et al. (1985) used whole body counting after $^{65}$Zn administration as their index of absorption. Fourteen days of prior supplementation with iron at a dosage of 50 mg failed to reduce the efficiency of zinc absorption in four subjects.

We can conclude from these data that high doses of oral iron will often reduce the efficiency of zinc absorption, but the duration of iron supplementation and the outcome variables used are important determinants of the final conclusions about effect.

**Zinc–iron interaction by direct competition.** The dietary context is of supreme importance in evaluating the existing evidence for a direct competitive interaction in the simultaneous administration of iron and zinc. Feeding studies in livestock and experimental animals with high Fe:Zn ratios in their rations have shown varying results in terms of zinc absorption and tissue uptake (Solomons, 1986a). High Fe:Zn ratios in test meals of foods and beverages fed to human subjects in single-meal absorption test formats have provided mixed results. With a meal of rice and meat sauce having either a 2.5:1 or 25:1 Fe:Zn ratio, Sandstrom et al. (1985) failed to find an effect of iron's inhibition. Similarly, the creation of a 3:1 Fe:Zn ratio with ferrous sulfate did not reduce the plasma uptake of zinc from a meal of Atlantic oysters (Solomons and Jacob, 1981). However, with extrinsically labeled turkey meat, Valberg et al. (1984) produced a 40% reduction in zinc absorption with a 10:1 Fe:Zn ratio. Recently, in our laboratory, Castillo-Duran and Solomons (1987) demonstrated the inhibition of plasma uptake of the intrinsic whole zinc in cooked beef with a 4:1 Fe:Zn ratio.

In the soluble state, and consumed alone, zinc and iron ions manifest a direct competition for intestinal absorption. This has been amply demonstrated, now, in numerous experiments in both animals and humans. This effect has been most dramatically and precisely shown in the elegant intestinal perfusion model in mice by Hamilton et al. (1978) in Ontario. The reflection of this experiment in human subjects is the administration of zinc and iron in different quantities and ratios in clear liquid solutions in a single swallow. Such experiments have been performed in several laboratories using various indices of absorption, including plasma zinc appearance (Solomons and Jacob, 1981; Aggett et al., 1983; Solomons et al., 1983a, b; Abu-Hamdan et al., 1984), whole body uptake of $^{65}$Zn (Aggett et al., 1983; Sandstrom et al., 1985); and fecal monitoring of radiozinc excretion (Payton et al., 1982; Valberg et al., 1984). In these studies, Fe:Zn ratios have varied from 1:1 to 22:1. Inhibition of zinc absorption was manifest in most of these studies (Solomons, 1986b). The first demonstration of the iron–zinc interaction using a zinc-tolerance test is illustrated in Fig. 4.1, from the report of Solomons and Jacob (1981). Surveying all of the aforementioned data, it was found that the simultaneous presence of at least 25 mg of total ionic species in the oral dose was more predictive of a competitive interaction and inhibition of zinc absorption than any

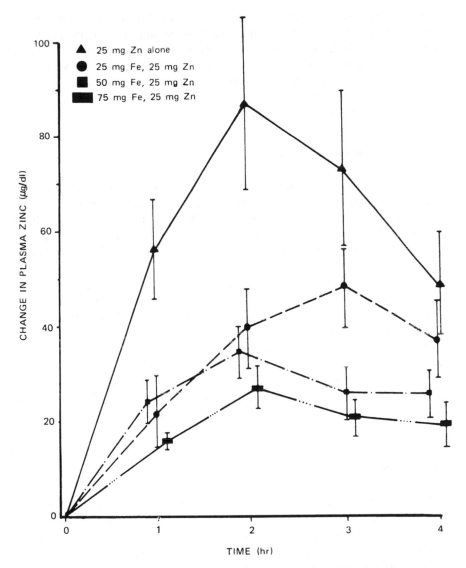

**FIG. 4.1** Iron–zinc interaction using zinc-tolerance test. (*From Solomons and Jacob, 1981.*)

specific Fe:Zn ratio. Interestingly, the converse interaction, i.e., zinc's inhibition of iron absorption, was demonstrated as well in the whole body counting uptake studies of [59]Fe by Aggett et al. (1983).

In settings in which iron and zinc are consumed in imbalanced proportions, away from meals, the chance of a biologically and nutritionally important competitive interaction in the human intestine is high. Chronic
dietary consumption of a regimen with a high Fe:Zn ratio may also adversely affect the zinc status of the consumer. Indeed, under some circumstances, medicinal supplementation with iron may also reduce the efficiency of absorption of dietary zinc.

## Iron–Manganese Interaction

Data in the experimental animal literature show inhibition of intestinal
manganese absorption in the presence of excessive dietary or intraluminal iron (Gruden, 1977; Kello and Kostial, 1977; King et al., 1980).
Comparable experimental observations in humans are not available, due,
in part, to the formidable technical problems in measuring manganese
absorption in human subjects.

Of intriguing interest to human nutrition, however, are findings of trace
mineral content of common infant formulas. In U.S. products, iron concentration ranged from 0.2 to 58.5 mg/liter (mean: 28 mg/liter), while
manganese concentration ranged from 0.2 to 7.8 mg/liter (mean: 1.1 mg/
liter) (Lonnerdal et al., 1983). The Fe:Mn ratios in these formulas ranged
from less than 1:1 to 425:1. In human milk, the Fe:Mn ratios are between
25:1 and 100:1. Thus, the potential for inhibition of manganese absorption
in formulas with the more extreme Fe:Mn ratios is signaled. This should
be pursued when appropriate technologies for manganese absorption
studies are developed.

## Zinc–Copper Interaction

The zinc–copper interaction plays a role in both normal nutrition and
disease and conforms to the Hill and Matrone dictum, since similar electron configurations exist in the two species. Feeding relatively increased
amounts of copper and normal amounts of zinc has no apparent consequence, but feeding zinc levels out of proportion to copper has important
biological consequences. A number of experiments in animals have
demonstrated that copper absorption is diminished when dietary levels of
zinc are high (Smith and Cousins, 1980; Fisher et al., 1981).

This phenomenon was evaluated in metabolic balance experiments in
humans. Greger et al. (1978), at Purdue, compared copper balance in
adolescent girls on a zinc intake of 11.5 and 14.7 mg per day. Although

there was no significant difference in apparent copper retention, fecal losses of copper were significantly greater with the higher zinc intake ($p <$ 0.04). Taper et al. (1980) at Virginia Polytechnic Institute provided 2 mg of copper to college women while adjusting the dietary zinc intakes to 8.0, 16.0, or 24.0 mg daily. No difference in apparent copper retention or fecal copper retention were seen with any level of zinc. In the same VPI unit, Burke et al. (1980) contrasted copper balance with a 2.3-mg intake of copper and either 7.8 or 23.3 mg of zinc in *elderly* subjects. In contradistinction to the results observed in the college-age subjects, a statistical difference was found between overall retention of copper: $0.94 \pm 0.68$ mg on the lower zinc dose, and $0.30 \pm 0.32$ mg at the higher level ($p < 0.05$).

The molecular basis for this zinc–copper interaction has been revealed through basic physiological research, much of it from the laboratories of Dr. Robert Cousins (Solomons and Cousins, 1984). He and his colleagues showed that high doses of zinc—admininstered enterally or parenterally—would increase the mucosal content of a sulfur-rich intracellular binding protein known as "thionein" (Richard and Cousins, 1976). This molecule is capable of binding a series of similar divalent metals, including $Zn^{2+}$, $Cu^{2+}$, $Hg^{2+}$, and $Cd^{2+}$. Zinc appears to operate at the level of genetic expression regulation of thionein, acting to increase the production of its mRNA (Menard et al., 1981). Figure 4.2 provides a schematic representation of zinc absorption by intestinal cells. The induction of the *metallo*thionein, i.e., thionein binding the available divalent metal ions (in this instance, zinc), by endogenous zinc is illustrated.

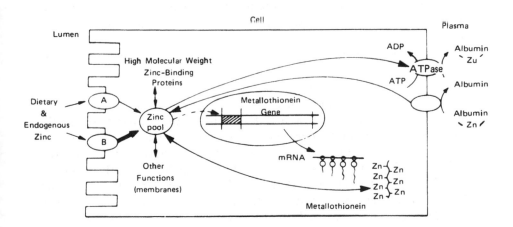

**FIG. 4.2** Schematic representation of zinc absorption by intestinal cells.

A *bystander* phenomenon occurs when copper in its usual dietary abundance confronts an intestinal mucosa primed with enriched levels of intracellular thionein. As thionein is not specific for zinc, much of the copper attempting to transit the enterocyte in the zinc-stimulated intestine will itself be bound and trapped on vacant sites in metallothionein chains. Thus, the diet ceases to resupply the body's copper reserves, and copper depletion can ensue.

One of the implications of the zinc–copper interaction in man is a toxicological one. Pharmacological doses of zinc can produce human copper deficiency and its clinical consequences: anemia and leukopenia. Pfeiffer and Jenny (1978) described a man who developed a frank anemia (Hgb = 6.2 g/dl) and neutropenia (700 white cells/mm$^3$) while chronically consuming 5 *grams* of zinc daily. His circulating copper level was 4 µg/dl, and his zinc level, 1,160 µg/dl. His hematological status normalized with 2 mg of copper and 60 mg of zinc in his diet. Patterson et al. (1985) reported a 57-year-old man with a 2-year history of a 450-mg daily intake of zinc who presented with exertional angina, fatigue, and shortness of breath. His hemoglobin level was 5.1 g/dl. He had 150 circulating neutrophils per mm$^3$. His bone marrow showed 21% ringed sideroblasts, constituting a sideroblastic anemia process. His serum zinc level was 320 µg/dl and his copper level, 4µg/dl. He normalized his hemoglobin status with the suspension of zinc supplementation.

At the dosages used therapeutically for the treatment of zinc deficiency, i.e., around 150 mg per day, copper deficiency anemia was described in patients with sickle cell disease (Prasad et al., 1978) and Crohn's disease (Porter et al., 1977). Duration of treatment had been from 1 to 2 years. The anemias resolved by providing supplemental copper and/or reducing the zinc dosage.

The antagonistic interaction of oral zinc and enteral copper has recently been turned to therapeutic advantage in the treatment of Wilson's disease (hepatolenticular degeneration). Wilson's disease is characterized by abnormal accumulation of copper in the liver, brain, and kidneys. Neurological damage and/or hepatic cirrhosis will develop unless the toxic accumulation of copper is controlled and reversed. Systemic chelation therapy with D(−)penicillamine is the traditional therapy for "decoppering" the Wilson's disease patient, but this drug has side effects. Dutch investigators at Utrecht reported the 14-year trajectory of a man with Wilson's disease, who was followed from age 30 to 44 years with 135 mg of zinc daily as the exclusive therapy. His symptoms of tremor and dysarthria regressed, and his handwriting improved. No deterioration in biochemical status was registered during the zinc-therapy period.

A more detailed study with oral zinc therapy in Wilson's disease was reported by Brewer et al. (1983). They developed a zinc-loading regimen of 150 mg of elemental zinc from zinc acetate daily. This consisted of doses of 25 mg of Zn at 7:00, 11:00, 15:00, and 19:00 hr, and 50 mg at bedtime. When copper balance studies were performed after from 2 to 7 weeks of this treatment, *negative* copper balance was observed in each of five Wilson's disease patients (see Table 4.4). In contradistinction to a patient's treatment with penicillamine (when he had 0.65 mg of copper in each stool and 480 µg of daily copper as urinary output on average), the oral zinc produced a 1.61-mg copper excretion per defecation, and urinary Cu output was 60 µg. It is felt that both dietary copper and copper from salivary and gastrointestinal secretions is blocked in its absorption by the priming of the mucosal thionein mechanism with oral zinc.

Cossack and Bouquet (1986) produced negative copper balance in two *juvenile* Wilson's patients with daily doses of oral zinc of 137 and 85 mg, respectively. Normally, in growing children a positive daily copper balance of about 0.20 mg is to be expected. Satisfactory maintenance of clinical, biochemical, and hepatic histological status was produced in two Dutch patients, previously decoppered with penicillamine, during 3 years of treatment with 135 mg of zinc as zinc sulfate. In one subject, a further decrease in hepatic copper content from 1460 to 890 µg/g dry weight was observed during 3 years of oral zinc therapy.

One of the common pathophysiological consequences of Wilson's disease is hyperuricemia and decreased renal excretion of uric acid. Umeki et al. (1986) treated a 12-year-old Japanese patient with 50 mg of zinc daily.

**TABLE 4.4** Copper Balance in Five Patients with Wilson's Disease Treated with Oral Zinc Therapy

| Age (yr) | Sex | Copper intake[a] | Copper excretion[a] | | Copper balance[a] |
|---|---|---|---|---|---|
| | | | Urinary | Fecal | |
| 25 | M | 1.06 | 0.08 | 1.42 | −0.44 |
| 31 | F | 0.92 | 0.01 | 1.07 | −0.16 |
| 32 | F | 1.12 | 0.04 | 1.11 | −0.03 |
| 33 | F | 1.16 | 0.06 | 1.52 | −0.42 |
| 34 | M | 1.19 | 0.10 | 1.11 | −0.02 |

[a]mg/day.
*Modified after Brewer et al. (1983).*

In addition to increasing fecal copper excretion and maintaining clinical remission of the Wilson's disease, oral zinc therapy reversed the hyperuricemia in this individual, and he returned to a state of normal uricosuria.

## Tin–Zinc Interaction

Consuming foods stored in unlacquered tin-alloy cans can provide substantial amounts of tin to the diet. Experiments in rodents revealed an antagonistic interaction between tin and zinc (Chmielnicka et al., 1981; Johnson and Greger, 1984; Johnson and Greger, 1985). The importance of the tin–zinc interaction in humans has been explored in metabolic balance studies (Johnson et al., 1981) and single-dose absorption tests (Solomons et al., 1983b; Valberg et al., 1984).

Johnson et al. (1981) fed diets containing 13.5 mg of zinc, and either 49.7 mg or <1 mg of tin in a metabolic balance setting. They noted a statistically significant increase in fecal zinc excretion and decrease in zinc retention with this dietary 4:1 Sn:Zn ratio.

With the plasma appearance test and an oral dose of 12.5 mg of zinc, simultaneous doses of 25, 50, and 100 mg of tin, constituting Sn:Zn ratios of 2:1, 4:1, and 8:1, produced no reduction in plasma uptake from the oral zinc sulfate (Solomons et al., 1983b). Using oral $^{65}$Zn either as the inorganic salt, zinc chloride, or intrinsically tagged to cooked turkey meat in a total amount of 4 mg of zinc and 36 mg of tin as stannous chloride, Valberg et al. (1984) found a significant decrease in zinc absorption ($p <$ 0.05). This represented a 9:1 Sn:Zn ratio (see Fig. 4.3). Thus, single-meal studies reveal conflicting findings. The convergence of the metabolic balance data (Johnson et al., 1981) and the results of Valberg et al. (1984), however, makes it prudent to consider excessive tin in the human diet a potential antagonist of dietary zinc bioavailability.

## Iron–Chromium Interaction

A curious and intriguing basis for an iron–chromium interaction of relevance to humans was developed by Sargent et al. (1979) at the Lawrence Berkeley Laboratory at the University of California. The setting was hemochromatosis, a familial iron storage disease characterized by bronze skin coloration, hypogonadism, and diabetes. Ordinarily, less than half of the binding sites on circulating transferrin (TF) are occupied with iron even in individuals with normal iron status. Hopkins and Schwarz (1964)

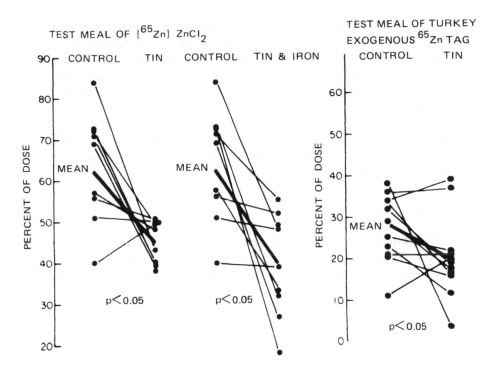

**FIG. 4.3** Decrease in zinc absorption with simultaneous zinc and tin intake. (*From Valberg et al., 1984.*)

first demonstrated competitive binding of both the $Fe^{3+}$ (ferric) and $Cr^{3+}$ (chromic) ions on TF. This was confirmed by Harris (1977). Given the lower circulating levels of serum chromium, relative to serum iron, it would stand to reason that only a fraction of the TF-binding sites need be available to support the usual circulation of $Cr^{3+}$ ions. Hence, only in extreme total body iron overload, as would be seen in untreated hemochromatosis (and possibly in extreme nutritional hemo*siderosis*), would iron compete with chromium for transport in the circulation.

The California investigation performed studies of $^{51}Cr$ retention after parenteral injection of radiolabeled $CrCl_3$ into normal subjects ($n = 5$), iron-depleted hemochromatotic subjects ($n = 3$), and untreated hemochromatotics ($n = 6$). The percent saturation of total iron-binding capacity in serum ranged from 75 to 93% in the active patients. It was 18, 22, and 60% in the treated patients. Consistent with the hypothesis of a displacement of chromium from TF in iron overload, the early component of total body and intravascular retention of $^{51}$Cr was only one-half that of normals in the iron-loaded hemochromatotics. The well-treated patients with normal iron stores had $^{51}$Cr kinetics that were indistinguishable from normals.

Since the etiology of the glucose intolerance ("bronze diabetes") of hemochromatosis is obscure, the authors suggest a role for chromium depletion. An additional speculation would be the existence of similar competitive chromium exclusion and chromium depletion in groups—such as the South African bantu—with severe hemo*siderosis* on a dietary basis. This remains to be evaluated by direct examination of iron-overloaded African subjects.

## Selenium–Iron Interaction

The relationship of selenium nutriture to iron intake in humans is speculative, but a series of recent findings converge to raise the question of a selenium–iron interaction to the investigative center stage. In recent years, we have come to understand the role of iron in the generation of intracellular free radicals (Slater, 1984; Halliwell and Gutteridge, 1986). Slivka et al. (1986) used the detection of volatile ethylene ($CH_2=CH_2$) from the reaction of 2-keto-4-methiolbutyric acid (KMB) with the hydroxyl radical ($\varepsilon$OH) as an indicator of .OH generation in vivo in rats. Oral iron produced a 15-fold increase in ethylene excretion over the group given KMB alone. Thus, the authors claim this as a direct, in vivo demonstration of hydroxyl radical generation from iron-dosing.

Simultaneously, the role of selenium as the trace element moiety in glutathione peroxidase has been defined (Condell and Tappel, 1982; Kraus et al., 1983). This enzyme catalyzes the donation of a proton from glutathione for the chemical reduction of intracellular peroxides, including hydrogen peroxide ($H_2O_2$), to a neutral species. As excessive iron promotes intracellular auto-oxidation, and as selenium participates in antioxidant protection, definition of the in vivo role of selenium from human dietaries in modulating cellular damage is required.

## DIETARY MODULATION OF MINERAL–MINERAL INTERACTION

It is suggested by some of the epidemiological studies cited that mineral-mineral interactions can and do occur in usual dietary contexts. The majority of experimental studies, however, employ *special* diets or even aqueous solutions of minerals as the dose. It is possible that other, *non-mineral* dietary constituents might confound the basic mineral–mineral interactions.

## Ascorbic Acid and Other Dietary Antioxidants

Ascorbic acid enhances the biological availability of nonheme (inorganic) iron in the diet. Since the competition of iron with zinc appears to be most intensive for the *ferrous* form of iron, a high ascorbic acid intake in a meal would potentiate the interaction. In fact, Solomons et al. (1983a) demonstrated the accentuation of iron–zinc competition by adding ascorbic acid to the solution containing the oxidized, *ferric* salt. In vivo conversion of the iron to its reduced, ferrous form by the ascorbate is the reasonable explanation.

Asayama et al. (1986) have described a unique, multiple interaction in rats that involves several elements (selenium, manganese, copper, and zinc) and the dietary context (vitamin E and selenium restriction). Selenium acts as an antioxidant by virtue of its role in glutathione peroxidase (Krause et al., 1983). Vitamin E is an antioxidant in cell membranes, which breaks the chain reaction of lipid peroxidation in unsaturated fatty acids. Enzymatic superoxide dismutase (SOD) activity converts an oxidant, the superoxide radical, to hydrogen peroxide. Two metalloenzyme forms of superoxide dismutase are found in mammalian systems: (1) Cu–Zn SOD in the cytosol of the cells; and (2) Mn SOD in the mitochondria. When rats were made vitamin E-deficient, selenium-deficient, or with combined deficiencies, alterations were observed among the dependent systems. In both groups with selenium-restricted diets as a common factor, GSH-peroxidase levels were reduced. Total SOD activity was maintained, but at the expense of a 75% reduction in activity of the manganoenzyme, but with a "compensatory" increase in the zinc–copper enzyme. The systemic vulnerability produced by lack of *dietary* antioxidant can influence the participation by trace metals in their endogenous, metalloenzyme antioxidant systems.

## Dietary Proteins

The nature of the protein in the diet is another consideration. The hemoglobin or myoglobins in diets containing red meats is an interesting example. It is well known that the mechanism for the absorption of iron in the "heme" form, i.e., as part of hemoglobin or myoglobin, is distinct from that of inorganic iron (Refsum and Schreiner, 1984). Thus, the potential of heme iron to impede zinc uptake has been debated. With heme iron in hemoglobin, Valberg et al. (1984) demonstrated a reduction in $^{65}$Zn absorption. With heme iron in the protein-free *hemin* form (heme chloride), no interaction of iron and zinc was observed (Solomons and Jacob, 1981). When minerals occur in the diet as part of the matrix of foods, a difference in mineral–mineral interaction behavior might be expected. This possibility must be examined on a case-by-case basis.

## CONCLUSIONS

Only in the past two decades have the conceptual consequences of the pioneering animal experiments of Pollack et al. (1965) and Hill and Matrone (1970) on mineral–mineral interactions been explored in relation to human health and nutrition. Much of the experience in laboratory animals and livestock has been reflected in observations in humans. In certain situations, mineral–mineral interactions are continuous and indispensable; the relationship of magnesium and calcium is a prime example (Salama and Scarpa, 1985). In other situations, conscious distortion of usual intake ratios of minerals, as adding abundant amounts of zinc to the diet, can be used therapeutically; this is exemplified in the novel approach of maintenance therapy for Wilson's disease (Hoogenraad et al., 1979; Brewer et al., 1983). One of the potential adverse consequences of heavy prenatal supplementation with iron is zinc depletion (Hambidge et al., 1983; Meadows et al., 1983; Krebs et al., 1987).

Herein, we have made logical connections between mammalian biology regarding manganese and iron, and selenium and iron in human dietary practices, and have *speculated* as to possible interactive situations that should be evaluated prospectively.

Individual genetic variation, duration and intensity of mineral intake, and specifically the composition of the diet's additional constituents can all modify the consequences of mineral–mineral interactions in people. The physiological bases for mineral–mineral interactions have been established in nonhuman species or in the test tube. The challenge to clinical investigation is to define further the operation of these physiological processes in human health, nutrition, and disease.

# REFERENCES

Abu-Hamdan, D. K., Mahajan, S. K., Migdal, S. D., Prasad, A. S., and McDonald, F. D. 1984. Zinc absorption in uremia: effects of phosphate binders and iron supplements. *J. Am. Coll. Nutr.* 3: 283.

Aggett, P. J., Crofton, R. W., Khin, C., Gvozdanovic, S., and Gvozdanovic, D. 1983. The mutual inhibitory effects on their bioavailability of inorganic zinc and iron. In *Zinc Deficiency in Human Subjects,* Prasad, A. S., Cadvar, A. O., Brewer, G. J., and Aggett, P. J., (Ed.). Alan R. Liss, New York.

Al-Jurf, A. S. and Chapman-Furr, F. 1985. Magnesium balance and distribution during total parenteral nutrition: effect of calcium additives. *Metabolism* 34: 658.

Altura, B. M. and Altura, B. T. 1985. New perspectives on the role of magnesium in the pathophysiology of the cardiovascular system. *Magnesium* 4: 226.

Altura, B. M., Altura, B. T., Gebrewold, A., Ising, H., and Gunther, T. 1984. Magnesium deficiency and hypertension: correlation between magnesium-deficient diets and microcirculatory changes in situ. *Science* 223: 1315.

Anderson, R. A. 1986. Chromium metabolism and its role in disease processes in man. *Clin. Physiol. Biochem.* 4: 31.

Anonymous. 1981. Nutritional influences on lead absorption in man. *Nutr. Rev.* 39: 363.

Asayama, K. Kooy, N. W., and Burr, I. M. 1986. Effect of vitamin E deficiency and selenium deficiency on insulin secretory reserve and free radical scavenging systems in islets: decrease of islet manganosuperoxide dismutase. *J. Lab. Clin. Med.* 107: 459.

Astrup, H. N. and Lyso, A. 1986. Nutritional Cu, Fe interrelation affected by energy level? The effect of iron and thyroxine on pigs given extra copper. *Acta Pharmacol. Toxicol.* 59 *Supp* 7: 152.

Bataille, P., Pruna., A., Finet, I., Leflon, P., Makdassi, R., Galy, C., Fievet, P., and Fournier, A. 1985. Magnesium renal wasting in calcium stone formers with idiopathic hypercalciuria contrasting with lower magnesium:calcium urinary ratio. *Proc. Eur. Dial. Transplant Assoc.-Eur. Ren. Assoc.* 71: 747.

Bengoa, J. and Wood, R. 1984. Magnesium. In *Absorption and Malabsorption of Mineral Nutrients*, Solomons, N. W. and Rosenberg, I. H. (Ed). p. 69. Alan R. Liss, Inc. New York.

Breskin, M. W., Worthington-Roberts, B. S., Knopp, R. H., Brown, Z., Plovie, B., Mottet, N. K., and Mills, J. L. 1983. First trimester serum zinc concentration in human pregnancy. *Am. J. Clin. Nutr.* 38: 943.

Brewer, G. J., Hill, G. M., Prasad, A. S., Cossack, Z. T., and Rabbani, P. 1983. Oral zinc therapy for Wilson's disease. *Ann. Intern. Med.* 99: 314.

Burke, D. M., DeMicco, F. J., Taper, L. J., and Ritchey, S. J. 1980. Copper and zinc utilization in elderly adults. *J. Gerontol.* 36: 558.

Campbell-Brown, M., Ward, R. J., Haines, A. P., North, W. R. S., Abraham, R., and McFayden, I. R. 1985. Zinc and copper in Asian pregnancies—is there evidence for a nutritional deficiency? *Br. J. Obstet. Gynecol.* 92: 975.

Castillo-Duran, C. and Solomons, N. W. 1987. Evaluation of zinc bioavailability using plasma response to a dietary portion of beef: effects of inorganic iron. *Fed. Proc.* 41:

Chmielnicka, J., Szymanska, J. A., and Sniec, J. 1981. Distribution of tin in the rat and disturbances in the metabolism of zinc and copper due to repeated exposure to $SnCl_2$. *Arch. Toxicol.* 47: 263.

Clydesdale, F. M. 1988. Mineral interactions in foods. In *Nutrient Interactions*, Erdman, J. and Bodwell, C. E. (Ed.), IFT Basic Symposium Series. Marcel Dekker, Inc., New York.

Condell, R. A. and Tappel, A. L. 1982. Amino acid sequence around the active-site selenocysteine of rat liver glutathione peroxidase. *Biochem. Biophys. Acta* 709: 304.

Cossack, Z. T. and Bouquet, J. 1986. The treatment of Wilson's disease in paediatrics: Oral zinc therapy versus penicillamine. *Acta Pharmacol. Toxicol.* 59, Supp. 7: 514.

Craig, W. J., Balbach, L., Harris, S., and Vyhmeister, N. 1984a. Plasma zinc and copper levels of infants fed different milk formulas. *J. Am. Coll. Nutr.* 3: 183.

Craig, W. J., Balbach, L., and Vyhmeister, N. 1984b. Zinc bioavailability and infant formulas. *Am. J. Clin. Nutr.* 39: 981.

Fischer, P. W. F., Giroux, A., and L'Abbe, M. 1981. The effect of dietary zinc on intestinal copper absorption. *Am. J. Clin. Nutr.* 34: 1976.

Flanagan, P. R., Haist, J., and Valberg, L. S. 1980. Comparative effects of iron deficiency induced by bleeding and a low-iron diet on the intestinal absorptive interactions of iron, cobalt, manganese, zinc, lead and cadmium. *J. Nutr.* 110: 1754.

Greger, J. L., Bula, E. N., and Gum, E. T. 1985. Mineral metabolism of rats fed moderate levels of varous aluminum compounds for short periods of time. *J. Nutr.* 115: 1708.

Greger, J. L. and Johnson, M. A. 1981. Effects of dietary tin on zinc, copper and iron utilization by rats. *Food Cosmet. Toxicol.* 19: 163.

Greger, J. L., Zahie, S. C., Abernothy, R. P., Bennett, O. H., and Huffman, J. 1978. Zinc, nitrogen, copper, iron and manganese balance in adolescent females fed two levels of zinc. *J. Nutr.* 108: 1449.

Gruden, N. 1977. Suppresion of transduodenal manganese transport by milk diet supplemented with iron. *Nutr. Metab.* 21: 305.

Haeger-Aronsen, B., Abdulla, M., and Fristedt, B. I. 1971. Effect of lead on delta-aminolevulinic and dehydratase activity in red cells. *Arch. Environ. Health* 23: 440.

Halliwell, B. and Gutteridge, J. M. C. 1986. Oxygen free radicals and iron in relation to biology and medicine: Some problems and concepts. *Arch. Biochem. Biophys.* 246: 501.

Hambidge, K. M., Krebs, N. F., Jacobs, M. A., Favier, A., Guyette, L., and Ickle, D. N. 1983. Zinc nutritional status during pregnancy: a longitudinal study. *Am. J. Clin. Nutr.* 37: 429.

Hamilton, D. L., Bellamy, J. E. C., Valberg, J. D., and Valberg, L. S. 1978. Zinc, cadmium and iron interaction during intestinal absorption in iron-deficient mice. *Can. J. Physiol. Pharmacol.* 56: 384.

Harris, D. C. 1977. Different metal-binding properties of the two sites of human transferrin. *Biochemistry* 16: 560.

Hill, C. H. and Matrone, G. 1970. Chemical parameters in the study of in vivo and in vitro interactions of transition elements. *Fed. Proc.* 29: 1474.

Hoogenraad, T. U., Loevoet, R., and De Ruyter Korver, E. G. 1979. Oral zinc sulphate as long-term treatment in Wilson's disease (hepatolenticular degeneration). *Eur. Neurol.* 18: 205.

Hopkins, L. L. and Schwarz, K. 1964. Chromium (III) binding to serum proteins, specifically siderophilin. *Biochem. Biophys. Acta* 90: 484.

Johnson, M. A., Baier, M., and Greger, J. L. 1981. Effect of dietary tin on zinc, copper, iron, manganese and magnesium metabolism of adult males. *Am. J. Clin. Nutr.* 35: 1332.

Johnson, M. A. and Greger, J. L. 1984. Absorption, distribution and endogenous excretion of zinc by rats fed various dietary levels of inorganic tin and zinc. *J. Nutr.* 114: 1843.

Johnson, M. A. and Greger, J. L. 1985. Tin, copper, iron and calcium metabolism of rats fed various dietary levels of inorganic tin and zinc. *J. Nutr.* 115: 615.

Kello, D. and Kostial, K. 1977. Influence of age and milk diet on manganese absorption from the gut. *Toxicol. Appl. Pharm.* 40: 277.

Kim, Y. and Linkswiler, H. 1980. Effect of level of calcium and of phosphorus and magnesium metabolism in young adult males. *Fed. Proc.* 39: 895.

King, B. D., Lassiter, J. W., Neathery, M. N., Miller, W. J., and Gentry, R. P. 1980. Effect of lactose, copper and iron on manganese retention and tissue distribution in rats fed dextrose-casein diets. *J. Animal Sci.* 50: 452.

Kraus, R. J., Foster, S. J., and Ganther, H. E. 1983. Identification of selenocysteine in glutathione peroxidase by mass spectroscopy. *Biochemistry* 22: 5853.

Krebs, N. F., Hambidge, K. M., Sibley, L., and English, J. 1987. Acute effects of iron therapy on zinc status during pregnancy. *Fed. Proc.* 46: 747.

Lappalainen, R. and Knuuttila, M. 1985. Mg content of healthy and chronically diseased human cancellous bone in relation to age and some physical and chemical factors. *Med. Biol.* 63. 144.

Levander, O. A. and Cheng, L. (Ed). 1980. *Micronutrient Interactions: Vitamins, Minerals, and Hazardous Elements.* New York Academy of Sciences, New York.

Lonnerdal, B., Keen, C. L., Ohtake, M., and Tamora, T. 1983. Iron, zinc, copper, and manganese in infant formulas. *Am. J. Dis. Child.* 137: 433.

Mahloudji, M., Reinhold, J. G., Haghasenass, M., Ronagy, H. A., Spivey-Fox, M. R., and Halsted, J. A. 1975. Combined zinc and iron supplementation of diets of 6- and 12-year-old school children in southern Iran. *Am. J. Clin. Nutr.* 28: 721.

Masters, D. G., Keen, C. L., Lonnerdal, B., and Hurley, L. S. 1986. Release of zinc from maternal tissues during zinc deficiency in the pregnant rat. *J. Nutr.* 116: 2148.

Meadows, N. J., Grainger, S. L., Ruse, W., Keeling, P. W. N., and Thompson, R. P. H. 1983. Oral iron and the bio-availability of zinc. *Br. Med. J.* 287: 1013.

Menard, M. P., McCormick, C. C., and Cousins, R. J. 1981. Regulation of intestinal metallothionein biosynthesis in rats by dietary zinc. *J. Nutr.* 111: 1353.

Mills, C. F. 1985. Dietary interactions involving the trace elements. *Annual Rev. Nutr.* 4: 173.

Morsy, F. A. and Shamoo, A. E. 1985. Trans-magnesium dependency of ATP-dependent calcium uptake into sarcoplasmic reticulum of skeletal muscle. *Magnesium* 4: 182.

Nielsen, F. H. 1985. The importance of diet composition in ultratrace element research. *J. Nutr.* 115: 1239.

Nielsen, F. H., Shuler, T. R., McLoed, T. G., and Zimmerman, T. J. 1984. Nickel influences iron metabolism through physiologic, pharmacologic and toxicologic mechanisms in the rat. *J. Nutr.* 114: 1280.

Patterson, W. P., Winkelmann, M., and Perry, M. C. 1985. Zinc-induced copper deficiency: megamineral sideroblastic anemia. *Ann. Intern. Med.* 103: 385.

Payton, K. B., Flanagan, P. R., Stinson, E. A., Chrodiker, D. R., Chamberlain, M. J., and Valberg, L. S. 1982. Technique for determination of human zinc absorption from measurement of radioactivity in a fecal sample of the body. *Gastroenterology* 83: 1264.

Pfeiffer, C. C. and Jenney, E. H. 1978. Excess oral zinc in man lowers copper levels. *Fed. Proc.* 37: 324.

Pollack, S., George, J. M., Reba, R. C., Kaufman, R. M., and Crosby, W. H. 1965. The absorption of nonferrous metals in iron deficiency. *J. Clin. Invest.* 4: 1470.

Porter, K. G., McMaster, D., Elmer, M. E., and Love, A. H. 1977. Anaemia and low serum-copper during zinc therapy. *Lancet* 2: 774.

Prasad, A. S., Brewer, C. J., Schoomaker, E. B., and Rabbani, P. 1978. Hypocupremia induced zinc therapy in adults. *J. Am. Med. Assoc.* 240: 166.

Prasad, A. S., Rabbani, P., Abassi, A., Bowersox, E., and Spivey-Fox, M. R. 1978. Experimental zinc deficiency in humans. *Ann Intern. Med.* 89: 483.

Refsum, S. B. and Schreiner, B. B. 1984. Regulation of iron balance by absorption and excretion: A critical review of a new hypothesis. *Scan. J. Gastroenterol.* 19: 867.

Richards, M. P. and Cousins, R. J. 1976. Metallothionein and its relationship to the metabolism of dietary zinc in rats. *J. Nutr.* 106: 1591.

Riordan, J. F. and Vallee, B. L. 1976. Structure and function of zinc metalloenzymes. In *Trace Elements in Human Health and Disease,* vol 1

*Zinc and Copper.* Prasad, A. S. (Ed.) p. 227 Academic Press, New York.

Rosenberg, I. H. and Solomons, N. W. 1982. Biological availability of minerals and trace elements: A nutritional overview. *Am. J. Clin. Nutr.* 35: 781.

Rosenberg, I. H. and Solomons, N. W. 1984. Physiological and pathophysiological mechanism in mineral absorption. In *Absorption and Malabsorption of Mineral Nutrients,* Solomons, N. W. and Rosenberg, I. H. (Ed.), p. 1. Alan R. Liss, Inc., New York.

Salama, G. and Scarpa, A. 1985. Magnesium permeability of sarcoplasmic reticulum. $Mg^{+2}$ is not counter-transported during ATP-dependent $Ca^{2+}$ uptake by sarcoplasmic reticulum. *J. Biol. Chem.* 160: 11697.

Salvioli, G. P., Faldella, G., Alessandroni, R., Lanari, M., and Benfenati, L. 1986. Plasma zinc concentrations in iron supplemented low birthweight infants. *Arch. Dis. Child.* 61: 346.

Sandstrom , B., Davidsson, L., Cederblad, A., and Lonnerdal, B. 1985. Oral iron, dietary ligands and zinc absorption. *J. Nutr.* 115: 411.

Sargent, T. 3rd, Lim, T. H., and Jenson, R. L. 1979. Reduced chromium retention in patients with hemochromatosis, a possible basis of hemochromatotic diabetes. *Metabolism* 28: 70.

Sheldon, W. L., Aspillaga, M. O., Smith, P. A., and Lind, T. 1985. The effect of oral iron supplementation on zinc and magnesium levels during pregnancy. *Br. J. Obstet. Gynecol.* 92: 892.

Shoback, D. M., Thatcher, J. G., and Brown, E. M. 1984. Interaction of extracellular calcium and magnesium in the regulation of cytosolic calcium and PTH release in dispersed bovine parathyroid cells. *Mol. Cell. Endocrinol.* 38: 179.

Shorr, E. and Carter, A. C. 1950. *Metabolic Interrelations,* p. 144. Josiah Macy Jr. Foundation, New York.

Simmer, K., Iles, C. A., James, C., and Thompson, R. P. H. 1987. Are iron-folate supplements harmful? *Am. J. Clin. Nutr.* 45: 122.

Slater, T. F. 1984. Free-radical mechanisms in tissue injury. *Biochem. J.* 222: 1.

Slivka, A., Kong, J., and Cohen, G. 1986. Hydroxy radicals and the toxicity of oral iron. *Biochem. Pharmacol.* 35: 553.

Smith, K. T. and Cousins, R. J. 1980. Quantitative aspects of zinc absorption by isolated, vascularly perfused rat intestine. *J. Nutr.* 110: 316.

Solomons, N. W. 1983. Competitive mineral:mineral interactions in the intestine: Implications for zinc absorption in humans. In *Nutritional*

*Bioavailability of Zinc,* ACS Symposium Series, p. 247. American Chemical Society. Washington, DC.

Solomons, N. W. 1986a. Competitive interaction of iron and zinc in the diet: Consequences for human nutrition. *J. Nutr.* 116: 927.

Solomons, N. W. 1986b. Zinc bioavailability to humans. In *Proceedings of the XIII International Congress of Nutrition,* Taylor, T. G. and Jenkins, N. K. (Ed.), p. 504. John Libbey, London.

Solomons, N. W. and Cousins, R. J. 1984. Zinc. In *Absorption and Malabsorption of Mineral Nutrients,* Solomons, N. W. and Rosenberg, I. H. (Ed.), p. 125. Alan R. Liss Inc., New York.

Solomons, N. W. and Jacob, R. A. 1981. Studies on the bioavailability of zinc in humans: Effect of heme and nonheme iron on the absorption of zinc. *Am. J. Clin. Nutr.* 34: 475.

Solomons, N. W., Marchini, J. S., Duarte-Favaro, R-M., Vannuchi, H., and Dutra de Oliveira, J. E. 1983b. Studies on the bioavailability of zinc in humans. VI. Intestinal interaction of tin and zinc. *Am. J. Clin. Nutr.* 37: 566.

Solomons, N. W., Pineda, O., Viteri, F. E., and Sandstead, H. H. 1983a. Studies on the bioavailability of zinc in humans: Mechanism of the intestinal interaction of nonheme iron and zinc. *J. Nutr.* 113: 337.

Spencer, H. 1979. Calcium and magnesium interaction in man. *Clin. Chem.* 25: 1043.

Spencer, H. 1986. Minerals and mineral interactions in human beings. *J. Am. Dietet. Assoc.* 86: 864.

Spencer, H. and Kramer, L. 1985. Osteoporosis: Calcium, flouride and aluminum interactions. *J. Am. Coll. Nutr.* 4: 121.

Spencer, H., Kramer, L., Norris, C., and Osis, D. 1984. Effect of calcium and phosphorus on zinc metabolism in man. *Am. J. Clin. Nutr.* 40: 1213.

Spencer, H., Kramer, L., Norris, C., and Osis, D. 1985. Inhibitory effect of zinc on intestinal absorption of calcium. *Clin. Res.* 33: 872.

Spencer, H., Schwartz, R., Norris, C., and Osis, D. 1985. Magnesium—28 studies and magnesium balances in man. *J. Am. Coll. Nutr.* 4: 316.

Taper, L. J., Hinners, M. L., and Ritchey, S. J. 1980. Effect of zinc intake on copper balance in adult females. *Am. J. Clin. Nutr.* 33: 1077.

Umeki, S., Ohga, R., Konishi, Y., Yasuda, T., Morimoto, K., and Terao, A. 1986. Oral therapy normalizes serum uric acid level in Wilson's disease patients. *Am. J. Med. Sci.* 292: 289.

Valberg, L. S., Flanagan, P. R., and Chamberlain, M. J. 1984. Effects of

iron, tin, and copper on zinc absorption in humans. *Am. J. Clin. Nutr.*
40: 536.

Van Caillie-Bertrand, M., Degenhart, H. J., Visser, H. K. A., Sinaasappel,
M., and Bouquet, J. 1985. Oral zinc sulphate for Wilson's disease. *Arch.
Dis. Child.* 60: 656.

Van Campen, D. R. 1966. Effects of zinc, cadmium, silver and mercury on
the absorption and distribution of copper-64 in rats. *J. Nutr.* 88: 125.

Walravens, P. A. and Hambidge, K. M. 1976. Growth of infants fed a zinc
supplemented formula. *Am. J. Clin. Nutr.* 29: 1114.

Walser, M. 1967. Magnesium metabolism. *Ergeb. Physiol. Biol. Chem. Exp.
Pharmakol.* 59: 185.

Watson, W. S., Hume, R., and Moore, M. R. 1980. Absorption of lead and
iron. *Lancet* 2: 236.

White, H. S. and Gynne, T. N. 1971. Utilization of inorganic elements by
young women eating iron-fortified foods. *J. Am. Dietet. Assoc.* 59: 27.

Yip, R., Reeves, J. D., Lonnerdal, B., Keen, C. L., and Dallman, P. R. 1985.
Does iron supplementation compromise zinc nutrition in healthy infants? *Am. J. Clin. Nutr.* 42: 683.

Zareba, G., Chmielnicka, J., and Kustra, J. 1986. Interaction of tin and
zinc in some processes of heme biosynthesis in rabbits. *Ecotoxicol. Environ. Safety* 11: 144.

# 5

# Protein-Iron Interactions: Influences on Absorption, Metabolism, and Status

**Elaine R. Monsen**

University of Washington
Seattle, Washington

Although iron is the most abundant heavy metal in the earth's crust, iron deficiency ranks as a major nutritional problem worldwide. Absorption of iron is controlled strictly, and excretion of iron is limited. Protein is one of the major participants in the tight regulation of iron: bioavailability of dietary iron is influenced by concomitantly ingested proteins; dietary patterns and protein status affect iron status; additionally, specialized proteins are instrumental in iron absorption and transport.

Food iron may be classified as either heme iron or nonheme iron. Humans absorb heme iron at a higher rate than nonheme iron. However, as food contains more nonheme iron, it is nonheme iron that makes the larger contribution to the body's iron pool. Absorption of nonheme iron is markedly influenced by a variety of factors found in foods. Most proteins are inhibitory to the absorption of nonheme iron. Cellular animal proteins are the exception, in that they enhance nonheme iron absorption. The controlling mechanism whereby proteins serve to either enhance or inhibit nonheme iron absorption is proposed to be the specific chelates of iron and protein/protein digestion products that are formed in the gut.

149

## NONCELLULAR PROTEIN INHIBITION OF NONHEME IRON ABSORPTION

Absorption of nonheme iron has been studied extensively by the use of extrinsic radioactive iron labels added to single foods or to meals (Cook et al., 1972; Hallberg, 1980). The rate of absorption is estimated by counting blood samples taken two weeks after the test dose. Thus, iron that enters the mucosal cells, passes into the plasma, and is metabolized is measured. In studies where subjects were fed meals composed of semisynthetic components (Monsen and Cook, 1979), egg albumen strongly inhibited absorption of nonheme iron (Fig. 5.1). Compared to a 700-kcal semisynthetic meal composed of 67 g dextrimaltose, 35 g corn oil, 37 g egg albumen, and 4.1 mg of iron, deletion of the egg albumen resulted in a 150% increase in absorption, while doubling the egg albumen to 74 g resulted in a 40% decrease in the amount of iron absorbed. Thus, removal of the noncellular animal protein resulted in over four times as much nonheme iron being absorbed as when the larger amount of egg albumen was presented. In contrast, deleting or doubling the other dietary components, i.e., dextrimaltose or corn oil, produced only modest changes in nonheme iron absorption.

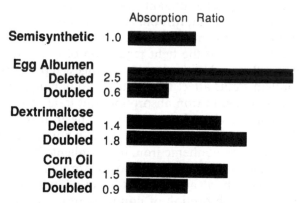

**FIG. 5.1** Absorption of nonheme iron from semisynthetic meals. Absorption ratios compare % absorption of nonheme iron from test meal to % absorption of nonheme iron from semisynthetic meal. Absorption from the semisynthetic meals set at 1.0 for the 27 subjects. (*From Monsen and Cook, 1979.*)

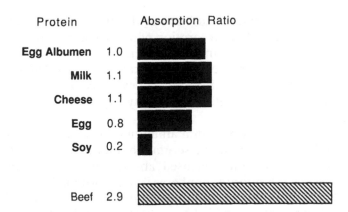

**FIG. 5.2** Absorption of nonheme iron when an equal quantity of noncellular animal protein was substituted for 25 g egg albumen or 100 g beef. These substitutions were made in 700-kcal meals composed of semisynthetic components or standard foods. These meals were fed to 33 adult subjects. Absorption of the egg albumen–containing meals set at 1.0. (*From Cook and Monsen, 1976.*)

Other noncellular animal proteins are also inhibitory to the absorption of nonheme iron (Fig. 5.2). When substituted in the semisynthetic meal for the egg albumen, milk and cheese were similarly inhibitory and dried whole egg was slightly more inhibitory than egg albumen (Cook and Monsen, 1976). The similarity of egg, egg albumen, milk, and cheese was seen whether or not the 700-kcal meals were composed of semisynthetic ingredients or standard foods. Absolute absorption was higher with the standard food meals (geometric mean of 10.0% from the standard food meal compared to 1.7% from the semisynthetic meal), yet the absorption ratio of the substituting protein source to egg albumen was constant, regardless of whether the semisynthetic meal or the standard food meal was served. As subjects consumed two to four test meals, individual variation was accounted for by comparing each individual's absorption of nonheme iron from the test meal to his or her absorption of nonheme iron from a reference dose or reference meal.

In subsequent studies, soy protein was reported to have an exceptionally inhibitory effect; nonheme iron was absorbed at only 20% of the reference level when soy protein was substituted for egg albumen in the semisynthetic meal (Cook et al., 1981; Lynch et al., 1985). In additional studies it was shown that the addition of 100 mg ascorbic acid to the soy protein meals increased the absorption of nonheme iron from 0.6 to 3.2% (Morck et al., 1982). The soy inhibition seemed to be modulated more by ascorbic acid than by the addition of meat to the isolated soy protein meals. Lynch and Covell (1987) provided evidence that iron in soybean flour is bound to phytoferritin, which may explain its poor absorption by humans.

Other studies with bovine serum albumin (Hurrell et al., 1988) showed that the addition of 30 g bovine serum albumin to a meal of 67 g corn starch and 35 g corn oil decreased absorption, while bovine serum albumin was not inhibitory to nonheme iron absorption in a meal of corn gruel and butter. In contrast, the bovine serum albumin increased absorption by 60% when it was added to a meal containing Egyptian baladi bread. Although inhibitory, bovine serum albumen did not exhibit as great an inhibitory effect as egg albumen.

## ENHANCEMENT OF NONHEME IRON ABSORPTION BY CELLULAR PROTEIN

In contrast to noncellular animal protein, cellular animal proteins appear to enhance the absorption of nonheme iron (Fig. 5.3). Approximately three times as much nonheme iron is absorbed when cellular animal proteins are substituted for 20 g protein from egg albumen in semisynthetic meals (Cook and Monsen, 1976). This effect was also observed in standard food meals which also contained 700 kcal, 29.4 g protein, 67.6 g carbohydrate, 34.9 g fat, 202 mg calcium, 414 mg phosphorus, and 4.1 mg iron, but in the form of beef, potatoes, cornmeal, bread, margarine, ice milk, and peaches. Substitution of the beef with liver produced a 21% increase in absorption, while substitution with either lamb or pork resulted in absorption similar to that seen with beef. Chicken and fish were somewhat lower in their enhancing effect in comparison to beef, with the absorption of nonheme iron being 17% lower with the substitution of chicken and 28% lower with the substitution of fish. When the noncellular animal protein, egg albumen, was substituted for beef, absorption of nonheme iron was reduced by 66%.

Other animal cellular proteins have also been reported to enhance absorption of nonheme iron. Addition of calf thymus gland increased absorption of nonheme iron from a meal of maize at a rate similar to the addition of beef (Bjorn-Rasmussen and Hallberg, 1979). In these studies the ratio of absorption of nonheme iron was again approximately three times

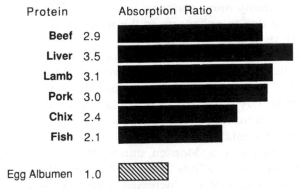

**FIG. 5.3** Absorption of nonheme iron where an equal quantity of cellular animal protein was substituted for 25 g egg albumen in the semisynthetic meal or 100 g beef in the standard food meal. These meals were fed to 44 adult subjects. Absorption of the egg albumen containing meals set at 1.0. Chicken represented by Chix. (*From Cook and Monsen, 1976.*)

higher in the presence of beef, chicken, fish, or calf thymus gland than it was in the presence of egg albumen.

The enhancing effect of cellular animal protein has been shown to be dose-related (Fig. 5.4) (Cook and Monsen, 1975; Bjorn-Rasmussen and Hallberg, 1979). In the semisynthetic meal, increasing the quantity of beef substituted on a protein basis for egg albumen caused a steady increase in the absorption of nonheme iron. When 100 g of beef was given in the semisynthetic meal, absorption was 3.6 times that seen when no beef was included. The standard food meal of the same macronutrient content

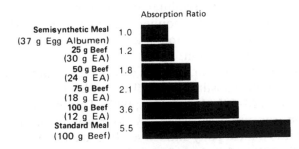

**FIG. 5.4** Absorption of nonheme iron from 700-kcal semisynthetic meals containing 29.4 g protein contributed by various combinations of egg albumen and beef. Comparison is also made to a standard food meal with the same total protein content. Absorption from the semisynthetic meals set at 1.0 for the 32 women subjects. (*From Cook and Monsen, 1975.*)

allowed considerably more nonheme iron to be absorbed. This may, in part, have been due to enhancers in the meal or to the absence of egg albumen with its known inhibiting effect.

Many studies have looked at the enhancement of nonheme iron absorption by the addition of meat (Hallberg and Rossander, 1984). Both raw and cooked meat appear to increase nonheme iron absorption (Monsen, unpublished). A water extract of cooked beef has been reported to have less stimulation than the residue (Bjorn-Rasmussen and Hallberg, 1979). Addition of a mixture of amino acids that mimic beef increased absorption of nonheme iron (Monsen, unpublished). However, individual amino acids have specific effects. Histidine has been shown to have no enhancing effect, while cysteine increased nonheme iron absorption 3.5 times (Layrisse et al., 1984). In further studies, cysteine-containing peptides produced during meat digestion were seen to enhance absorption of nonheme iron (Taylor et al., 1986). In contrast, oxidizing these products of meat digestion to cysteine-containing peptides reduced the absorption of nonheme iron by 63%. Thus, it appears that free sulfhydryl groups aid absorption of nonheme iron.

## HEME IRON ABSORPTION

Heme iron plays an important role in iron status. Approximately 40–60% of the iron in meat, fish, and poultry is in the form of heme (Cook and Monsen, 1976; Lynch et al., 1985; Martinez-Torres et al., 1986). Heme in meat, fish, and poultry is readily absorbed by humans as an intact protoporphyrin molecule. However, when isolated hemoglobin was administered in meals containing no cellular animal protein, absorption of heme iron was depressed along with nonheme iron absorption (Hallberg et al., 1979). Thus, cellular animal protein appears to enhance absorption of both heme and nonheme iron.

Another protein factor that has been reported to enhance heme iron absorption is soy flour. Substituting soy flour for beef increased absorption of heme iron from 30–60% (Lynch et al., 1985). This increase in the absorption of heme iron was seen when soy flour was incorporated with the ground beef into a meat patty and fed solely as a patty or in a mixed meal composed of a patty, white bun, french fries, and a vanilla milk shake. In other studies, large quantities of heme blocked absorption of nonheme iron; correspondingly, large quantities of nonheme iron in a meal blocked absorption of heme (Hallberg et al., 1979). This observation supports the concept that nonheme iron released from heme forms a common pool in the mucosal cell. As this pool becomes saturated, further absorption would decline sharply.

Interesting studies have been conducted where heme has been destroyed during extensive cooking and, with this destruction, decreased absorption of heme iron has been observed (Martinez-Torres et al., 1986). In these studies, a 26% reduction in absorption was observed from meat fried for 10 minutes and held in a 350° F oven for 60 minutes compared to meat that was fried for only 10 minutes. Meals with similarly overcooked meat showed a 56% reduction in the absorption of heme iron.

Fortification of foods with bovine hemoglobin is currently under investigation in Chile (Stekel et al., 1986). Initial studies show higher serum ferritin levels in females over 10 years of age who received daily three or four 10-g cookies fortified with bovine hemoglobin compared to similar subjects receiving unfortified cookies.

Other inhibitors and enhancers of nonheme absorption have been shown to have no effect on the absorption of heme. For example, ascorbic acid does not increase heme absorption, nor do bran, tea, or desferrioxamine appear to appreciably affect the absorption of heme iron.

## SPECIALIZED PROTEINS INFLUENCING IRON ABSORPTION

Extensive work by Lonnerdal and co-workers have illustrated the key role of human lactoferrin in absorption of iron from breast milk (Lonnerdal, 1984). Lactoferrin, a glycoprotein synthesized in the mammary gland, is a major protein of human milk. It is highly resistant to proteolysis. Receptors have been identified in the small intestine which accept human lactoferrin and rhesus monkey lactoferrin but not bovine lactoferrin. This mechanism may explain why human infants absorb over 50% of the iron in breast milk, yet absorb iron poorly from cow's milk. Iron supplementation of cow's milk formulas compensates effectively.

Another protein that has been observed to be specific for iron is uteroferrin. This specialized protein has been shown to transport iron in the placenta of pigs (Roberts et al., 1986). Uteroferrin is synthesized and secreted by the glandular epithelial cells of maternal uterus of the pig. It is readily taken up by the fetus and efficiently incorporated into fetal hemoglobin.

## DIETARY PATTERNS: METABOLIC EFFECTS OF THE QUANTITY AND SOURCE OF DIETARY PROTEIN

Calorie and protein deficiency in humans as exhibited in times of famine has been shown to increase iron stores (Murray and Murray, 1987). Similarly, weanling rats fed a protein-deficient diet exhibited higher levels of storage iron (Enwonwu et al., 1972). The effect of a short supply of protein and/or calories in either growing or adult animals may be to conserve iron. As the ability to form red blood cells is diminished by the protein-calorie deficit, more absorbed iron would be diverted to storage.

Long-term dietary patterns are reflected in iron status. Concern has arisen regarding vegetarian regimens for three reasons: (1) the lack of heme, which is known to be absorbed at a high rate by humans: (2) the lack of meat, which increases absorption of nonheme iron; and (3) excessive use of foods, such as bran, which are inhibitory to nonheme iron absorption (Hallberg et al., 1987). Differentiation needs to be made, however, between long-term vegetarians with careful diet selection and new vegetarians where diet selection may be neither well controlled nor well designed. The effects of long-term intake have been studied in Seventh Day Adventist women who had been vegetarians for 19 or more years (Anderson et al., 1981). In these 56 women, 12.5 mg iron was consumed per day, 92% of which was from plant sources. Hemoglobin was satisfactory at 131 g/liter, and transferrin saturation was adequate at 0.38.

In contrast, new vegetarians have been shown to have considerably lower ferritin levels than omnivores (Helman and Darnton-Hill, 1987). Unsupplemented omnivores had serum ferritin levels of 70 µg/liter while new vegetarians had 45 µg/liter. Thus, estimated iron stores were 36% lower for the new vegetarians. It is interesting to note that approximately 22% of the new vegetarians were taking vitamin or mineral supplements, but few of these supplements contained iron.

A recent study compared 51 premenopausal women whose usual self-selected diets differed by the predominant protein source (Worthington-Roberts et al., 1988), being either red meat (at least five times per week), fish/poultry (with no more than one serving of red meat per week), or lacto-ovo plants (with no red meat use and not more than one serving of fish/poultry per week). No differences were seen in length of menses or menstrual cycle (Table 5.1). Similar daily intakes of 1800 kcal, 160 mg ascorbic acid, and 12 mg iron were recorded for the three groups. However, statistically different intakes of protein, carbohydrate, and fat were reported: Red meat eaters had the highest protein and fat and lowest carbo-

**TABLE 5.1** Iron Status and Dietary Patterns of Premenopausal Women

| | Predominant protein source | | |
|---|---|---|---|
| | Red meat | Fish/ Poultry | Lacto-ovo vegetarian |
| Subjects, *n* | 16 | 15 | 20 |
| Age, years | 31 | 28 | 29 |
| Menstrual cycle | | | |
| Interval, days | 29 | 30 | 29 |
| Flow, days | 4.5 | 4.6 | 4.7 |
| Diet intake/day | | | |
| Iron, mg | 13 | 12 | 12 |
| Ascorbic acid, mg | 144 | 169 | 169 |
| kcal | 1788 | 1785 | 1814 |
| Protein, g* | 76 | 72 | 64 |
| Carbohydrate, g* | 197 | 209 | 249 |
| Fat, g | 78 | 70 | 64 |
| Hemoglobin g/l | 140 | 130 | 130 |
| Serum ferritin µg/liter* | 30.5 | 15.6 | 19.1 |

*$p < 0.05$; difference between protein sources indicated by analysis of variance.
*Source: Worthington-Roberts et al. (1988).*

hydrate intakes; the reverse was seen in the lacto-ovo vegetarians, whose reported intakes were lowest in protein and fat and highest in carbohydrate. Iron status was adequate in all subjects, although hemoglobin levels and serum ferritin were statistically higher in the red meat eaters. Interestingly, the lowest serum ferritin levels were seen in the fish/poultry group. This suggests that long-term dietary patterns impact iron stores either through influences on absorption or metabolism of iron. While meat consumption is associated with higher iron stores, a well-designed and carefully selected lacto-ovo vegetarian diet can support adequate hemoglobin levels and modest iron stores.

## IRON STORES: A REGULATORY OF IRON ABSORPTION

Opportunity for excretion of iron from the healthy body is limited to exfoliated skin, sloughed gastrointestinal cells, modest gastrointestinal blood loss, and menstrual losses. With such restricted avenues for ridding the body of iron, it is apparent that absorption must be strictly controlled. As iron stores increase, absorption of both heme and nonheme iron

**TABLE 5.2**  Characteristics of Theoretical Subjects at Selected Levels of Iron Stores and Estimated Obligatory Iron Absorption to Maintain or Achieve Stated Iron Stores

100 mg—*Iron-deficient*
        Absorption up to maximum (~25%)
        Anemia may develop
300 mg—*Average iron stores for 12–50 year old woman*
        11 mg iron average daily intake
        1.4 mg iron absorbed/day required for balance
        13% absorption of 11-mg intake allows 1.4 mg absorbed
     —*5% of 12–50-year-old women require*
        >2.4 mg iron absorbed/day for balance
        22% absorption of 11-mg intake needed to absorb 2.4 mg iron
1000 mg—*Average iron stores of adult man*
        17 mg iron average daily intake
        1.0 mg iron absorbed per day for balance
        to achieve 1000-mg store, additional absorption of 0.06 mg/day suffices
        6% absorption of 17-mg intake is needed

declines. Correspondingly, lowered iron stores increase rate of iron absorption up to a maximum approximately 25% of dietary iron. This rate has been recorded in subjects with depleted iron stores from blood donations (Monsen et al., 1983) or frequent blood withdrawals (Mahalko, 1983). A similar rate of absorption is expected in iron-deficient individuals with iron stores ≤ 100 mg (Table 5.2).

The average woman during her reproductive years will have iron stores of approximately 300 mg. Average iron intake reported in NHANES II for women 12–50 years of age is 11 mg/day (NCHS, 1983). To meet her average needs, a woman would need to absorb 13% of the ingested 11 mg. If menstrual losses were above average, absorption would need to increase and could reach 25% of ingested iron (Table 5.2).

The average adult man consumes 17 mg iron/day (NCHS, 1983) and has iron stores approximating 1000 mg. If 6% of ingested iron were absorbed, an adult man could maintain iron balance plus store 0.06 mg iron/day; this modest retention would allow 1000 mg iron stores to be built up over 45 years (Table 5.2).

The impact of iron stores on absorption of dietary iron was incorporated in a model proposed to calculate available dietary iron. Depending on iron stores, absorption of heme iron was estimated to range from 15–35% and nonheme iron from 2–20% (Monsen et al., 1978). Thus, absorption of dietary iron would be expected to be within the range 5–25% from a mixed diet. This range of absorption is compatible with the Recommended Dietary Intakes suggested by Herbert (1987).

## CONCLUSION

Protein interacts with iron at many levels, impacting absorption, metabolism, and nutritional status of iron. Iron readily oxidizes and precipitates from solution unless it is chelated. In the gut, protein and products of its digestion are able to chelate iron freed from foods (Fig. 5.5). The specific iron chelates formed may either be excreted or be brought to the mucosal brush border, where the iron is released and absorbed at a rate governed by an individual's iron stores. Noncellular proteins tend to depress nonheme iron absorption, while cellular animal proteins act as enhancers. Specific amino acids, peptides, and products of protein digestion appear to be potent factors in controlling iron absorption.

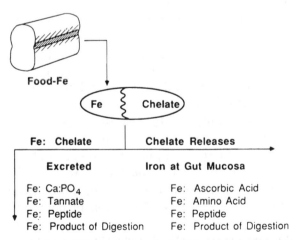

**FIG. 5.5** A proposed mechanism for nonheme iron absorption suggests that during digestion iron is released from food. The free iron forms chelates with a variety of substances, some of which bind tightly and are excreted along with the bound iron, while other substances release the iron at the mucosal surface.

# REFERENCES

Anderson, B. M., Gibson, R. S., and Sabry, J. H. 1981. The iron and zinc status of long-term vegetarian women. *Am. J. Clin. Nutr.* 34: 1042.

Bjorn-Rasmussen, E. and Hallberg, L. 1979. Effect of animal proteins on the absorption of food iron in man. *Nutr. Metab.* 23: 192.

Cook, J. D., Layrisse, M., Martinez-Torres, C., Walker, R., Monsen, E. R., and Finch, C. A. 1972. Food iron absorption measured by an extrinsic tag. *J. Clin. Invest.* 51: 805.

Cook, J. D. and Monsen, E. R. 1975. Food iron absorption. I. Use of a semi-synthetic diet to study absorption of nonheme iron. *Am. J. Clin. Nutr.* 28: 1289.

Cook, J. D. and Monsen, E. R. 1976. Food iron absorption in human subjects III. Comparison of the effects of animal protein on non-heme iron absorption. *Am. J. Clin. Nutr.* 29: 859.

Cook, J. D., Morck, T. A., and Lynch, S. R. 1981. The inhibitory effect of soy products on nonheme iron absorption in man. *Am. J. Clin. Nutr.* 34: 2622.

Enwonwu, C. O., Monsen, E. R., and Jacobson, K. 1972. Absorption of iron in protein-calorie deficient rats and immediate effects of refeeding an adequate protein diet. *Am. J. Digestive Dis.* 17: 959.

Hallberg, L. 1980. Food iron absorption. In *Methods in Hematology*. Cook, J. D. (Ed.), p. 116. Churchill, London.

Hallberg, L., Bjorn-Rasmussen, E., Howard, L., and Rossander, L. 1979. Dietary heme iron absorption: possible mechanisms for the absorption promoting effect of meat and for the regulation of iron absorption. *Scand. J. Gastroenterol.* 14: 769.

Hallberg, L. and Rossander. L. 1984. Improvement of iron nutrition in developing countries: comparison of adding meat, soy protein, ascorbic acid, citric acid, and ferrous sulphate on iron absorption from a simple Latin American-type of meal[1-3]. *Am. J. Clin. Nutr.* 39: 577.

Hallberg, L., Rossander, L., and Skanberg, A-B. 1987. Phytates and the inhibitory effect of bran on iron absorption in man. *Am. J. Clin. Nutr.* 45: 988.

Helman, A. D., and Darnton-Hill, I. 1987. Vitamin and iron status of new vegetarians. *Am. J. Clin. Nutr.* 45: 785.

Herbert, V. 1987. Recommended dietary intakes (RDI) of iron in humans. *Am. J. Clin. Nutr.* 45: 679.

Hurrell, R. F., Lynch, S. R., Trinidad, P. T., Dassenko, S. A., and Cook, J. D. 1988. Iron absorption in humans: bovin serum albumin compared with beef muscle and egg white. *Am. J. Clin. Nutr.* 47: 102.

Layrisse, M., Martinez-Torres, C., Leets, I., Taylor, P., and Ramirez, J. 1984. Effect of histidine, cysteine, glutathione, or beef on iron absorption in humans. *J. Nutr.* 114: 217.

Lonnerdal, B. 1984. Iron in breast milk. In *Iron Nutrition in Infancy and Childhood*, Nestle Nutrition Workshop Series, Vol. 4, Stekel, A. (Ed.), p. 95. Vevey/Raven Press, New York.

Lynch, S. R., and Covell, A. M. 1987. Iron in soybean flour is bound to phytoferritin. *Am. J. Clin. Nutr.* 45: 866.

Lynch, S. R., Dassenko, S. A., Morck, T. A., Beard, J. L., and Cook, J. D. 1985. Soy protein products and heme iron absorption in humans. *Am. J. Clin. Nutr.* 41: 13.

Mahalko, J. R., Sandstead, H. H., Johnson, L. K., and Milne, D. B. 1983. Effect of a moderate increase in dietary protein on the retention and excretion of calcium, copper, iron, magnesium, phosphorus, and zinc by adult males. *Am. J. Clin. Nutr.* 37: 8.

Martinez-Torres, C., Leets, I., Taylor, P., Ramirez, J., Camacho, M., and Layrisse, M. 1986. Heme, ferritin and vegetable iron absorption in humans from meals denatured of heme iron during cooking. *Am. J. Clin. Nutr.* 116: 1720.

Monsen, E. R. and Cook, J. D. 1979. Food iron absorption in human subjects. V. Effects of the major dietary constituents of the semisynthetic meal. *Am. J. Clin. Nutr.* 32: 804.

Monsen, E. R., Critchlow, C. W., Finch, C. A., and Donohue, D. M. 1983. Iron balance in superdonors. *Transfusion* 23: 221.

Monsen, E. R., Hallberg, L., Layrisse, M., Hegsted, D. M., Cook J. D., Mertz, W., and Finch, C. A. 1978. Estimation of available dietary iron. *Am. J. Clin. Nutr.* 31: 134.

Morck, T. A., Lynch, S. R., and Cook, J. D. 1982. Reduction of the soy induced inhibition of nonheme iron absorption. *Am. J. Clin. Nutr.* 36: 219.

Murray, M. J. and Murray, A. B. 1987. Mobilization of storage iron as a determinant of refeeding infection. *Trace Elements in Man and Animals*, 6th Conference, p. 16.

National Center for Health Statistics. 1983. *Dietary Intake Source Data: United States, 1976–1980.* Vital and Health Statistics Series 11-No 231,

DHHS Pub. No. (PHS) 83-1681. U. S. Government Printing Office, Washington, DC.

Roberts, R. M., Raub, T. J. and Bazer, F. W. 1986. Role of uteroferrin in transplacental iron transport in the pig. *Federation Proc.* 45: 2513.

Stekel, A., Monckeberg, F., and Beyda, V. 1986. Combatting iron deficiency in Chile: a case study. International Nutritional Anemia Consultative Group (INACG), Step.

Taylor, P. G., Martinez-Torres, C., Romano, E. L., and Layrisse, M. 1986. The effect of cysteine-containing peptides released during meat digestion on iron absorption in humans. *Am. J. Clin. Nutr.* 43: 68.

Worthington-Roberts, B. S., Breskin, M. W., and Monsen, E. R. 1988. Iron status of premenopausal women in a university community and its relationship to habitual dietary sources of protein. *Am. J. Clin. Nutr.* 47: 275.

# 6

# Vitamin–Mineral Interactions

**Bo Lönnerdal**

University of California
Davis, California

The growing use of nutrient supplements as well as the popularity of low-calorie but often nutritionally inadequate diets gives rise to concern about interactions between vitamins and minerals. Such interactions are very different in their biochemistry but can occur by two fundamental mechanisms: (1) one micronutrient directly affects the absorption of another micronutrient, and (2) deficiency or excess of one micronutrient within the organism affects the metabolism of another micronutrient. Examples of the first type of interaction are the interactive effects of ascorbic acid on iron absorption and the potential interaction between negatively charged folates and positively charged trace elements, such as zinc. The second type of interaction is exemplified by the effect of zinc deficiency on retinol binding protein (RBP) and vitamin A metabolism and the effect of high doses of ascorbic acid on copper metabolism. The effects of negative micronutrient interactions need to be considered when single nutrient supplements are used. In addition, the need for supplementing populations at risk for one micronutrient deficiency with other micronutrients that would be affected by the primary deficiency must be assessed.

## ASCORBIC ACID-IRON INTERACTIONS

One of the more classical examples of a direct interaction between a vitamin and a mineral is the enhancing effect of ascorbic acid on non-heme iron absorption. It is believed that this stimulatory effect of ascorbic acid on iron absorption is exerted by both its reducing capacity, thereby keeping iron in the more easily absorbed ferrous (+II) form, and its chelating properties, keeping iron in a soluble, absorbable form (Hallberg, 1981). Numerous studies have shown that low iron bioavailability from various diets can be overcome by ascorbic acid supplementation. Recently, Gillooly et al. (1983) showed that, compared to several organic acids including malic acid, citric acid, and tartaric acid, ascorbic acid was most efficient in increasing iron absorption from various vegetables. The same authors (Gillooly et al., 1984) also showed that ascorbic acid above a certain level had a beneficial effect on iron absorption from cow's milk formula and, in particular, from soy formula. Thus, there is a need for supplementation with ascorbic acid beyond the nutritional requirement for vitamin C as such. In adults it is not necessary to use pure ascorbic acid supplements to enhance iron absorption. Hallberg and Rossander (1984) recently showed that by combining local food items rich in vitamin C, such as cauliflower, with other Latin-American food items with low iron bioavailability, a significant positive effect on iron absorption was observed. The relative importance of ascorbic acid acting as a reducing agent or as a chelator has not been assessed in detail. It appears likely, however, that the reducing capacity is a necessary prerequisite since iron in the ferric form is poorly absorbed. The importance of ascorbic acid as a reducing agent is supported by the fact that ascorbic acid was efficient in small doses, while other organic acids with chelating capacity but without reduction potential required considerably higher levels to observe a positive effect on iron absorption (Gillooly et al., 1983).

The interaction of ascorbic acid and iron also occurs at other levels. For example, it is well known that hemoglobin synthesis is impaired during ascorbic acid deficiency (Banerjee and Chakrabarty, 1965) and that iron supplementation cannot overcome this impairment (Table 6.1). In order to transport iron into the cell, conversions between the ferric and the ferrous forms are needed (Sirivech et al., 1974). Recently, Bridges and Hoffman (1986) showed that ascorbate stabilizes ferritin in isolated cells and that the pool of chelatable iron was increased, indicative of reduction of ferric iron in the ferritin core. Thus, in ascorbic acid deficiency, mobilization of storage iron is impaired and, similarly, when excess ascor-

**TABLE 6.1**  Hematologic Picture of Guinea Pigs

|  | Scorbutic guinea pig (5)[a] | Iron-treated scorbutic guinea pig (7) | Iron-treated normal guinea pig (8) |
|---|---|---|---|
| Hemoglobin, g/100 ml | 9.3 ± 0.63[b] | 5.9 ± 0.93 | 12.7 ± 0.55 |
| Red blood cell, 10⁶/mm | 3.7 ± 0.18 | 2.5 ± 0.40 | 5.0 ± 0.24 |
| Packed cell vol. (P.C.V.), % | 31.8 ± 1.83 | 20.0 ± 2.93 | 40.7 ± 1.38 |

[a]Number of animals.
[b]Values are expressed as mean ± SEM.
*Source: Banerjee and Chakrabarty (1965).*

bic acid is given, removal of iron from hemochromatosis patients with desferrioxamine is markedly improved. However, the detailed function of ascorbic acid in cellular iron metabolism needs to be studied further.

## ASCORBIC ACID–COPPER INTERACTIONS

Intake of high levels of ascorbic acid is known to have a negative effect on copper metabolism in experimental animals. Although some studies demonstrate a further reduction of copper status in animals fed copper-deficient diets (Carlton and Henderson, 1965; Hill and Starcher, 1965; Hunt and Carlton, 1965; Hunt et al., 1970), it is also obvious that copper deficiency can be induced in animals fed diets adequate in copper (Carlton and Henderson, 1965; Milne and Omaye, 1980; Smith and Bidlack, 1980). In several of these studies, serum concentrations of copper and ceruloplasmin, the major copper-binding protein in serum, were reduced, and soft tissue (liver) copper levels were lower in ascorbic acid–supplemented animals as compared to animals fed a control diet or a low copper diet (Table 6.2). Using a primate model, the adult cynomolgus monkey, Milne et al. (1981) fed a diet marginal in copper and no vitamin C for four weeks. Plasma ascorbate levels at 4 weeks were significantly lower than initial values, while serum copper and ceruloplasmin values showed little change. After this time, animals were fed the same diet with either a low ( 1 mg/kg body weight/day) or a high ascorbic acid level (25 mg/kg body weight/day) for four weeks. When fed the higher level of ascorbic acid, plasma ascorbate increased to the initial level, while serum copper and ceruloplasmin were reduced about 20% after 4 weeks. An in-

LÖNNERDAL

**TABLE 6.2**  Effect of Vitamin C on Copper and Iron in Plasma and Liver

| Treatment | $n^a$ | Plasma | | | Liver | |
|---|---|---|---|---|---|---|
| | | $CP^b$ (IU) | Cu (µg/ml) | Fe (µg/ml) | Cu (µg/g) | Fe (µg/g) |
| Low vitamin C | (7) | $71.2 \pm 9.2^c$ | $1.05 \pm 0.15$ | $1.47 \pm 0.20$ | $162.0 \pm 15.1$ | $310.0 \pm 25.0$ |
| High vitamin C | (7) | $35.2 \pm 2.6$ | $0.54 \pm 0.05$ | $2.38 \pm 0.29$ | $62.2 \pm 14.4$ | $736.9 \pm 103.8$ |
| Statistical significance | | $<0.001$ | $<0.001$ | $<0.001$ | $<0.001$ | $<0.01$ |

[a]Number of animals.
[b]Ceruloplasmin. IU = µm product/minute/liter plasma.
[c]Values are expressed as mean ± SEM.
*Source: Milne and Omaye (1980).*

teresting finding was that during the period of copper depletion, serum cholesterol levels increased significantly. This increase was more pronounced for animals receiving the higher level of ascorbic acid. It should be noted that although serum copper levels were reduced in these animals, the change was not significantly different. This is in contrast to earlier studies in which much more dramatic reductions in plasma copper and ceruloplasmin were observed. As pointed out by the authors, however, in earlier studies growing animals were used while Milne et al. (1981) used adult monkeys with a presumably lower requirement for copper. The finding of increased serum cholesterol levels in this study is consistent with similar findings in copper-deficient animals (Lei, 1977; Allen and Klevay, 1978). The higher levels of serum cholesterol in monkeys fed the higher level of ascorbic acid is also in agreement with the hypercholesterolemia due to ascorbic acid described by Klevay (1976).

Recently, Finley and Cerklewski (1983) showed that high intake of ascorbic acid (1500 mg/day) in humans consuming a regular diet with normal copper concentration (1.8 mg/day) also can lead to low serum copper and ceruloplasmin. The study was performed with only male subjects to avoid sex-related differences in copper metabolism caused by different estrogen levels. In addition, termination of ascorbic acid supplementation caused an increase in serum copper and ceruloplasmin levels and a reversal to initial levels by 20 days.

The mechanism behind the interaction of ascorbic acid and copper has not been studied in detail. It is believed, however, that high levels of ascorbic acid reduce copper (+II) to copper(+I), which is less absorbable at the initial uptake phase so that the absorptive step is impaired (Van Campen and Gross, 1968). Serum ceruloplasmin is known to closely follow serum copper levels. Both of these parameters show lower than normal circulating levels of copper; however, high levels of serum ascorbic acid may reduce ceruloplasmin activity in the assay. The hypercholesterolemia observed is then a secondary consequence induced by the copper deficiency and not the high levels of ascorbic acid per se. It should also be noted that ascorbic acid and copper may interact at several levels. More recently, Hunt et al. (1970) have shown that the activity of cytochrome oxidase, a copper-dependent enzyme, is low in hearts of copper-deficient rabbits and that a diet supplemented with ascorbic acid further reduces the activity of this enzyme (Table 6.3). DiSilvestro and Harris (1981) observed that when L-ascorbic acid was given "postabsorptive," i.e., after allowing for absorption and transport of copper, a beneficial effect on restoration of copper-dependent enzymes in tissues was obtained (Table 6.4). Thus, ascorbic acid may also be needed in the oxidoreduction of copper when incorporated into proteins.

**TABLE 6.3** Cytochrome Oxidase Activities in Rabbit Tissues after 12 Weeks on Experimental Diets

| Diet | Liver[a] (3)[b] | Heart[a] (3) |
|------|-----------------|--------------|
| Control | $89.4 \pm 2.2$[c] | $37.9 \pm 2.9$ |
| Copper-deficient + ascorbic acid | $43.1 \pm 3.0$ | $11.9 \pm 2.1$ |
| Copper-deficient | $41.3 \pm 4.1$ | $29.1 \pm 3.1$ |

[a]Each value is the rate constant/min per mg nitrogen ($\times 10^3$ for liver, $\times 10^2$ for heart).
[b]Number of animals.
[c]Values are expressed as mean $\pm$ SEM.
*Source: Hunt et al. (1970).*

**TABLE 6.4** Effect of L-Ascorbate Administration on the Copper-induced Activation of Aortic Lysyl Oxidase

| Treatment | Lysyl oxidase activity[a] (8)[b] |
|-----------|----------------------------------|
| Copper-deficient | $0.4 \pm 0.2$[c] |
| + L-Ascorbate | $0.8 \pm 0.5$ |
| + CuSO$_4$ | $41.0 \pm 0.5$* |
| + CuSO$_4$, L-Ascorbate (0 min)[d] | $34.2 \pm 4.9$** |
| + CuSO$_4$, L-Ascorbate ($-75$ min) | $29.7 \pm 2.1$** |
| + CuSO$_4$, L-Ascorbate ($+75$ min) | $62.3 \pm 4.9$*** |
| Copper-fed | $65.4 \pm 4.3$ |

[a]Each value is cpm $^3$H$_2$O/hr/g ($\times 10^{-3}$).
[b]Number of animals.
[c]Values are expressed as mean $\pm$ S.D. Values not sharing a common superscript letter are significantly different at **$p < 0.05$ or ***$p < 0.01$.
[d]Ascorbate time release in relation to CuSO$_4$ dose.
*Source: DiSilvestro and Harris (1981).*

Common use of high dose ascorbic acid supplements combined with the marginal copper content of the U.S. diet makes this a relevant nutritional concern.

## ZINC–VITAMIN E INTERACTIONS

In experimental animals the severity of some signs of zinc deficiency is affected by the polyunsaturated fatty acid (PUFA) levels in their diet (Bettger et al., 1979, 1980). The pathology observed was correlated to the degree of peroxidation of PUFAs. Because vitamin E serves a protective role in inhibiting peroxidative processes, Bettger et al. (1980) studied the interaction of zinc and vitamin E in the chick. After feeding a low-zinc diet (5μg/g), severe skin lesions and gross joint abnormalities were observed (Table 6.5). In addition, lipid peroxidation as measured by thiobarbituric acid (TBA)-reactive substances was demonstrated in zinc-deficient chicks. Supplementation of the zinc-deficient diets with antioxidants, such as vitamin E, significantly decreased the severity of the dermal and joint lesions. Furthermore, high levels of dietary vitamin E decreased the formation of TBA-reactive products, indicating an inhibitory effect on lipid peroxidation. The authors concluded that there is a physiological interaction between vitamin E and zinc in this species and suggested that zinc protects against peroxidative damage and promotes membrane integrity.

In order to investigate whether vitamin E can alleviate some of the signs of zinc deficiency, we supplemented a zinc-deficient diet with large doses of vitamin E and fed this to pregnant rats (Hurley et al., 1983). The teratogenicity of zinc deficiency is well documented (Hurley, 1981). Through its protective effect on peroxidation of essential fatty acids, we hypothesized that vitamin E might reduce the teratogenic effects of zinc deficiency. However, vitamin E did not affect pregnancy outcome as determined by number of malformations, resorption sites, or live fetuses. Furthermore, no effect on tissue zinc levels were found. The lack of effect in our study differs from the findings of Bettger et al. (1980) but may be explained by the low placental transfer of vitamin E. The fetus may not have received any benefit from the additional vitamin E given to the zinc-deficient pregnant dam. Martin and Hurley (1977) have shown that high levels of vitamin E are not teratogenic, most likely because of low transfer into the fetus.

The mechanism of zinc and vitamin E interaction is believed to be at the membrane level. Chvapil (1973) showed that high levels of zinc will reduce lipid peroxidation and described a role for zinc in the structure and integrity of biological membranes. Bettger et al. (1978) showed that erythrocytes from zinc-deficient rats are more fragile and that extracellular zinc can improve the integrity of the membranes under conditions of peroxidative stress. Since vitamin E serves a dual role in protecting membranes against peroxidation and maintaining membrane structure, zinc and vitamin E may act synergistically to preserve cell membrane integrity. Since only a few studies have investigated this synergism, further research is needed on this interaction.

Another possible interaction between zinc and vitamin E was recently reported by Bunk et al. (1987). These authors suggested that intestinal absorption and/or plasma transport of vitamin E are impaired during zinc

**TABLE 6.5**  Effect of Dietary Vitamin E Level on Zinc Status, Growth Rate, Leg Abnormalities, and Dermal Lesions

| Dietary supplement | | Plasma zinc[b] (ppm) | Body weight[b] (g) | Leg score[b] | Dermal score[b] |
|---|---|---|---|---|---|
| Zinc (ppm) | Vitamin E (IU/kg) | | | | |
| 5 | none[a] | 0.4 ± 0.1* | 112 ± 4* | 3.2 ± 0.1* | 2.6 ± 0.2* |
| 5 | 25 | 0.4 ± 0.1* | 109 ± 5* | 2.6 ± 0.2** | 2.5 ± 0.1* |
| 5 | 525 | 0.4 ± 0.1* | 103 ± 6* | 1.2 ± 0.2*** | 0.6 ± 0.2** |
| 100 (F.R.)[c] | none[a] | 1.3 ± 0.1** | 113 ± 4* | 0 | 0 |
| 100 (F.R.) | 25 | 1.4 ± 0.1** | 110 ± 4* | 0 | 0 |
| 100 (F.R.) | 525 | 1.4 ± 0.1** | 110 ± 5* | 0 | 0 |
| 100 | none[a] | 1.3 ± 0.1** | 503 ± 6** | 0 | 0 |
| 100 | 25 | 1.3 ± 0.1** | 501 ± 9** | 0 | 0 |
| 100 | 525 | 1.3 ± 0.1** | 510 ± 12** | 0 | 0 |

[a]All diets contained 5% corn oil which is estimated to supply 30 IU of vitamin E per kilogram of diet.
[b]Values are expressed as mean ± SEM. Values having different superscripts are significantly different by Student's $t$-test ($p <$ 0.05).
[c]Feed restricted so that the weights of these chicks matched those of the chicks fed the 5 ppm zinc diet ad libitum.
*Source: Bettger et al. (1980).*

**TABLE 6.6**   Zinc Concentrations of Maternal Tissues from Control and Vitamin E–Supplemented Zinc-Adequate and Zinc-Deficient Rats

| Experimental group | Plasma (μg/ml) | Liver (μg/g[a]) | Kidney (μg/g[a]) |
|---|---|---|---|
| Control | $1.05 \pm 0.07$[b] | $24.79 \pm 1.11$ | $20.99 \pm 0.69$ |
| Control + E | $1.30 \pm 0.05$[c] | $24.09 \pm 0.70$ | $24.04 \pm 0.70$ |
| Zn-deficient | $0.40 \pm 0.04$[c] | $23.46 \pm 0.07$ | $16.35 \pm 0.86$[c] |
| Zn-deficient + E | $0.67 \pm 0.11$[c,d] | $24.87 \pm 0.92$ | $18.51 \pm 1.44$ |

[a]Concentrations are on a wet weight basis.
[b]Values are expressed as mean $\pm$ SEM.
[c]Significantly different from control group, $p < 0.05$ using Student's $t$-test.
[d]Significantly different from Zn-deficient group, $p < 0.05$ using Student's $t$-test.
*Source: Hurley et al. (1983).*

deficiency. In their study, they found significantly lower plasma vitamin E levels in zinc-deficient rats ($4.0 \pm 1.2$ μg/ml) than in pair-fed ($9.2 \pm 0.7$ μg/ml) or ad libitum fed controls ($9.5 \pm 0.9$ μg/ml). In addition, plasma retinol, triglyceride, and cholesterol levels were lower in zinc-deficient rats than in controls. It was suggested that vitamin E, which is transported by plasma lipoproteins or lipids, are poorly absorbed during zinc deficiency. It is interesting to note that in our study (Hurley et al., 1983), we found higher plasma zinc levels in animals fed supplemented vitamin E, regardless of whether they were zinc-deficient or zinc-adequate (Table 6.6). It is therefore possible that transport of zinc/vitamin E across biological membranes is dependent on the levels of these two nutrients in the membrane.

## ZINC–VITAMIN A INTERACTIONS

A negative effect of zinc deficiency on the metabolism of vitamin A was shown in 1956 by Stevenson and Earle. These authors found that swine fed zinc-deficient diets had low levels of serum vitamin A and that these levels could not be increased by large supplements of retinyl acetate. Subsequently, studies in other species have documented that zinc-deficient animals exhibit low serum vitamin A levels despite adequate levels of vitamin A in the diet; hepatic stores of vitamin A were normal or elevated. Smith and co-workers (1973) showed that rats fed a zinc-deficient diet had markedly reduced levels of plasma vitamin A as compared to rats fed a diet adequate in zinc. They therefore suggested that mobilization of vitamin A from the liver was impaired by zinc deficiency. The authors recognized the possibility that reduced food intake, which occurs in zinc-deficient animals, might have caused the low serum vitamin A levels. They therefore designed another series of experiments (Smith et al., 1974) in which control animals were pair-fed the same amount of diet as the zinc-deficient animals consumed ad libitum (Table 6.7). Lower plasma vitamin A levels were found in the pair-fed rats than in control rats fed ad libitum showing an effect of reduced food intake independent of zinc deficiency. However, the zinc-deficient group had even lower plasma vitamin A levels than the pair-fed rats and, in addition, reduced levels of plasma retinol-binding protein (RBP). The lower level of plasma RBP was much more pronounced than the general decrease in plasma protein observed in the zinc-deficient group, suggesting a specific effect on RBP. This finding is consistent with the low level of liver RBP in zinc-deficient, but not in zinc-adequate, pair-fed animals. Another confounding parameter, however, was recognized: pair-fed animals had reduced growth rate, which could affect vitamin A utilization. In their next study (Smith et al., 1976), these authors showed that a severely food-restricted group fed a zinc-adequate diet had plasma vitamin A levels similar to zinc-deficient rats. It was concluded that absorption of vitamin A and its transport to the liver was not impaired since liver vitamin A levels were normal, that zinc deficiency or growth depression decreases plasma vitamin A levels, and that zinc deficiency per se may affect liver RBP synthesis.

Studies in humans fail to present a clear picture of the mechanisms

**TABLE 6.7**  Effect of Zinc Deficiency on Plasma Proteins and Retinol-Binding Protein (RBP) in Rats

| Treatment | Total plasma proteins (7)[a] (mg/ml) | Plasma RBP (7) (ng/mg protein) |
|---|---|---|
| Zinc-adequate (ad libitum) | 69.0 ± 1.1[b] | 49.7 ± 2.0 |
| Zinc-adequate (pair fed) | 68.1 ± 2.3 | 35.9 ± 1.2 |
| Zinc-deficient | 53.1 ± 1.8 | 13.9 ± 2.3 |
| Difference (%)[c] | 22 | 61 |

[a]Number of animals.
[b]Values are expressed as mean ± SEM.
[c]Difference is a comparison of the zinc-adequate, pair-fed group with the zinc-deficient group.
*Source: Smith et al. (1974).*

behind low serum vitamin A levels in zinc deficiency. In several conditions manifested by low zinc status in humans, such as alcoholic cirrhosis and protein–energy malnutrition, zinc supplementation sometimes, but not always, resulted in improved vitamin A status (Smith, 1982). It is possible that there is a threshold level of plasma zinc below which an effect can be expected and that no effect will be found above this level. It appears obvious that if zinc status is normal but vitamin A status impaired, no effect of zinc supplementation should be expected.

The mechanisms underlying the impaired vitamin A metabolism found in zinc deficiency and/or food restriction remain speculative. Smith (1982) has suggested that a severe zinc deficiency is required to affect vitamin A metabolism at the cellular level. Under such conditions, liver RBP synthesis may be impaired and retinene reductase activity reduced. This enzyme is zinc-dependent, and low activity of retinene reductase may be responsible for both the impaired dark adaptation found in zinc-deficient humans and retinal dysfunction described in zinc-deficient cats (Jacobson et al., 1986).

Both zinc deficiency and vitamin A deficiency during pregnancy can cause teratogenic effects. Since the American diet can be considered to be marginal in both zinc and vitamin A, particularly during pregnancy, there has been concern about the consequences of consumption of such diets during pregnancy. Duncan and Hurley (1978) found that outcome of pregnancy in rats depended on the intake of both zinc and vitamin A; there were significant interactions between the two nutrients. Teratogenic indices, such as affected implantation sites and proportion of malformed fetuses, were used. Recently, Baly et al. (1984) showed in pregnant rhesus monkeys fed a marginally zinc-deficient diet that concentration of plasma vitamin A was positively correlated to the concentration of plasma zinc and also that there may be a threshold level of plasma zinc above which no effect on plasma vitamin A will occur. Since vitamin A is a component of prenatal vitamin supplements, we were interested in the possibility that supplemental vitamin A could overcome some of the negative effects of feeding zinc-deficient diets on pregnancy outcome and vitamin A metabolism (Peters et al., 1986). Teratogenic effects were observed in the groups fed the lowest level of dietary zinc (0.5 µg/g diet), but no beneficial effect on pregnancy outcome was observed when vitamin A supplements were given. Maternal plasma vitamin A levels were low in the zinc-deficient groups and plasma zinc levels were significantly correlated to plasma vitamin A levels (Table 6.8). Both dams and fetuses showed increased levels of vitamin A in the liver, as compared to the control groups. Neither plasma nor liver vitamin A levels were affected by supplemental

174     LÖNNERDAL

**TABLE 6.8** Vitamin A and Trace Elements in Fetal Plasma and Liver

| Retinyl acetate intake (µg/g) | Zinc intake (µg/g) | Plasma Vitamin A[e] (µg/100 ml) | Liver Vitamin A (µg/g) | Liver Zinc (µg/g) | Liver Copper (µg/g) | Liver Iron (µg/g) |
|---|---|---|---|---|---|---|
| 4 | 100 | 33.3 ± 10.9[a] | 10.3 ± 1.2[a] | 62.5 ± 5.9[a] | 12.2 ± 0.7[a] | 209.6 ± 10.2[a] |
| 4 | 4.5 | 29.3 ± 6.7[a] | 14.9 ± 1.5[abc] | 35.8 ± 2.6[b] | 15.8 ± 1.1[a] | 300.9 ± 11.2[b] |
| 4 | 0.5 | 27.6 ± 6.2[a] | 19.8 ± 5.4[cd] | 27.6 ± 9.7[b] | 11.8 ± 1.5[a] | 325.7 ± 36.9[b] |
| 8 | 100 | 34.3 ± 10.5[a] | 12.2 ± 0.9[ab] | 61.6 ± 8.4[a] | 13.3 ± 1.6[a] | 198.0 ± 17.6[a] |
| 8 | 4.5 | 31.6 ± 10.7[a] | 15.8 ± 1.8[bcd] | 47.8 ± 4.4[b] | 14.5 ± 1.7[a] | 276.5 ± 26.3[b] |
| 8 | 0.5 | 27.8 ± 2.9[a] | 21.5 ± 1.2[d] | 38.4 ± 4.5[b] | 12.3 ± 1.0[a] | 305.9 ± 28.7[b] |

[a-d]Differing superscripts denote significantly different means at the $p < 0.05$ level.
[e]Values are expressed as mean ± SEM. Each value represents the mean of analyses from 7 to 10 liters.
Source: Peters et al. (1986).

vitamin A. Thus, vitamin A supplements could not improve the terato-genic effects of zinc deficiency and did not help in increasing circulating vitamin A levels. Our results are therefore in agreement with the earlier observations by Smith (1982) that the underlying mechanism behind the low plasma vitamin A levels is an impairment in liver vitamin A mobi-lization.

## VITAMIN A-IRON INTERACTIONS

It has been known for quite some time that vitamin A deficiency causes impaired hematopoiesis (Koessler et al., 1926). The anemia usually ob-served is different from that seen in iron deficiency and in rats has been described as hypochromic, microcytic (Amine et al., 1970). Emphasizing that this interaction may be a problem in humans, Mejia et al. (1977) reported that low plasma retinol levels in children were correlated to low hemoglobin, serum iron, and transferrin saturation values. These obser-vations were made when iron intake was adequate; no correlation was found when dietary iron was low. In a study of experimental vitamin A deficiency in human volunteers, Hodges et al. (1978) found that hemo-globin values decreased in a pattern similar to that of plasma vitamin A and that during repletion with vitamin A, hemoglobin values increased with plasma vitamin A. In their paper, the authors also reviewed many previous studies showing or indicating the effect of vitamin A deficiency on serum iron and hemoglobin.

Studies on vitamin A deficiency in experimental animals showed that liver and spleen iron increased concomitantly with the decrease in serum iron and hemoglobin (Mejia et al., 1979a). In order to study the mech-anisms behind these observations, the same authors (Mejia et al., 1979b) followed the absorption of $^{59}Fe$ in vitamin A-deficient rats. No difference in iron absorption was found between vitamin A-deficient and control rats. Iron turnover and $^{59}Fe$ clearance from plasma were not affected by vitamin A status either. The incorporation of $^{59}Fe$ into red blood cells, however, was significantly decreased in vitamin A-deficient rats com-pared to control rats (Table 6.9). Consistent with this observation was the increased accumulation of $^{59}Fe$ in the liver of vitamin A-deficient animals. Thus, it appears that the mechanism of interaction between vitamin A and iron is an impairment in the mobilization of iron from liver and/or incor-poration of iron into the erythrocyte.

**TABLE 6.9**  Distribution of $^{59}$Fe in Young Rats Sufficient and Deficient in Vitamin A Six Days Postinjection

| Diet | Plasma[a] (%) | RBC[b] (%) | Spleen[c] (%) | Liver[c] (%) |
|------|------------|---------|------------|-----------|
| Control | $1.2 \pm 0.1^d$ | $73 \pm 1^d$ | $0.38 \pm 0.01^d$ | $4.95 \pm 0.34^d$ |
| Vitamin A–deficient | $15.5 \pm 1.0^e$ | $41 \pm 1^e$ | $0.67 \pm 0.16^e$ | $8.54 \pm 1.1^e$ |
| Control, food restricted | $0.9 \pm 0.1^d$ | $70 \pm 2^d$ | $0.29 \pm 0.01$ | $6.70 \pm 0.3^e$ |

[a]Percentage of total dose injected estimated in plasma after correction for changes in volume.
[b]Percentage of total dose injected estimated in red blood cells after correction for changes in volume.
[c]Percentage of total dose injected.
[d]Values are expressed as mean $\pm$ SEM. Number of animals is 6 or greater. Values differing in superscript are significant at $p < 0.05$.
*Source: Mejia et al. (1979b).*

## RIBOFLAVIN–IRON INTERACTIONS

Impaired iron metabolism due to riboflavin deficiency is manifested by low hemoglobin levels, hypoplasia of the bone marrow, and anemia (Mookerjee and Hawkins, 1960; Lane et al., 1964; Alfrey and Lane, 1970). Induced riboflavin deficiency in humans was accompanied by an anemia, which was reversed when riboflavin status was restored (Table 6.10). Subsequent studies in rats showed that the activity of NADH-FMN oxidoreductase was low in riboflavin-deficient animals and that this may be the underlying mechanism for the abnormal iron metabolism observed (Sirivech et al., 1977). Studies in Gambia on the effect of iron supplementation of anemic human subjects showed that iron given with riboflavin supplements was more efficient in restoring hematological parameters than iron given alone (Powers et al., 1983, 1985).

Recent studies have attempted to further delineate the sequence of events leading to impaired iron status and, ultimately, anemia. Powers (1986) studied the relative effects of riboflavin deficiency on iron metabolism in weanling and adult rats. Riboflavin-deficient diets were fed for 35 and 49 days, respectively, and impaired riboflavin status was assessed by the activation constant for erythrocyte glutathione reductase. Ariboflavinosis was evident after 14 days in weanling rats and after 28 days in adult rats, demonstrating higher vulnerability of the rapidly growing weanling rat. Using an in vitro system with mucosal homogenate, ferritin, and NADH, it was shown that NADH-FMN oxidoreductase ac-

**TABLE 6.10** Hematological Effects of Induced Riboflavin Deficiency

| Patient | Reticulocytes | | | | | Hemoglobin | | | |
|---|---|---|---|---|---|---|---|---|---|
| | Initial (%) | Height of deficiency (%) | (day) | Riboflavin reversal (%[a]) | (day[b]) | Initial (g/100 ml) | Height of deficiency (g/100 ml) | Riboflavin reversal (g/100 ml) | (day[c]) |
| V.T. | 1.6 | 0.4 | 43 | 3.2 | 16 | 11.0 | 8.0 | 9.4 | 21 |
| S.H. | 1.0 | 0.2 | 50 | 2.0 | 21 | 12.1 | 9.0 | 9.3 | 21 |
| C.R. | 0.6 | 0.3 | 49 | 3.0 | 10 | 12.7 | 8.5 | 8.8 | 10 |
| R.C. | 1.4 | 0.2 | 46 | 4.4 | 10 | 10.4 | 6.4 | 6.6 | 10 |
| B.P. | 2.0 | 0.0 | 32 | 7.8 | 23 | 14.0 | 8.0 | 10.3 | 23 |
| G.H. | 0.7 | 0.0 | 23 | 6.4 | 30 | 15.0 | 7.3 | 12.6 | 30 |

[a]Maximal reticulocyte response.
[b]Duration of reticulocytosis in days.
[c]Last day recorded before termination of study from any cause.
Source: Lane et al. (1964).

tivity (defined as iron released/min/mg mucosal protein) was dramatically reduced (2.9% of control rats) by day 28 in riboflavin-deficient weanling rats as compared to control rats. In adult rats, riboflavin deficiency also decreased the enzyme activity, but the decrease was not as pronounced (23.3% of control rats). In both groups of animals, rate of iron mobilization was strongly correlated to riboflavin status. Some effects of plasma iron levels were found; however, all levels were high and showed a high degree of variability, indicating problems with hemolysis of samples. Iron absorption may have been impaired, since ferritin stores were low in riboflavin-deficient animals. Adelakan and Thurnham (1986) showed this impairment using $^{59}$Fe to assess iron uptake. Fasted riboflavin-deficient and control weanling rats were intubated with ferric chloride. The concentration of $^{59}$Fe in plasma was analyzed every 30 minutes after dosing until animals were killed four hours later. Slower iron uptake and lower net iron appearance in plasma were found in riboflavin-deficient animals. More unabsorbed $^{59}$Fe remained in the gastrointestinal tract of the riboflavin-deficient animals, supporting the idea of impairment in the uptake phase of iron. Liver ferritin accumulation was lower in riboflavin-deficient animals but not in pair-fed or weight-matched control animals.

Taken together, these studies show that NADH-FMN oxidoreductase activity is low in riboflavin-deficient animals and that iron absorption is impaired. NADH-FMN oxidoreductase is involved in the mobilization of iron from ferritin:

$$
\begin{array}{ccc}
\text{NADH} \diagdown \quad \diagup \text{FMN} \diagdown & \diagup \text{Fe(II)} \\
\qquad \text{Enzyme} & \\
\text{NAD}^+ \diagup \quad \diagdown \text{FMNH}_2 & \diagdown \text{Ferritin-Fe(III)}
\end{array}
$$

Thus, in riboflavin-deficient animals release of iron (temporarily stored in mucosal ferritin) may be reduced significantly and more iron than normal would be lost by mucosal sloughing. The findings of Zaman and Verwilghen (1977) support the theory of impairment in the absorptive mechanism of the mucosa. They showed by intravenous administration of $^{59}$Fe that iron uptake into liver ferritin is not affected by riboflavin deficiency.

## FOLATE–ZINC INTERACTIONS

Folate supplements were reported by Milne et al. (1984) to exert a negative effect on zinc absorption. These authors studied eight men who were fed diets with varying zinc levels and either "adequate" (150 μg folacin/day) or "high" (400 μg folic acid/every other day) folate diet. The "high" level in this case is equivalent to the U.S. RDA for folate, which often is regarded as high. Each level of zinc (3.5, 7.5, or 33.5 mg/day) was fed for 4 weeks. The folic acid supplement given in this study was pteroylmonoglutamic acid. It was hypothesized that this form of folate would chelate zinc and make it less available for absorption. They found that during periods of low or control levels of zinc (3.5 or 7.5 mg/day), urinary zinc excretion was reduced and fecal zinc excretion was increased in folate-supplemented individuals compared to unsupplemented men (Table 6.11). A recent study by Simmer et al. (1987) also found an effect of folate supplements on zinc absorption. Plotting a plasma uptake curve after oral administration of 50 mg zinc in fasted individuals, these investigators found a lower area under the curve and a lower peak concentration for plasma zinc when folate was given as a supplement. This effect was observed 24 hours after folate administration, thus arguing against a direct inhibitory effect of folate by chelation of zinc in the lumen. The method for assessing zinc absorption by using nonphysiological doses of zinc has been questioned (Valberg et al, 1985); the authors acknowledged this by emphasizing that the effect observed could also have been due to a slower uptake of zinc and that net uptake may not necessarily have been affected.

Even though a direct effect of folate supplements on zinc absorption is debatable, the metabolic pathways of zinc and folate are definitely interrelated. Patients with megaloblastic anemia due to folate deficiency exhibit low levels of red blood cell zinc (Fredricks et al., 1964). In addition,

**TABLE 6.11** Effect of Folate Supplementation on Fecal and Urinary Losses of Zinc

| | Zn (mg/day) | | |
|---|---|---|---|
| | Low Folate (4)[a] | Folate supplement (4) | $p$ |
| Control | | | |
| Diet | 8.07 ± 0.28[b] | 8.48 ± 0.72 | NS[c] |
| Feces | 5.83 ± 1.20 | 7.30 ± 1.21 | <0.008 |
| Urine | 0.45 ± 0.11 | 0.28 ± 0.10 | 0.001 |
| Balance | 1.79 ± 1.21 | 0.90 ± 1.64 | NS |

[a]Number of subjects.
[b]Values are expressed as mean ± SD.
[c]NS: $p > 0.05$.
*Source: Milne et al. (1984).*

**TABLE 6.12** Effect of Zinc Depletion on Plasma Zinc, Hematological Parameters, and Serum Folate Levels

| | Before zinc-deficient diet (6)[a] | After zinc-deficient diet (6) |
|---|---|---|
| Plasma zinc ($\mu$g/100 ml) | $100.8 \pm 24.0$[b] | $50.8 \pm 21.0$ ($p < 0.005$)[c] |
| Serum folate (ng/ml: >5) | | |
| PteGlu supplement (2.27 $\mu$mol) | | |
| 0 hr | $9.1 \pm 4.8$ | $10.7 \pm 5.2$ |
| 1 hr | $51.2 \pm 12.3$ | $41.5 \pm 6.3$[d] |
| 2 hr | $36.4 \pm 8.4$ | $39.6 \pm 8.1$[e] |
| PteGlu$_7$ supplement (2.27 $\mu$mol) | | |
| 0 hr | $12.8 \pm 3.2$ | $13.9 \pm 4.6$ |
| 1 hr | $45.8 \pm 8.8$ | $29.3 \pm 8.7$[d] ($p < 0.01$)[c] |
| 2 hr | $36.0 \pm 7.3$ | $28.7 \pm 6.1$[e] ($p < 0.05$)[c] |

[a]Number of subjects
[b]Values are expressed as mean ± SD.
[c]Significant difference before and after development of zinc depletion
[d]Significant difference between PteGlu and PteGlu$_7$ ($p < 0.01$)
[e]$p < 0.05$. All other comparisons (before and after zinc depletion, or between PteGlu and PteGlu$_7$ in the same tests) were not significantly different ($p > 0.05$).
Source: Tamura et al. (1978).

Williams and Mills (1973) have shown that zinc-deficient rats have lower levels of liver folate. In agreement with the latter observation, Tamura et al. (1978) showed that zinc-depleted humans have decreased intestinal absorption of pteroylheptaglutamate, but not of the monoglutamate (Table 6.12). It is likely that reduced uptake of folate will lead to lower liver folate levels. The impairment of folate absorption during zinc deficiency is possibly because folate conjugase (pteroylpolygammaglutamyl hydrolase), a brush border membrane enzyme necessary for cleavage of the polygammaglutamate part of folate, is a zinc-dependent enzyme (Chandler et al., 1986).

The common use of folate supplements during pregnancy, regardless of folate status, raises concerns about interaction with zinc nutriture. Zinc deficiency during pregnancy has a negative effect on the outcome of pregnancy. Mukherjee et al. (1984) found that low levels of plasma zinc and high levels of plasma folate were associated with complications in pregnancy. Thus, there is a potential for both high dietary folate interacting with zinc metabolism and low zinc status interfering with folate metabolism.

## CONCLUSION

It is evident that vitamins and essential trace elements interact at several levels. The obvious concern is in developing countries, in which riboflavin, vitamin A, or zinc deficiencies are common. The interactive effects of vitamin/mineral deficiencies can lead to further impairments in nutritional status. Hopefully, the growing concern about zinc status in industrialized nations will lead to studies on possible secondary nutritional problems induced by suboptimal zinc status. Vitamin supplements are commonly used in relief programs in developing countries and by various groups in developed countries. This raises the possibility of negative interactions with other nutrients, such as the trace elements. It is apparent that both suboptimal and excessive levels should be avoided. This is exemplified by ascorbic acid, which in moderate levels enhances iron absorption but in high levels has a negative effect on copper metabolism. Our knowledge of vitamin/mineral interactions is limited. It is apparent that further research on the mechanisms behind these interactions is sorely needed.

## REFERENCES

Adelakan, D. A. and Thurnham, D. I. 1986. The influence of riboflavin deficiency on absorption and liver storage of iron in the growing rat. *Br. J. Nutr.* 56: 171.

Alfrey, C. P. and Lane, M. 1970. The effect of riboflavin deficiency on erythropoiesis. *Sem. Haem.* 7: 49.

Allen, K. G. D. and Klevay, L. M. 1978. Copper deficiency and cholesterol metabolism in the rat. *Atherosclerosis* 31: 259.

Amine, E. K., Corey, J., Hegsted, D. M., and Hayes, K. C. 1970. Comparative hematology during deficiencies of iron and vitamin A in the rat. *J. Nutr.* 100: 1033.

Baly, D. L., Golub, M. S., Gershwin, M. E., and Hurley, L. S. 1984. Studies on marginal zinc deprivation in rhesus monkeys. III. Effects on vitamin A metabolism. *Am. J. Clin. Nutr.* 40: 199

Banerjee, S. and Chakrabarty, A. S. 1965. Utilization of iron by scorbutic guinea pigs. *Blood* 25: 839.

Bettger, W. J., Fish, T. J., and O'Dell, B. L. 1978. Effects of copper and zinc status of rats on erythrocyte stability and superoxide dismutase activity. *Proc. Soc. Exp. Biol. Med.* 158: 279.

Bettger, W. J., Reeves, P. G., Savage, J. E., and O'Dell, B. L. 1980. Interaction of zinc and vitamin E in the chick. *Proc. Soc. Exp. Biol. Med.* 163: 432.

Bridges, K. R. and Hoffman, K. E. 1986. The effects of ascorbic acid on the intracellular metabolism of iron and ferritin. *J. Biol. Chem.* 261: 14273.

Bunk, M. J., Dnistrian, A., Schwartz, M. K., and Rivlin, R. S. 1987. Dietary zinc deficiency impairs plasma transport of vitamin E. *Am. J. Clin. Nutr.* 45: 865.

Carlton, W. W. and Henderson, W. 1965. Studies in chickens fed a copper deficient diet supplemented with ascorbic acid, reserpine and diethylstilbesterol. *J. Nutr.* 85: 67.

Chandler, C. J., Wang, T. Y., and Halsted, C. H. 1986. Pteroylpolyglutamate hydrolase from human jejunal brush borders: purification and characterization. *J. Biol. Chem.* 261: 928.

Chvapil, M. 1973. New aspects in the biological role of zinc: a stabilizer of macromolecules and biological membranes. *Life Sci.* 13: 1041.

DiSilvestro, R. A., and Harris, E. D. 1981. A postabsorption effect of L-ascorbic acid on copper metabolism in chicks. *J. Nutr.* 111: 1964.

Duncan, J. and Hurley, L. S. 1978. An interaction between zinc and vitamin A in pregnant and fetal rats. *J. Nutr.* 108: 1431.

Finley, E. B. and Cerklewski, F. L. 1983. Influence of ascorbic acid supplementation on copper status in young adult men. *Am. J. Clin. Nutr.* 37: 553.

Fredricks, R. E., Tanaka, K. R., and Valentine, W. N. 1964. Variations of human blood cell zinc in disease. *J. Clin. Invest.* 43: 304.

Gillooly, M., Bothwell, T. H., Torrance, J. D., MacPhail, A. P., Derman, D. P., Bezwoda, W. R., Mills, W., Charlton, R. W., and Mayet, F. 1983. The effects of organic acids, phytates and polyphenols on the absorption of iron from vegetables. *Br. J. Nutr.* 49: 331.

Gillooly, M., Torrance, J. D., Bothwell, T. H., MacPhail, A. P., Derman, D., Mills, W., and Mayet, F. 1984. The relative effect of ascorbic acid on iron absorption from soy-based and milk-based infant formulas. *Am. J. Clin. Nutr.* 40: 522.

Hallberg, L. 1981. Bioavailability of dietary iron in man. *Ann. Rev. Nutr.* 1: 123.

Hallberg, L. and Rossander, L. 1984. Improvement of iron nutrition in developing countries: comparison of adding meat, soy protein, ascorbic acid, citric acid, and ferrous sulphate on iron absorption from a simple Latin American-type of meal. *Am. J. Clin. Nutr.* 39: 577.

Hill, C. H. and Starcher, B. 1965. Effect of reducing agents on copper deficiency in the chick. *J. Nutr.* 85: 271.

Hodges, R. E., Sauberlich, H. E., Canham, J. E., Wallace, D. L., Rucker, R. B., Mejia, L. A., and Mohanram, M. 1978. Hematopoietic studies in vitamin A deficiency. *Am. J. Clin. Nutr.* 31: 876.

Hunt, C. E. and Carlton, W. W. 1965. Cardiovascular lesions associated with experimental copper deficiency in the rabbit. *J. Nutr.* 87: 385.

Hunt, C. E., Carlton, W. W., and Newberne, P. M. 1970. Interrelationships between copper deficiency and dietary ascorbic acid in the rabbit. *Br. J. Nutr.* 24: 61.

Hurley, L. S., Dungan, D. D., Keen, C. L., and Lonnerdal, B. 1983. The effects of vitamin E on zinc deficiency teratogenicity in rats. *J. Nutr.*

Hurley, L. S., Dungan, D. D., Keen, C. L. and Lonnerdal, B. 1983. The effects of Vitamin E on zinc deficiency teratogenicity in rats. *J. Nutr.* 113: 1875.

Jacobson, S. G., Meadows, N. J., Keeling, P. W. N., Mitchell, W. D., and Thompson, R. P. H. 1986. Rod mediated retinal dysfunction in cats with zinc depletion: comparison with taurine depletion. *Clin. Sci.* 71: 559.

Klevay, L. M. 1976. Hypercholesterolemia due to ascorbic acid. *Proc. Soc. Exp. Biol. Med.* 151: 579.

Koessler, K. K., Mauer, S., and Loughlin, R. 1926. The relation of anemia, primary and secondary, to vitamin A deficiency. *J. Am. Med. Assoc.* 87: 476.

Lane, M., Alfrey, C. P., Jr. , Mengel, D. E., Doherty, M.A., and Doherty, J. 1964. The rapid induction of human riboflavin deficiency with galactoflavin. *J. Clin. Invest.* 43: 357.

Lei, K. Y. 1977. Cholesterol metabolism in copper deficient rats. *Nutr. Rep. Int.* 15: 597.

Martin, M. M. and Hurley, L. S. 1977. Effects of large amounts of vitamin E during pregnancy and lactation. *Am. J. Clin. Nutr.* 30: 1629.

Mejia, L. A., Hodges, R. E., Arroyave, G., Viteri, F., and Torun, B. 1977. Vitamin A deficiency and anemia in Central American children. *Am. J. Clin. Nutr.* 30: 1175.

Mejia, L. A., Hodges, R. E., and Rucker, R. B. 1979a. Clinical signs of anemia in vitamin A-deficient rats. *Am. J. Clin. Nutr.* 32: 1439.

Mejia, L. A., Hodges, R. E., and Rucker, R. B. 1979b. Role of vitamin A in the absorption, retention and distribution of iron in the rat. *J. Nutr.* 109: 129.

Milne, D. B., Canfield, W. K., Mahalko, J. R., and Sandstead, H. H. 1984. Effect of oral folic acid on zinc, copper, and iron absorption and excretion. *Am. J. Clin. Nutr.* 39: 535.

Milne, D. B. and Omaye, S. T. 1980. Effects of vitamin C on copper and iron metabolism in the guinea pig. *Int. J. Vitam. Nutr. Res.* 50: 301.

Milne, D. B., Omaye, S. T., and Amos, W. H., Jr. 1981. Effect of ascorbic acid on copper and cholesterol in adult cynomolgus monkeys fed a diet marginal in copper. *Am. J. Clin. Nutr.* 34: 2389.

Mookerjee, S. and Hawkins, W. W. 1960. Haematopoiesis in the rat in riboflavin deficiency. *Br. J. Nutr.* 14: 239.

Mukherjee, M. D., Sandstead, H. H., Ratnaparkhi, M. V., Johnson, L., K., Milne, D. B., and Stelling, H. P. 1984. Maternal zinc, iron , folic acid, and protein nutriture and outcome of human pregnancy. *Am. J. Clin. Nutr.* 46: 496.

Peters, A. J., Keen, C. L., Lonnerdal, B., and Hurley, L. S. 1986. Zinc-vitamin A interaction in pregnant and fetal rats: supplemental vitamin A does not prevent zinc-deficiency-induced teratogenesis. *J. Nutr.* 116: 1765.

Powers, H. J. 1986. Investigation into the relative effects of riboflavin deprivation on iron economy in the weanling rat and the adult. *Ann. Nutr. Metab.* 30: 308.

Powers, H. J., Bates, C. J., Prentice, A. M., Lamb, W. H., Jepson, M., and Bowman, H. 1983. The relative effectiveness of iron and iron with riboflavin in correcting a microcytic anemia in men and children in rural Gambia. *Hum. Nutr.: Clin. Nutr.* 37C: 413.

Powers, H. J., Bates, C. J., and Lamb, W. H. 1985. Haematological response to supplements of iron and riboflavin to pregnant and lactating women in rural Gambia. *Hum. Nutr.: Clin. Nutr.* 39C: 117.

Simmer, K., Iles, C. A., James, C., and Thompson, R. P. H. 1987. Are iron-folate supplements harmful? *Am. J. Clin. Nutr.* 45: 122.

Sirivech, S., Driskell, J., and Frieden, E. 1977. NADH-FMN oxidoreductase activity and iron content of organs from riboflavin-deficient and iron-deficient rats. *J. Nutr.* 107: 739.

Sirivech, S., Frieden, E., and Osaki, S. 1974. The release of iron from horse spleen ferritin by reduced flavins. *Biochem. J.* 143: 311.

Smith, J. C., Jr. 1982. Interrelationship of zinc and vitamin A metabolism in animal and human nutrition: a review. In *Clinical, Biochemical and Nutritional Aspects of Trace Elements,* Current Topics in Nutrition and Disease, Vol. 6, Prasad, A. S. (Ed.), p. 239. Alan R. Liss, Inc., New York.

Smith, C. H. and Bidlack, W. R. 1980. Interrelationships of dietary ascorbic acid and iron on the tissue distribution of ascorbic acid, iron and copper in female guinea pigs. *J. Nutr.* 110: 1398.

Smith, J. C., Jr., McDaniel, E. G., Fan, F. F., and Halsted, J. A. 1973. Zinc: A trace element in vitamin A metabolism. *Science* 181: 954.

Smith, J. C., Jr., Brown, E. D., McDaniel, E. G., and Chan, W. 1976. Alterations in vitamin A metabolism during zinc deficiency and food and growth restriction. *J. Nutr.* 106: 569.

Smith, J. E., Brown, E. D., and Smith, J. C., Jr. 1974. The effect of zinc deficiency on the metabolism of retinol binding protein in the rat. *J. Lab. Clin. Med.* 84: 692.

Stevenson, J. W. and Earle, I. P. 1956. Studies on parakeratosis in swine. *J. Anim. Sci.* 15: 1036.

Tamura, T., Shane, B., Baer, M. T., King, J. C., Margen, S., and Stokstad, E. L. R. 1978. Absorption of mono- and polyglutamyl folates in zinc depleted men. *Am. J. Clin. Nutr.* 31: 1984.

Valberg, L. S., Flanagan, P. R., Brennan, J., and Chamberlain, M. J. 1985. Does the oral zinc-tolerance test measure zinc absorption? *Am. J. Clin. Nutr.* 41: 37.

Van Campen, D. R. and Gross, E. 1968. Influence of ascorbic acid on the absorption of copper by rats. *J. Nutr.* 95: 617.

Williams, R. B. and Mills, C. F. 1973. Relationships between zinc deficiency and folic acid status of the rat. *Proc. Nutr. Soc.* 32: 2A.

Zaman, Z. and Verwilghen, L. 1977. Effect of riboflavin deficiency on activity of NADH-FMN oxidoreductase (ferrireductase) and iron content of rat liver. *Biochem. Soc. Trans.* 5: 306.

# 7

# Lipid–Vitamin–Mineral Interactions in the Diet and in the Tissues

**Harold H. Draper**

University of Guelph
Guelph, Ontario, Canada

## INTRODUCTION

The interactions of primary importance among lipids, vitamins, and minerals are those affecting the oxidative stability of polyunsaturated fatty acids (PUFA) and other unsaturated lipids. Oxidative deterioration of PUFA in foods and in the body shares certain commonalities with respect to the initial products of oxidation, catalysis by metal ions, and inhibition by antioxidants, but there are important mechanistic differences arising from the involvement of enzymes in oxygen free radical metabolism in vivo, as well as in the biological consequences of lipid oxidation in living and nonliving systems. The chief significance of autoxidation of food lipids lies in the accompanying deterioration in flavor and nutritional quality rather than in the toxicity of the products formed. In contrast, lipid oxidation in vivo can result in pervasive biological damage. To protect against such damage, living organisms have developed a complex antioxidant defense system. Lack of one component of this system can result in tissue damage, disease, and death. The focus of research in this field in recent years has been on factors affecting the generation and metabolism of the free radicals of molecular oxygen that catalyze lipid peroxidation in vivo and on the biological impact of its products.

## LIPID–VITAMIN–MINERAL INTERACTIONS IN FOODS

### Products of Lipid Oxidation

Oxidation of fatty acids in edible oils and in foods is a complex process that is influenced by numerous factors, including the degree of unsaturation of the fatty acids involved, the nature of the catalysts, chelators, and antioxidants present, and conditions of temperature, moisture, and time. This process has been reviewed in detail elsewhere (Frankel, 1984) and will be discussed here only to the extent necessary to illustrate the role of vitamins and minerals.

Autoxidation (i.e., oxidation involving atmospheric oxygen) affects primarily the polyunsaturated fatty acids (PUFA) and is classically described in terms of a sequence of four steps: initiation, oxygen uptake, propagation, and termination (Table 7.1). The C–H bond energy of saturated fatty acids is sufficiently high to render them stable to oxidation under most circumstances, but bond energy at the α-position (allylic) to double bonds decreases progressively with the degree of unsaturation. Propagation rate constants correspondingly increase: oleic <1: linoleic 62; linolenic 120; arachidonic 180 (Frankel, 1984). The high rate constants for the more highly unsaturated fatty acids of the n-3 series in fish oils, such as eicosapentaenoic acid and docosahexaenoic acid, create a major problem with respect to their oxidative stability in foods.

Fatty acid hydroperoxides in foods undergo a series of complex reactions, including decomposition, further oxidation, and condensation. Decomposition produces aldehydes, ketones, hydrocarbons, esters, furans, and lactones, many of which produce off-flavors. Oxidation yields secondary products, including epoxides, keto derivatives, dihydroperoxides, and endoperoxides. Condensation leads to dimers and polymers, some of which decompose to form volatile compounds on heating. Complex interactions occur between lipid oxidation products, proteins, and amino acids in foods during processing and cooking, which can affect the availability of amino acids as well as the forms in which the oxidation products are absorbed from the intestine.

**TABLE 7.1** Steps in the Autoxidation of Polyunsaturated Fatty Acids

| | $\cdot OH, {}^1O_2, Fe^{3+}/Fe^{2+}(?)$ | |
|---|---|---|
| Initiation: | $RH \xrightarrow{\hspace{3cm}} R\cdot$ | |
| $O_2$ Uptake: | $R\cdot + O_2 \xrightarrow{\hspace{3cm}} ROO\cdot$ | |
| Propagation: | $ROO\cdot + RH \longrightarrow ROOH + R\cdot$ | |
| Termination: | $ROO\cdot + AH \longrightarrow ROOH + A\cdot$ | |
| | $ROO\cdot + ROO\cdot \longrightarrow$ Nonradicals | |

## Catalysis of Lipid Oxidation by Metals

Transition metals catalyze lipid oxidation by mechanisms which, despite extensive investigation, remain unclear. Their role in oxygen radical reactions has been reviewed by Aust et al. (1985). Iron and copper are the most important catalysts in food systems, although other elements, including cobalt, are also active. These ions catalyze lipid peroxidation in two ways: by accelerating the decomposition of hydroperoxides to form alkoxy radicals, which propagate chain reactions, and by increasing the generation of partially reduced dioxygen radicals, which initiate oxidation.

The general reactions proposed for the decomposition of alkyl hydroperoxides by metal ions are illustrated for iron:

$$ROOH + Fe^{2+} \rightarrow RO\cdot + OH^- + Fe^{3+}$$

$$RO\cdot + RH \rightarrow ROH + R$$

Heme compounds are stronger catalysts of ROOH degradation than inorganic iron salts, although the latter may be more important in some foods because of their high concentrations. The general order of activity is hematin > hemoproteins > $Fe^{2+}$ > $Fe^{3+}$. Myoglobin, hemoglobin, and the cytochromes, as well as the nonheme iron proteins ferritin and transferrin, are lipid hydroperoxide decomposers in foods.

Superoxide ($O_2^-$), a major form of oxygen radicals produced enzymatically in biological systems, is also formed nonenzymatically by autoxidation of ferrous salts (Aust et al., 1985):

$$Fe^{2+} + O_2 \rightarrow Fe^{3+} + O_2^-$$

Superoxide dismutates rapidly to form hydrogen peroxide, which, in the presence of an active iron catalyst, generates the highly reactive hydroxyl radical. This process is illustrated in three reactions, referred to collectively as the Haber-Weiss reaction:

$$2O_2^- + 2H^+ \rightarrow H_2O_2 + O_2$$

$$O_2^- + Fe^{3+} \rightarrow Fe^{2+} + O_2$$

$$Fe^{2+} + H_2O_2 \rightarrow \cdot OH + OH^- + Fe^{3+}$$

Although there is evidence that the ratio of inorganic ferric and ferrous ions is important in the initiation of lipid peroxidation (Aust et al., 1985; Braughler et al., 1986), hydroxyl radicals generated in the third, so-called Fenton reaction are generally considered to be the most active initiators:

$$RH + \cdot OH \rightarrow R \cdot + H_2O$$

Chelators may depress, enhance, or have no effect on metal catalysis of lipid oxidation. For example, EDTA promotes the Haber-Weiss reaction by maintaining $Fe^{3+}$ in a soluble form, whereas desferrioxamine inhibits it by blocking the reduction of $Fe^{3+}$ to $Fe^{2+}$ (Halliwell and Gutteridge, 1985).

## Role of Singlet Oxygen in Lipid Oxidation

Much attention has been given to factors affecting the initiation of autoxidation, i.e., the formation of PUFA free radicals. Atmospheric oxygen, which is in the triplet state, is very reactive with free radicals but is relatively unreactive with stable molecules (Frankel, 1985). Singlet oxygen ($^1O_2$), on the other hand, is highly reactive with stable compounds. This oxygen species is generated in foods in the presence of "sensitizers," such as chlorophyll and hematoporphyrins, which absorb light energy and transfer it to $O_2$ to produce $^1O_2$ by a non–free radical reaction. It is also produced from water by ionizing radiation. Singlet oxygen may be an important initiator of lipid peroxidation in foods containing such sensitizers.

## Role of Antioxidants in Lipid Oxidation

Oxidation of PUFA in foods is accompanied by oxidation of other lipid nutrients, including the unsaturated retinoids β-carotene and vitamin A

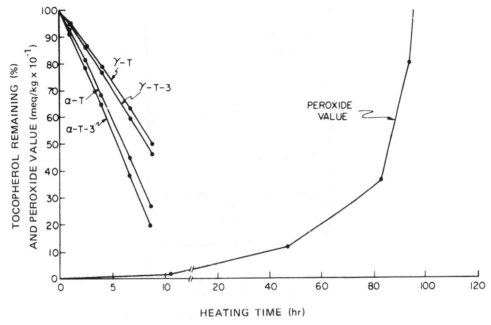

**FIG. 7.1** Depletion of α-tocopherol (α-T), γ-tocopherol (γ-T), and the corresponding tocotrienols (α-T-3 and γ-T-3) in treated corn oil. (*From Chow and Draper, 1974.*)

and the natural antioxidant vitamin E. Depletion of α-tocopherol, γ-tocopherol, and their unsaturated homologs α-tocotrienol and γ-tocotrienol in heated corn oil is illustrated in Fig. 7.1. Note that the accumulation of peroxides is prevented until the antioxidants are exhausted, whereupon the peroxide value rises exponentially. β-Carotene and other carotenoids in foods scavenge singlet oxygen and thereby serve as antioxidants. The stabilizing action of vitamin E on PUFA is, under some circumstances, also partially attributable to its scavenging of this species. The importance of any singlet oxygen generated nonphotochemically in foods is unclear but appears to be minimal.

The antioxidant action of vitamin E is generally depicted in terms of a chain-breaking reaction with a fatty acid hydroperoxy radical generated in the oxygen uptake reaction (Fig. 7.2). In biological systems the resulting hydroperoxide (ROOH) is, in part, reduced enzymatically by peroxidases to stable hydroxy fatty acids (ROH). In food systems, and to some unknown extent in vivo, metal-catalyzed decomposition of fatty acid hydroperoxides yields the alkoxy radical RO · (Fig. 7.2), and the quenching

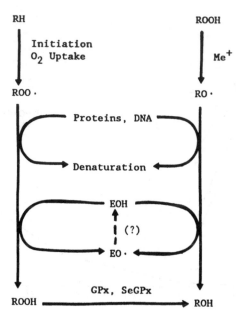

FIG. 7.2   General scheme for the metabolism of fatty acid peroxides formed in the tissues. Abbreviations: RH, a polyunsaturated fatty acid; ROO·, peroxy free radical; RO·, alkoxy free radical; ROOH, hydroperoxide; ROH, hydroxy acid; GPx, glutathione peroxidase; EOH, tocopherol; EO·, tocopheroxy radical.

of this radical by vitamin E is also probably important in the antioxidant action of this vitamin.

   Oxidation of α-tocopherol by lipoxy radicals, weak chemical oxidizing agents such as ferricyanide, and superoxide yields a dimer, oxydimer, trimer, and p-quinone (Csallany and Draper, 1970). The proportions of these products vary according to circumstances, mild conditions of oxidation favoring dimer formation, heat and stronger oxidizing agents, such as ferric chloride, favoring quinone formation. The initial formation of a tocopheroxy radical is followed by an intramolecular shift of electrons to form a carbon radical at the 5-methyl position and dimerization via a condensation of carbon radicals:

$$EOH + ROO \cdot \ (RO \cdot) \rightarrow EO \cdot \ + ROOH \ (ROH)$$

$$EO \cdot \ \rightleftharpoons EOH\text{--}5\text{--}CH_2 \cdot$$

$$2\ EOH\text{--}5\text{--}CH_2 \cdot \ \rightarrow EOH\text{--}CH_2\text{--}CH_2\text{--}EOH$$

$$EO \cdot \ + ROO \cdot \ (RO \cdot) \rightarrow E{=}O + ROOH \ (ROH)$$

A further reaction between EO ˙ and the dimer results in formation of a trimer containing an ether bond. Small amounts of an oxydimer are also formed in peroxidized oils. α-Tocopheryl quinone (often cited as the only oxidation product of α-tocopherol) is formed by divalent oxidation involving rupture of the chroman ring.

γ-Tocopherol, the predominant form of vitamin E in cereal oils, and γ-tocotrienol, the predominant form in latex lipids (the richest natural source of vitamin E), lack a methyl group at the 5-position and hence their oxidation involves a different mechanism from that of α-tocopherol. In these cases, a dimeric carbon–carbon bond is formed directly between two chroman rings at the 5-position. A second dimer is produced by a reaction between a carbon radical and a tocopheroxy radical (Chow and Draper, 1970).

Generally analogous reactions apply to synthetic phenolic food antioxidants. Although for specific purposes a synthetic antioxidant may be more effective than the natural antioxidant vitamin E in preventing lipid oxidation in foods, it is important to realize that the common food antioxidants, such as BHA, BHT, and propyl gallate, are not active as biological antioxidants. Since the vitamin E requirement is determined largely by the intake of PUFA (or, more specifically, their concentration in the tissues), an increase in the ingestion of such acids increases the vitamin E requirement. For example, while synthetic antioxidants are effective in stabilizing fish oil concentrates (currently being recommended for the prevention of heart attacks), the ingestion of such low-tocopherol oils places a stress on the vitamin E requirement for control of lipid peroxidation in vivo.


## Biological Effects of Oxidized Food Lipids

The deleterious effects of the autoxidation of dietary lipids are associated mainly with organoleptic deterioration and destruction of nutrients (PUFA, β-carotene, vitamin A, and vitamin E). Chronic feeding of oils to rodents at various levels of oxidation results in food refusal at high peroxide values, but if adequate nutrients are provided, produces little gross pathology or histopathology. The high molecular weight dimers and polymers produced by extensive oxidation are poorly absorbed. The liver contains an active mitochondrial aldehyde oxidase of low specificity which metabolizes aldehydic decomposition products of alkyl hydroperoxides. Although fatty acid hydroperoxides are highly toxic when administered intravenously in the free form (producing, among other effects, hemolysis), they are not absorbed from the intestine. Oral administration of $^{14}$C-methyl linoleate hydroperoxide to the rat leads to the appearance of the reduced hydroxyester in the lymph, probably as a consequence of the action of glutathione peroxidases in the intestinal wall (Bergan and Draper, 1970).

The lipoxidation product in foods that has evoked the greatest toxicological interest is malondialdehyde (MDA), which is mutagenic in bacterial and human cell cultures and has been reported to produce skin and liver tumors when applied to the skin of mice. However, recent work has indicated that there is little, if any, free MDA present in foods (Piche and Draper, unpublished). The only significant forms of MDA found in the in vitro digestion products of meats and fish were adducts of lysine (Fig. 7.3). These products are excreted in the urine partially in unchanged form and partially after N-acetylation at the α-amino position. N-acetylation is a common detoxification reaction in the liver. These observations serve to allay concern over the possible carcinogenicity of MDA in the diet but do not necessarily reflect its toxicity when formed by peroxidation of nuclear membrane lipids in proximity to DNA.

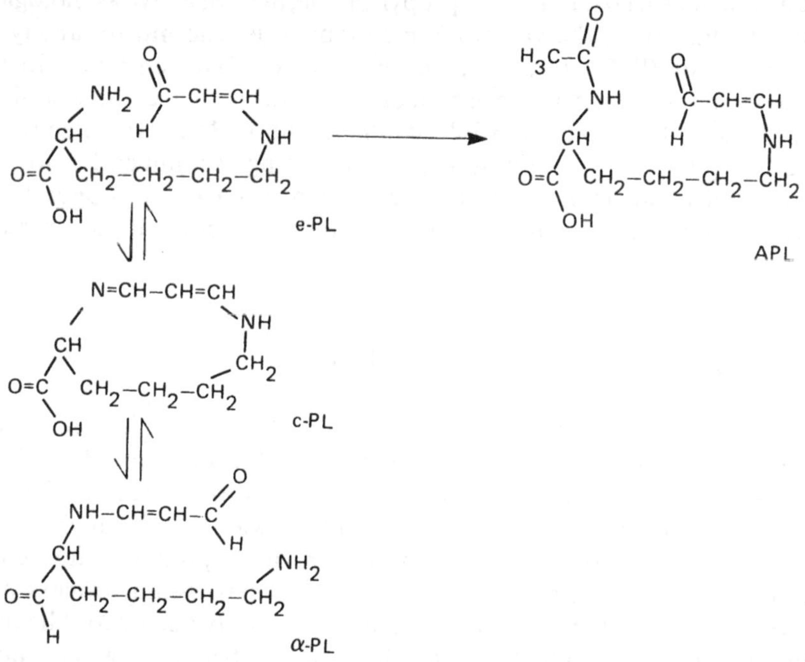

FIG. 7.3   Adducts of MDS with lysine in food and urine. e-Propenal lysine (e-PL) is the main form of MDA released by in vitro digestion of animal foods. It is excreted in human and animal urine in part as the N-acetylated compound APL. In aqueous solution, e-PL forms an equilibrium with c-PL and α-PL.

# LIPID–VITAMIN–MINERAL INTERACTIONS IN VIVO

Inorganic elements play both prooxidant and antioxidant roles in lipid peroxidation in vivo. As chelates with proteins of still uncertain identity, metals catalyze lipid oxidation by reactions similar to those described for dietary lipids. As constituents of enzymes that inhibit the initiation and propagation of lipid free radical reactions, minerals also have an essential role in the cellular antioxidant defense system. Comprehensive reviews of work in this field have been published by Halliwell and Gutteridge (1985, 1986) and by Aust et al. (1985).

## Role of Metals in the Cellular Formation of Oxygen Radicals

Although most of the $O_2$ consumed by aerobic cells is reduced to $H_2O$ by cytochrome C oxidase without the appearance of intermediates, other components of the electron transport chain "leak" electrons, which reduce $O_2$ univalently to form superoxide ($O_2^-$). The main sources of electron leakage appear to be the NADH-coenzyme Q complex and the reduced forms of coenzyme Q (ubiquinone). However, various other sources of superoxide have been identified, including oxyhemoglobin, xanthine oxidase, and aldehyde oxidase.

Superoxide is not notably reactive in aqueous media but in the presence of an active iron catalyst it generates the highly reactive hydroxyl radical ($\cdot OH$). This radical is capable of oxidizing numerous macromolecules, including DNA and proteins. It is also strongly implicated in the initiation of PUFA autoxidation. The mechanism of $\cdot OH$ generation involves the reduction of complexed $Fe^{3+}$ to $Fe^{2+}$ by $O_2^-$ followed by $Fe^{2+}$-catalyzed reduction of $H_2O_2$ in the Fenton reaction.

Inorganic iron salts appear not to exist in significant concentrations in the free state in the tissues, and the major heme proteins are not active catalysts of $\cdot OH$ formation in the presence of $O_2^-$ and $H_2O_2$. The identity of the iron catalysts involved in $\cdot OH$ generation is unclear, but a number of substances that bind iron loosely, including ATP and citrate, as well as some that bind iron tightly, such as ferritin, are active catalysts. The overall transaction catalyzed by iron is described by the sum of the second and third steps in the Haber-Weiss reaction:

$$O_2^- + H_2O_2 \xrightarrow{Fe^{2+}-complex} \cdot OH + OH^- + O_2$$

Copper ions can also catalyze this reaction but are efficiently bound to proteins and appear not to be important catalysts under physiological conditions.

## Nonenzymatic Role of Metals in Peroxide Decomposition

Except for differences in the nature of the complexes involved and the minor role of inorganic ions in vivo, nonenzymatic metal catalysis of hydroperoxide decomposition and the resulting propagation of lipid free radical chain reactions in the tissues is similar to that described for food lipids (Fig. 7.2). Alkoxy and peroxy radicals arising from decomposition of fatty acid peroxides are major initiators of additional peroxidation. Contamination with salts of iron, copper, cobalt, and other elements during processing and handling makes inorganic ions a more important source of metal catalysis of lipid peroxidation in foods than in living systems.

The urinary excretion of malondialdehyde (MDA), a decomposition product of lipid peroxides, is a useful indicator of peroxidation in the diet and in the tissues. Excretion of MDA derivatives increases in response to a number of factors which have been determined by other criteria (formation of conjugated dienes and lipofuscin pigments, pentane evolution) to enhance lipid peroxidation in vivo. A list of factors is shown in Table 7.2. Anion exchange chromatography of rat and human urine revealed a number of MDA derivatives (Fig. 7.4). The main compounds were shown to be lysine adducts (Fig. 7.3). Although these adducts are mainly of dietary origin, their presence in rat urine during fasting or consumption of

**TABLE 7.2**  Factors that Increase Urinary MDA

Vitamin E deficiency
Iron administration
Dietary MDA
High PUFA diet
High tissue PUFA
Lipolytic hormones
Fasting
Adriamycin
Carbon tetrachloride administration

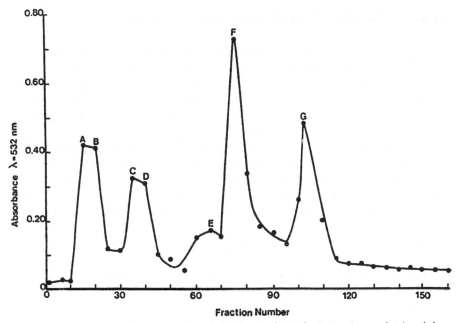

**FIG. 7.4**  Elution profile of malondialdehyde (MDA) derivatives obtained by anion exchange chromatography of urine from rats fed a vitamin E–deficient diet containing cod liver oil. A and F are the lysine derivatives e-PL and APL (see Fig. 7.3). G is free MDA. B and E are MDA adducts with ethanolamine and serine, respectively (see Fig. 7.5). C and D are unidentified. (*From Hadley and Draper, unpublished.*)

a peroxide-free diet indicates that they are also formed in vivo. Adducts with the phospholipid bases serine and ethanolamine (Fig. 7.5) also were identified, providing direct evidence for peroxidative decomposition of the polyunsaturated fatty acids in phospholipid-rich cell membranes.

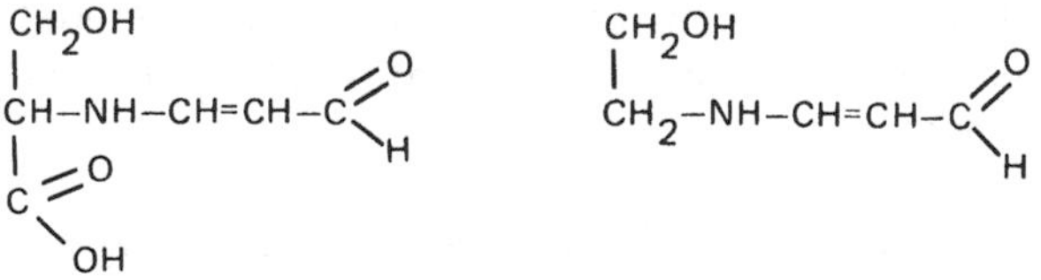

**FIG. 7.5**  Structures of malondialdehyde adducts with serine (left) and ethanolamine (right) excreted in urine.

## Role of Vitamin E in Lipid Peroxidation In Vivo

The role of vitamin E as a biological antioxidant in the tissues (where it functions as a de facto chemical antioxidant) resembles its role in foods. However, there are several noteworthy differences. α-Tocopherol is the predominant form of the vitamin in vivo, whereas γ-tocopherol is the predominant form in most diets. Although the antioxidant activity of these two forms in foods is comparable, α-tocopherol has about ten times the antioxidant activity of γ-tocopherol as a nutrient. This discrepancy is due, at least in part, to the much faster turnover of γ-tocopherol in the blood, arising from its weak binding to lipoproteins (Chow et al., 1971). When the two forms of vitamin E are fed in amounts that sustain comparable levels in the plasma, they are equally effective in inhibiting lipid peroxidation in vivo.

The oxidation products of α-tocopherol formed in the body are analogous to those formed under controlled conditions in oils and fats, a further testimony to the role of the vitamin as a biological antioxidant. At physiological concentrations, none of the oxidation products of α-tocopherol has significant biological activity. α-Tocopheryl-p-quinone in large doses displays vitamin E activity in animals which is attributable to its reduction to the hydroquinone. Under normal conditions the small amount of hydroquinone generated is rapidly catabolized by conjugation with glucuronic acid and is excreted via the bile and urine (Chow and Draper, 1967).

## Enzymatic Role of Metals in Superoxide Metabolism

Metal-containing superoxide dismutases constitute the main defense against intracellular superoxide-mediated generation of $\cdot OH$ radicals, lipid peroxidation, and their biological sequelae. One dismutase (CuZnSOD) has two protein subunits containing $Cu^{2+}$ and $Zn^{2+}$, respectively. This enzyme catalyzes the dismutation of superoxide by the following mechanism:

$$E - Cu^{2+} + O_2^- \rightarrow E - Cu^+ + O_2$$

$$E - Cu^{2+} + O_2^- + 2H^+ \rightarrow E - Cu^{2+} + H_2O_2$$

$$\text{Net: } O_2^- + O_2^- + 2H^+ \rightarrow H_2O_2 + O_2$$

The copper ion is the active site on the enzyme; zinc is not involved in catalysis but is necessary to stabilize the enzyme. Copper deficiency, but not zinc deficiency, results in a decrease in enzyme activity. CuZnSOD is present mainly in the cytosol, although some activity has been reported in lysosomes and between the mitochondrial membranes.

A manganese-containing enzyme (MnSOD) performs the same function in the mitochondrial matrix and, to a lesser extent, extramitochondrially. The enzyme from higher organisms has four protein subunits with a Mn ion at the active site of each unit.

## Enzymatic Role of Se and Fe in Peroxide Metabolism

Hydrogen peroxide, an important product of the superoxide dismutase reaction, is a potential source of hydroxyl radicals generated in the iron-catalyzed Fenton reaction. There are also several enzymes, including D-amino acid oxidase and urate oxidase, which produce $H_2O_2$. In contrast to superoxide, $H_2O_2$ crosses cell membranes readily and, under some conditions, can be a cause of serious biological damage following its decomposition by iron chelates to form hydroxyl radicals. This property also can have beneficial results, such as the destruction of bacterial and tumor cells by $H_2O_2$ produced by macrophages.

Hydrogen peroxide is removed by two sets of cellular enzymes, the peroxidases and the catalases. The most important of these is a selenium-containing glutathione peroxidase (SeGPx). This ubiquitous enzyme catalyzes the reduction of $H_2O_2$ by reduced glutathione (GSH):

$$H_2O_2 + 2GSH \rightarrow H_2O + GSSG$$

Selenium deficiency enhances the requirement for vitamin E and produces a specific cardiomyopathy (Keshan Disease) in humans. SeGPx is made up of four protein units, each with a molecule of selenocysteine at its active site. GSH is regenerated by glutathione reductase, which catalyzes the reduction of GSSG by NADPH formed in the pentose shunt:

$$GSSG + NADPH + H^+ \rightarrow 2GSH + NADP^+$$

SeGPx also catalyzes the reduction of lipid hydroperoxides by GSH, although this reaction is catalyzed mainly by a nonselenium glutathione peroxidase (Fig. 7.2):

$$ROOH + 2GSH \rightarrow ROH + GSSG + H_2O$$

The reduction product, an hydroxy fatty acid, is dehydrated to form a fatty acid with an additional double bond, which is metabolized in the normal lipolytic pathway. These two glutathione peroxidases serve to prevent the decomposition of fatty acid hydroperoxides to peroxy and alkoxy radicals, which catalyze lipid free radical chain reactions (Fig. 7.2).

The $Fe^{3+}$-containing heme enzyme catalase activates the conversion of $H_2O_2$ to $H_2O$ and $O_2$:

$$2H_2O_2 \rightarrow 2H_2O + O_2$$

Catalase activity is concentrated in the peroxisomes; hence $H_2O_2$ generated in the mitochondria, endoplasmic reticulum, and cytosol cannot be metabolized in this way and is removed by SeGPx. There are also major tissue differences with respect to the prevalence of SeGPx and catalase.

## Roles of Vitamin C

Ascorbic acid has an ambivalent role in lipid oxidation as both a prooxidant and an antioxidant. Its ability to reduce $Fe^{3+}$ to $Fe^{2+}$ in hydroxylase enzymes may explain its role in collagen synthesis and many of the symptoms of scurvy. However, this same ability can catalyze the production of hydroxyl radicals via the Fenton reaction. Ascorbic acid is toxic to iron-overload patients. The antioxidant action of ascorbate is due, in part, to its ability to scavenge $O_2^{\bar{\cdot}}$, $\cdot OH$, and $^1O_2$. This action may be important in tissues, such as the lens, which are low in SOD activity, and the lungs, which are exposed to gaseous oxidants. While it has been reported that ascorbic acid can reduce the tocopheroxy free radical, thereby regenerating vitamin E from its univalently oxidized form, this reaction has not been demonstrated in vivo and is unlikely to be of physiological importance. A liver microsomal enzyme has been reported to catalyze the reduction of the tocopheroxy radical by reduced glutathione (McCay et al., 1986).

These seemingly conflicting roles of ascorbate are not fully understood, but they appear to be at least partially dependent upon its concentration at specific sites. The extent of generation of $Fe^{2+}$ catalysts by vitamin C in vivo is unknown, but it seems unlikely that an essential nutrient present at concentrations required to prevent symptoms of deficiency would have any toxicological effects.

## The Extracellular Antioxidant Defense System

There is little SOD, peroxidase, or catalase activity in extracellular fluids, which therefore must depend on some other antioxidant defense system. Vitamin E is the main lipid-soluble chain-breaking antioxidant in plasma, but it accounts for only a small fraction of total antioxidant activity (Halliwell and Gutteridge, 1985). β-Carotene and certain other carotenoids are efficient scavengers of singlet oxygen; however, this oxygen species is not prevalent in extracellular fluids, and β-carotene, a nonessential nutrient of uncertain availability, could not be relied upon as an integral component of an antioxidant system.

Halliwell and Gutteridge (1985) have proposed that control of oxygen free radical generation in plasma is effected mainly by efficient binding of metal catalysts. Under physiological conditions, the iron-binding protein transferrin and the related protein lactoferrin may act as inhibitors of the iron-catalyzed Haber-Weiss reaction. Ceruloplasmin, a copper-containing protein, has antioxidant activity (e.g., in the prevention of iron-catalyzed hemolysis), which may be due to its capacity to act as a ferroxidase, i.e., to oxidize $Fe^{2+}$ to $Fe^{3+}$. This protein has been credited with a major part of the antioxidant activity of human plasma. Uric acid, which binds iron and copper ions, is another active antioxidant.

These observations indicate that in plasma and other extracellular fluids the antioxidant defense system is designed mainly to prevent the generation of free radical forms of stable oxygen by sequestration of metal catalysts, whereas within cells it depends mainly on enzymatic inactivation of such radicals and the products of their reactions. The profound role of inorganic elements in both of these systems is evident from the foregoing discussion.

## Free Radicals in Clinical Disease

Apart from the free radical–induced membrane damage associated with the major diseases of dietary antioxidant deficiency (heart and skeletal myopathies, sterility, hemolytic anemia, etc.) and with an overload of oxidants (iron, ozone, $NO_2$, adriamycin, etc.), numerous other forms of tissue injury recently have been shown to be influenced by oxygen radicals. Indeed, the last five years has seen an explosion of interest in the role of free radicals and the ameliorative effects of antioxidants in clinical disease. For example, it has been reported that administration of a megadose of vitamin E before cardiopulmonary surgery significantly reduces postoperative trauma and tissue damage. Surgery releases metal catalysts, which stimulate peroxidation of lysosomal membranes and thereby the release of hydrolytic enzymes. Injury to the brain may release iron catalysts into the cerebrospinal fluid, which is low in iron-binding capacity, causing peripheral neurological damage. Desferrioxamine, a chelator of iron and an inhibitor of lipid peroxidation, modifies some drug toxicities and the inflammation of some diseases. CuZnSOD injected intravenously into animals reduces the inflammatory response to a range of antigens.

The role of oxygen free radicals in biology and medicine has been the subject of several recent national and international conferences. Fuller discussions of developments in this field can be found in publications on free radical mechanisms in tissue injury (Slater, 1984), free radicals in biology and medicine (Halliwell and Gutteridge, 1985, 1986), the toxicology of molecular oxygen (DiGuiseppi and Fridovich, 1984), autoxidation in food and biological systems (Simic and Karel, 1980), free radicals, lipid peroxidation, and cancer (McBrien and Slater, 1982), free radicals in molecular biology, aging, and disease (Armstrong et al., 1984), and the physiology of free radicals (Taylor et al., 1986). A new journal, *Free Radicals in Biology and Medicine,* is devoted to rapid dissemination of research findings. These events reflect the pace of developments in the field and the important role of nutrition in the prevention and management of diseases associated with free radical pathology.

## REFERENCES

Armstrong, D., Sohal, R. S., Cutler, R. G., and Slater, T. F. (Ed.). 1984. *Free Radicals in Molecular Biology, Aging and Disease.* Raven Press, New York.

Aust, S. D., Morehouse, L. A., and Thomas, C. E. 1985. Role of metals in oxygen radical reactions. *J. Free Radicals in Biol. Med.* 1: 3.

Bergan, J. G. and Draper, H. H. 1970. Absorption and metabolism of 1-$^{14}$C-methyl linoleate hydroperoxide. *Lipids* 5: 976.

Braughler, J. M., Duncan, L. A., and Chase, R. L. 1986. The involvement of iron in lipid peroxidation. *J. Biol. Chem.* 261: 10282.

Chow, C. K., Csallany, A. S., and Draper, H. H. 1971. Relative turnover rates of the tocochromanols in rabbit plasma. *Nutr. Rep. Intern* 4: 45.

Chow, C. K., Draper, H. H., Csallany, A. S., and Chiu, M. 1967. The metabolism of $C^{14}$-α-tocopheryl quinone and $C^{14}$-α-tocopheryl hydroquinone. *Lipids* 2: 390.

Chow, C. K. and Draper, H. H. 1974. Oxidative stability and antioxidant activity of tocopherols in corn and soybean oils. *Intern. J. Vitamin Nutr. Res.* 44: 396.

Csallany, A. S. and Draper, H. H. 1970. Oxidation products of α-tocopherol formed in autoxidizing methyl linoleate. *Lipids* 5: 1.

DiGuiseppi, J. and Fridovich I. 1984. The toxiology of molecular oxygen. *CRC Critical Rev. Toxicol.* 12: 315.

Frankel, E. N. 1984. Lipid oxidation: mechanism, products, and biological significance. *J. Am. Oil Chem. Soc.* 61: 1908.

Halliwell, B. and Gutteridge, J. M. C. 1985. *Free Radicals in Biology and Medicine.* Clarendon Press, Oxford University.

Halliwell, B. and Gutteridge, J. M. C. 1986. Oxygen free radicals and iron in biology and medicine: some problems and concepts. *Arch. Biochem. Biophys.* 246: 501.

McBrien, D. C. H. and Slater, T. F. (Ed.). 1982. *Free Radicals, Lipid Oxidation and Cancer.* Academic Press, New York.

McCay, P. B., Lai, E. K., Powell, S. R., and Breuggemann, G. 1986. Vitamin E functions as an electron shuttle for glutathione-dependent "free radical reductase" activity in biological membranes. *Fed. Proc.* 45: 451 (Abstract).

Simic, M. G. and Karel, M. (Ed.). 1980. *Autoxidation in Food and Biological Systems.* Plenum Press, New York.

Slater, T. F. 1984. Free-radical mechanisms in tissue injury. *Biochem. J.* 222: 1.

Taylor, A.E., Matalon, S., and Ward, P. A. 1986. *Physiology of Oxygen Radicals.* Am. Physiol. Soc., Bethesda, MD.

# 8

# Effect of Variations in Dietary Protein, Phosphorus, Electrolytes, and Vitamin D on Calcium and Zinc Metabolism

**Janet L. Greger**

University of Wisconsin
Madison, Wisconsin

Alterations in dietary levels of protein, phosphorus, vitamin D, calcium, and electrolytes, such as sodium and chloride, can produce a complex set of interactions. Each of these nutrients can influence the utilization of many other nutrients directly and indirectly. Sometimes the biological effect of one of these nutrients can be counteracted or modified by another.

Variations in dietary levels and forms of proteins, peptides, and amino acids can affect the absorption and excretion of a variety of minerals. The difference between the Recommended Dietary Allowance (RDA) for protein and typical intakes of protein by many Americans is large. USDA officials estimated that the average American consumed 165% of the RDA for protein daily in 1977 (Science and Education Administration, 1980). Moreover types of proteins consumed, i.e., vegetable versus animal sources, vary greatly among individuals. Thus there are many practical implications to studies on the effects of protein on mineral metabolism.

## EFFECT OF ISOLATED PROTEIN ON CALCIUM METABOLISM

Several investigators have noted that human subjects tended to absorb calcium somewhat more efficiently when levels of isolated protein (>90 g protein daily versus ≃50 g protein daily) were elevated and dietary calcium intakes were ≥800 mg daily (Hegsted et al., 1981; Linkswiler et al., 1981; Allen, 1982). However, this effect was small in comparison to the effect of elevated dietary protein levels on urinary calcium losses. It is well established that the ingestion of additional (>50 g) protein, if dietary levels of phosphorus are kept constant, will cause hypercalciuria in humans (Linkswiler et al., 1981; Allen, 1982). Thus, high intakes of protein, but levels still consistent with actual protein intakes of some Americans, tend to be associated with reductions in calcium retention by males and females of different ages (Hegsted et al., 1981; Hegsted and Linkswiler, 1981; Schuette and Linkswiler, 1982; Lutz, 1984). The hypercalciuric effect is not transient and is still oberved after 2 months of dietary treatment (Hegsted and Linkswiler, 1981).

The increase in urinary calcium losses when elevated protein levels are ingested can be related to increases in glomerular filtration rate (GFR) and decreases in fractional tubular reabsorption of calcium (FTRCa) (Linkswiler et al., 1981). Table 8.1 summarizes the results of several studies conducted by Linkswiler and her students. When protein intakes were increased by 2- to 3-fold, the GFR was increased by 10–15%; the amount of plasma ultrafiltrable calcium was not changed significantly; and the calcium load reabsorbed by tubular cells was reduced by about 1%. These decreases in calcium reabsorption and increases in GFRs are large enough to account for much of the increase in urinary calcium losses.

Alteration in dietary protein levels may affect kidney function by several different mechanisms. These mechanisms relate to change in renal sulfate and acid excretion and to changes in circulating insulin levels. However, the addition of isolated protein to the diet does not alter circulating levels of immunoreactive parathyroid hormone or 1,25-dihydroxyvitamin D (Schuette et al., 1981). Gollaher et al. (1984) found that removing the parathyroid glands from rats did not alter the calciuretic response to an amino acid infusion.

Investigators have noted that increased urinary sulfate and even elevated serum sulfate levels were correlated to increased urinary calcium excretion (Whiting and Draper, 1980; Schuette et al., 1981; Zemel et al., 1981; Cole and Zlotkin, 1983). However, Zemel et al. (1981) noted that the increase in urinary calcium losses and the decrease in rate of fractional tubular reabsorption of calcium were only 43 and 42%, respectively, as

**TABLE 8.1** Function and Handling of Calcium in Response to Changes in Dietary Protein

| Protein intake (g/day) | Glomerular filtration rate (ml/min) | Ultrafiltrable Ca (mg/dl) | Fractional tubular reabsorption of Ca (%) |
|---|---|---|---|
| | Young men | | |
| 47 | 105 | 6.75 | 98.4 |
| 142 | 116 | 6.95 | 97.4 |
| | Young women | | |
| 46 | 91 | 5.70 | 98.5 |
| 123 | 103 | 5.64 | 97.6 |
| | Older men and women | | |
| 47 | 84 | 6.02 | 98.5 |
| 112 | 101 | 6.00 | 97.8 |

*Source: Linkswiler et al. (1981).*

great when subjects were fed a low protein diet supplemented with sulfur amino acids to the level in a high protein diet as when the high protein diet was fed (Table 8.2). Urinary sulfate excretion was similar when these subjects were fed the high protein diet or the low protein diet supplemented with sulfur amino acids.

Schuette and Linkswiler (1982) have demonstrated that subjects' excretion of sulfate in urine was highly correlated to renal acid excretion. The catabolism of excess sulfur amino acids to sulfate and concomitant

**TABLE 8.2** Urinary Excretion of Calcium by Adult Males Fed Varying Levels of Protein and Sulfur Amino Acids

| Dietary treatments | Diet N (g/day) | Urine Ca (mg/day) | Fractional reabsorption of Ca (%) |
|---|---|---|---|
| Low protein | 8.3 | 182[a] | 90.0[c] |
| Low protein with SAA[d] | 8.8 | 267[b] | 97.1[b] |
| High protein | 24.3 | 380[c] | 95.9[a] |

[a-c]Means ($n = 8$) in a column not sharing a common superscript letter are significantly ($p < 0.01$) different.
[d]Supplemented with sulfur amino acids to level present in high protein diet.
*Source: Zemel et al. (1981).*

release of hydrogen acids is believed to be the primary source of increased acid production when dietary protein levels are increased.

Lutz (1984) demonstrated that the addition of sodium bicarbonate to a high protein diet, without presumably changing urinary sulfate levels, reduced urinary calcium losses by 32%. The amount of bicarbonate (5.85 g/day) consumed was sufficient to lower renal acid excretion below that which occurred when a low protein diet was consumed. The idea that renal acid excretion is related to urinary calcium excretion is not new. Farquharson et al. (1931a,b) studied it in 1931. They noted that an increase in protein or sulfur amino acid intake resulted in increased renal $NH_4^+$ excretion. Moreover, this component of renal acid excretion was more related to urinary calcium excretion than acid excretion in the form of titratable acidity (Farquharson et al. 1931a,b; Schuette et al., 1981).

Some of the effect of dietary protein may also be mediated by insulin (Allen, 1982). Wood and Allen (1983) found that infusion of arginine (a known insulin secretagogue) increased urinary calcium excretion by 575% in streptozotocin diabetic rats. However, carbohydrate ingestion will also cause a rise in serum insulin levels and urinary calcium excretion (Lennon and Piering, 1970).

These human balance studies do not provide data on tissue changes, particularly bone alterations. Studies with animal models can provide this type of data. However, differences in bone metabolism in rats and humans may lessen the usefulness of data collected with rats. Rats excrete more calcium in urine but generally do not experience changes in the size or calcium content of bones when the protein content of their diets is increased (Graves and Wolinsky, 1980; Calvo et al., 1982; Yuen and Draper, 1983). Changes in protein intake also affected phosphorus metabolism. Graves and Wolinsky (1980) reported that rats excreted more phosphorus in urine when protein intake was elevated. Howe and Beecher (1983) observed that young rats had increased concentrations of phosphorus in bone when fed elevated levels of protein, but mature rats had depressed length and strength of femurs when fed elevated levels of protein.

## EFFECT OF ISOLATED PROTEIN ON ZINC METABOLISM

Changes in dietary protein levels will affect the metabolism of a number of minerals. Only the effects of dietary protein on zinc metabolism will be discussed here.

Ingestion of additional protein also significantly increases urinary excretion of zinc (Allen et al., 1979; Greger and Snedeker, 1980: Colin et al.,

1983; Mahalko et al., 1983). Snedeker and Greger (1981) demonstrated that subjects excreted less zinc in urine when fed a low protein diet (0.83 mg Zn/day) than when fed a low protein diet supplemented with sulfur amino acids to the level present in a high protein diet or a high protein diet (1.32 and 1.10 mg Zn/day, respectively). Ingestion of high amounts of histidine has also been found to increase urinary excretion of zinc (Henkin et al, 1975). However, the amount of zinc lost in urine is fairly small. Thus the effect of dietary protein on overall zinc balance in humans generally does not reflect the effect of dietary protein on urinary zinc losses.

The effect of dietary protein on zinc absorption in human subjects is controversial. Greger and Snedeker (1980) observed that human subjects absorbed zinc more efficiently when fed a high protein (24.1 g N daily) diet than a low protein diet (8.1 g N daily) when phosphorus levels were maintained at a moderate level (1010 mg daily). The difference was not observed when high levels of phosphorus (2525 mg P daily) were fed. However, serum zinc levels were elevated when subjects were fed high protein diets and, to a lesser extent, when subjects were fed low protein diets supplemented with sulfur amino acids (Greger and Snedeker, 1980; Snedeker and Greger, 1981). Similarly, Spencer et al. (1982) observed higher zinc retention in a subject fed high levels of zinc (146–157 mg/d) when the protein intake was elevated from 27 to 90 g protein/day. Other investigators have observed either improvements in zinc absorption or retention (Price et al., 1970; Price and Bunce, 1972) or no changes (Colin et al., 1983; Mahalko et al., 1983) when dietary protein levels were elevated. The differences in the data may reflect the fact that investigators have allowed dietary zinc and phosphorus levels to increase when protein was added to the diet (Price et al., 1970; Sandstead et al., 1979; Spencer et al., 1982; Colin et al., 1983; Sandstrom et al., 1987).

The effect of variations in dietary protein levels on absorption and tissue retention of zinc has been somewhat more consistent in animal studies. This may reflect the fact that isolated proteins are fed and dietary phosphorus and zinc levels are allowed to vary less with changes in protein intake in animal studies than in many human studies. Investigators have noted lower apparent absorption of zinc and/or lower tissue zinc levels in rats fed protein-deficient diets (Van Campen and House, 1974, Filteau and Woodward, 1982). Johnson and Evans (1984) and Johnson and Canfield (1985) observed that diabetic rats fed a high protein diet had higher bone zinc levels and absorbed more zinc than rats fed lower levels of protein. Others have observed increased levels of zinc in bone or soft tissues when more protein was added to the diet, but the tissue affected differed among studies (Snedeker and Greger, 1983; Wallwork et al., 1983; Sherman et al., 1985).

Several individual amino acids, particularly histidine, cysteine, and tryptophan, have been found to increase absorption and/or retention of zinc in tissue (Smith et al., 1978; Wapnir et al., 1981; Hsu and Smith, 1983; Snedeker and Greger, 1983; Greger and Mulvaney, 1985; Wapnir and Stiel, 1986). Snedeker and Greger (1983) found that rats fed diets containing 15% protein that were supplemented with cysteine and histidine to equal the amounts of sulfur amino acids and histidine in high protein diets (45% protein) absorbed zinc as efficiently and retained as much zinc in bone as rats fed a high protein (45% protein) diet (Table 8.3). The amino acids may improve absorption by forming soluble complexes with zinc and thus prevent the formation of insoluble hydroxides. However, Wapnir and Stiel (1986) observed that zinc absorption in the jejunum and/or ileum of rats was improved more by amino acids than by related homologues (i.e., tryptophan vs. tryptophol, proline vs. pyroglutamate, and 1-cysteine vs. N-acetyl-1-cysteine). They believe that these data demonstrate that zinc absorption is enhanced by mediated and nonmediated transport systems for amino acids.

The negative effect of soy products on mineral absorption is largely attributable to the phytate content of these products (Fordyce et al., 1987). However, differences in the amino acid content of soy proteins, particularly in regard to sulfur amino acids, may also affect mineral bioavailability from diets containing soy proteins (Greger and Mulvaney, 1985).

**TABLE 8.3**  Absorption and Tissue Zinc Levels in Rats in Response to Changes in Dietary Protein, Cysteine, and Histidine

| Dietary treatments | Tibia Zn (µg/g) | Liver Zn (µg/g) | Apparent absorption Zn (%) |
|---|---|---|---|
| Low protein | $238 \pm 4^a$ | $5 \pm 10^a$ | $35 \pm 1^a$ |
| High protein | $303 \pm 6^c$ | $43 \pm 2^b$ | $41 \pm 1^b$ |
| Low protein w/histidine[d] | $260 \pm 6^b$ | $13 \pm 6^a$ | $35 \pm 1^a$ |
| Low protein w/cysteine[e] | $306 \pm 6^c$ | $28 \pm 5^{ab}$ | $38 \pm 1^{ab}$ |
| Low protein w/histidine and cysteine[d,e] | $325 \pm 9^c$ | $43 \pm 3^b$ | $37 \pm 1^a$ |

[a-c]Mean $\pm$ SEM ($n = 6$). Means in a column not sharing a common superscript letter are significantly ($p < 0.01$) different.
[d]Supplemented with histidine to level in high protein diet.
[e]Supplemented with cysteine to level of sulfur amino acids in high protein diet.
*Source: Snedeker and Greger (1983).*

Dietary proteins can also affect mineral utilization when minerals are "trapped" inside protein or peptide complexes that are resistant to proteolysis. Lönnerdal (1987) suggested that incomplete hydrolysis of casein in cow's milk by infants led to decreased absorption of zinc. Browning reaction products (amino acid–carbohydrate complexes) in toasted and other heat-treated foods are resistant to hydrolysis in the gut and can complex metals; they have been shown to reduce zinc absorption (Lykken et al., 1986). Similarly when protein–phytate mineral complexes are incompletely digested, mineral absorption is decreased (Erdman et al., 1980).

## EFFECT OF PHOSPHORUS ON CALCIUM METABOLISM

The hypocalciuric effect of dietary phosphorus is well established (Spencer et al., 1978a; Linkswiler et al., 1981). In fact, phosphate supplements have sometimes been used to reduce urinary calcium levels and crystalluria in individuals prone to kidney stone formation (Goldsmith et al., 1969; Burdette et al., 1976). Zemel and Linkswiler (1981) reported that the conservation of urinary calcium was not attributable to changes in glomerular filtration rate but reflected improved fractional tubular reabsorption of calcium in the kidneys. Schuette et al. (1981) noted that the increase in total renal acid excretion that occurred when monobasic potassium phosphate was added to the diet reflected increases in titratable acidity due to organic acids as well as that buffered by phosphate. Lau et al. (1979) noted that the hypocalciuric effect of phosphorus varied with the form of phosphorus given. Neutral phosphate supplements had 30% more effect on urinary calcium levels than acidic phosphate supplements.

Most investigators have observed that ingestion of additional orthophosphate had only small effects on calcium absorption (Spencer et al., 1978a; Greger et al., 1981; Hegsted et al., 1981). Thus the overall effect of ingesting additional phosphorus is generally either no change in calcium balance or an improvement in calcium balance among human subjects (Table 8.4).

These results might appear to have considerable implications for the prevention and treatment of osteoporosis. However, a positive balance in regard to calcium does not necessarily mean that the extra calcium is retained in bone. Hulley et al. (1971) observed that although the ingestion of additional phosphorus prevented the hypercalciuria induced by bed rest in human subjects, the additional phosphorus did not delay bone loss induced by bed rest. Alternatively, Goldsmith et al. (1967) reported that phosphate supplement decreased the healing time of patients with frac-

**TABLE 8.4** Calcium and Phosphorus Excretion and Retention by Adult Males Fed Various Levels of Calcium and Phosphorus

| Diet | | Calcium | | | Phosphorus | | |
| Ca | P | Fecal losses | Urinary losses | Apparent retention | Fecal losses | Urinary losses | Apparent retention |
|---|---|---|---|---|---|---|---|
| | | ------- | | | (mg/day) ------- | | |
| 780[d] | 843 | 677[a] | 240[b] | −137 | 406[a] | 481[a] | −44[a] |
| 780 | 2442 | 689[a] | 147[a] | −56 | 957[b] | 1382[c] | 103[b] |
| 2382 | 2443 | 2222[b] | 212[b] | −52 | 1314[c] | 1062[b] | 66[b] |

[a-c]Means (n = 9) in columns with different superscripts are significantly (p < 0.01) different.

*Source: Greger et al. (1981).*

tures. Furthermore, Goldsmith et al. (1976) noted that phosphate supplements (1.0 g/day) given to seven postmenopausal women with osteoporosis reduced urinary calcium losses, improved calcium balance, and increased the density of the midradius slightly. However, bone-forming surfaces decreased and bone-reabsorbing surfaces increased in these patients during phosphate treatment. Moreover, Dudley and Blackburn (1970) observed extraskeletal calcifications in seven out of nine patients treated with neutral phosphate for 9 to 97 months.

The significance of hypocalciuric effect of phosphorus becomes even more debatable when data from animal studies are considered. Of course, differences in bone metabolism between rats and humans make interpretation of data difficult. Elevation of phosphorus intakes of rats has been reported to reduce urinary calcium levels, to increase fecal calcium levels, to reduce retention of calcium in bone, and to induce nephrocalcinosis (Draper and Scythes, 1981; Howe and Beecher, 1981; Yuen and Draper, 1983; Greger et al., 1987a). The last effect was often related to reduced magnesium utilization when high levels of calcium and phosphorus were fed (Woodard and Jee, 1984). However, the interaction is probably more complex. Greger et al. (1987b) observed that rats fed recommended levels of magnesium and calcium as dibasic calcium phosphate developed kidney calcification, even though tissue magnesium levels were normal. The effect was only seen when casein, not lactalbumin, was the protein source. Moreover, Yuen and Draper (1983) reported the ingestion of additional dietary protein prevented phosphorus-induced kidney calcinosis.

## EFFECT OF PHOSPHORUS ON ZINC METABOLISM

Generally the ingestion of high amounts of orthophosphates has not been found to significantly depress zinc absorption in humans (Spencer et al., 1979; Snedeker et al., 1981; Bour et al., 1984) or animals fed moderate levels of calcium (Greger, 1982; Zemel and Bidari, 1983). However, the ingestion of phosphorus as polyphosphates or with high levels of calcium has often been found to decrease zinc absorption (Greger, 1982; Zemel and Bidari, 1983; Bour et al., 1984). Generally zinc appears to be less sensitive than iron to the ingestion of elevated levels of phosphorus and calcium (Greger, 1982).

**TABLE 8.5** Protein, Phosphorus, Calcium, Sodium, Potassium, and Zinc Content of Selected Foods

| Food | Serving | Protein (g) | P | Na | K (mg) | Ca | Zn |
|------|---------|-------------|-----|-----|-----|-----|-----|
| Milk, whole | 1 cup | 8.0 | 227 | 120 | 371 | 290 | 0.9 |
| Cheese, cheddar | 1 oz | 7.1 | 145 | 176 | 28 | 204 | 0.9 |
| Egg | 1 | 6.2 | 92 | 70 | 66 | 29 | 0.7 |
| Hamburger, cooked | 85 g | 20.6 | 165 | 40 | 382 | 9 | 3.7 |
| Peanut butter | 2 T | 7.8 | 122 | 182 | 200 | 18 | 0.8 |
| Apple, raw | 1 | 0.3 | 40 | 1 | 152 | 10 | 0.1 |
| Carrot, cooked | ½ cup | 0.7 | 24 | 26 | 172 | 26 | 0.2 |
| Spinach, cooked | ½ cup | 2.7 | 34 | 45 | 292 | 84 | 0.6 |
| Bread, French | 25 g | 2.3 | 21 | 145 | 23 | 11 | 0.3 |
| Bread, whole wheat | 23 g | 2.4 | 52 | 121 | 63 | 23 | 0.4 |

*Source: Leveille et al. (1983).*

## DIET MILIEU

The preceding discussion has been somewhat theoretical. As demonstrated in Table 8.5, most high protein foods, i.e., meat, milk products, and legumes, are also rich sources of phosphorus and zinc. Foods that are rich in calcium generally contain at least moderate amounts of phosphorus, but the converse is not true. Moreover, foods that differ in their protein, calcium, and phosphorus content also differ in their sodium, potassium, and chloride content. Thus addition of a high protein food to a diet will drastically change the phosphorus, zinc, and potassium content of the diet and perhaps the calcium and sodium content of the diet.

Much of the controversy that has surrounded the effects of dietary protein on calcium and zinc utilization is attributable to the fact that nutrients do not occur in isolation in real foods. One good example involves protein, phosphorus, and calcium. As would be expected, simultaneous changes in dietary protein and phosphorus levels, as occurs when meat or milk products are added to the diet, usually result in smaller changes in urinary calcium excretion than the addition of isolated protein, as shown in Table 8.6 (Spencer et al., 1978b; Hegsted et al., 1981; Schuette and Linkswiler, 1982). Thus Schuette and Linkswiler (1982) observed that the addition of fairly large quantities of meat (i.e., 520 g/day) to diets that contained about 600 mg Ca/day resulted in reductions in calcium balance that were insignificant.

**TABLE 8.6** Effects of Simultaneous Changes in Dietary Protein and Phosphorus Levels on Calcium Excretion and Retention by Adult Males

| Diet | | Absorbed Ca | Urinary Ca losses | Ca retention |
|---|---|---|---|---|
| Protein (g/day) | P (mg/day) | ------------------(mg/day)------------------ | | |
| 50 | 1010 | 181[ab] | 156[b] | 24[a] |
| 50 | 2525 | 100[a] | 93[a] | 8[a] |
| 150 | 1010 | 218[b] | 334[d] | −166[b] |
| 150 | 2525 | 175[ab] | 200[c] | −25[a] |

[a-d]Means ($n$ = 8) in a column not sharing a common superscript letter are significantly ($p < 0.05$) different.

*Source: Hegsted et al. (1981).*

Although Matkovic et al. (1979) attributed increased bone density in a population of Yugoslavians to high intakes of calcium in dairy products, individuals in this group were estimated to consume twice as much protein and 40–50% more phosphorus than another subpopulation of Yugoslavians. Moreover Recker and Heaney (1985) observed that bone remodeling rates (both accretion and resorption) in postmenopausal women were reduced less by milk than by calcium carbonate supplements, even though the improvement in calcium balance was proportional to the amount of supplemental calcium consumed in both cases.

## EFFECT OF ELECTROLYTES

The metabolism of calcium and sodium are intertwined in a variety of ways (Blaustein, 1977). Muldowney et al. (1982) noted that ingestion of additional sodium (120 mEq/day) as sodium chloride increased urinary calcium excretion (384 vs. 278 mg/day) and lowered serum calcium levels (9.68 vs. 9.80 mg/dl) in subjects with hypercalciuria. Silver et al. (1983) reported that moderate sodium restriction was an effective means of reducing hypercalciuria in patients who were calcium stone formers. Castenmiller et al. (1983) found that even normal subjects excreted significantly more calcium and potassium in urine when sodium chloride was added to their diet but the effect was less than that observed in patients with hypercalciuria.

Calcium intake may also affect an animal's sensitivity to salt. Calcium infusions will cause naturesis in animals (Suki et al., 1969). Doris (1985) observed that Wistar Kyoto (WKY) rats given 0.5% saline for fluid consumption developed hypertension, but the hypertensive response disappeared after 3 months in rats fed additional calcium (35,000 µg/g diet). Similarly McCarron et al. (1985) observed increased dietary calcium (20,000 vs. 1000 µg/g diet) intakes prevented hypertension in spontaneously hypertensive rats (SHR) fed diets containing 10,000 µg Na/g diet as sodium chloride. However, McCarron et al. also reported that SHR rats fed "restricted" levels of both calcium (1000 µg/g) and sodium (2500 µg/g) had higher blood pressures than rats fed excess sodium (10,000 µg/g) and restricted levels of calcium (1000 µg/g). These conclusions could be questioned because the investigators supplemented the high sodium diets but not the low sodium diets with additional (5000 µg/g diet) potassium.

Muldowney et al. (1982) and Castenmiller et al. (1985) fed sodium as sodium chloride; they did not consider the potential importance of the chloride ion or acid-base balance. Several investigators have noted that

**TABLE 8.7** Effect of Dietary Anions on Urinary Excretion of Calcium and Acid

| Dietary treatment[d] | Urinary Ca (μmol/day) | Urinary acid excretion | | |
|---|---|---|---|---|
| | | Titratable acid | NH$_4^+$ (mEq/day) | Net acid excretion |
| Carbonate diet | 15[a] | −1.07[a] | 0.87[a] | −0.20[a] |
| Chloride diet | 101[c] | −0.13[b] | 3.22[c] | 3.08[c] |
| Sulfate diet | 53[b] | 0.13[b] | 2.35[b] | 2.48[b] |

[a-c]Means ($n = 7$) in a column with different superscripts are significantly ($p < 0.05$) different.
[d]Magnesium and most calcium salts in diets were supplied as salts indicated.
Source: Whiting and Cole (1986).

adding sodium, as sodium bicarbonate not sodium chloride, reduced urinary calcium losses (Goulding et al., 1984; Lutz, 1984). Greger et al. (1987a) found that the ingestion of sodium chloride produced a larger and more sustained increase in urinary calcium losses than sodium bicarbonate, especially in rats fed high (10,000 μg/g) levels of calcium.

Ingestion of additional chloride, without changes in sodium intake, has been shown to affect urinary calcium excretion. Whiting and Cole (1986) found that the substitution of chloride or sulfate for carbonate in the diets fed rats increased urinary calcium losses; the calcium losses paralleled changes in acid excretion (Table 8.7). However, chloride tended to have a greater effect on urinary calcium losses because chloride was better absorbed than sulfate. Others have reported that the addition of ammonium chloride to diets induced metabolic acidosis, increased serum ionized calcium levels, increased urinary calcium losses, and sometimes reduced bone density and calcium and phosphorus levels in rats (Gafter et al., 1980; Petito and Evans, 1984; Kunkel et al., 1986) and increased urinary calcium and phosphorus losses in humans (Lemann et al., 1966). The complexity of this interaction is demonstrated by studies by Bushinsky et al. (1982). Rats made acidotic by adding ammonium chloride (1.5%) to their drinking water did not respond to ingestion of a low calcium diet (2000 μg/g diet) with increases in circulating levels of 1,25-dihydroxy vitamin D, although circulating levels of immunoreactive parathyroid hormone were elevated in animals fed the low calcium diet.

## EFFECT OF VITAMIN D

The active form of vitamin D in the body, 1,25-dihydroxyvitamin $D_3$ ($1,25(OH)_2D_3$), is formed by hydroxylation of vitamin D in the liver and kidneys. Both $1,25(OH)_2D_3$ and $1,25(OH)_2$ D-dependent calcium-binding proteins have been identified in a variety of tissues. Thus some actions of 1,25-dihydroxyvitamin D may be unappreciated (DeLuca and Schnoes, 1983).

There are three main sites of action of $1,25(OH)_2D_3$: villus cells in gut, distal renal tubule cells, and osteoblasts. Vitamin D acts, perhaps by several mechanisms, to increase the intestinal transport of calcium and phosphorus and functions with parathyroid hormone to mobilize calcium from bone and to increase the efficiency of calcium reabsorption from the kidney filtrate. The mechanisms of action of vitamin D have been reviewed previously and will not be discussed here (DeLuca and Schnoes, 1983).

Sowers et al. (1986) and Omdahl et al. (1982) have suggested that Americans consume considerably less vitamin D daily than the 400 IU suggested in the RDAs. A dietary insufficiency of vitamin D may not be of practical significance because human skin produces vitamin D when irradiated with ultraviolet light. Moreover, Sowers et al. (1986) found little evidence of inadequate vitamin D nutritional status in adults as indicated by circulating 25-OH vitamin D levels.

Omdahl et al. (1982) observed elderly subjects had lower circulating 25-OH vitamin D levels than young adults. The data are consistent with results from studies in the United Kingdom and Ireland that suggest that marginal status in regard to vitamin D among elderly individuals, especially women, may be related to bone loss (McKenna et al., 1985; Nordin et al., 1985). Differences in sunlight exposure, medications used, and food fortification and enrichment standards in the United States and the United Kingdom make it difficult to predict whether marginal nutritional status in regard to vitamin D, particularly among the elderly, is a problem in the United States. Moreover, the toxicity of vitamin D necessitates moderate use of the vitamin in fortified foods or supplements (Committee on Nutritional Misinformation, 1975; Goldman et al., 1985).

## OTHER EFFECTS OF CALCIUM

The effects of ingesting excess phosphorus on calcium utilization are parallel to the effects of ingesting excess calcium on phosphorus utilization. Generally individuals excrete significantly less phosphorus in the urine, absorb slightly less phosphorus, and perhaps have slightly improved phosphorus retention when dietary calcium levels are increased, as shown in Table 8.4 (Spencer et al., 1978; Greger et al., 1981; Zemel and Linkswiler, 1981).

Ingestion of excess calcium will affect the utilization of magnesium and trace elements. For example, Weaver and Evans (1986) demonstrated that rats fed high levels of calcium (25,000 µg/g) rather than normal levels of calcium (5000 µg/g) developed symptoms of magnesium deficiency, including kidney calcification, even though dietary magnesium levels were at recommended levels (500 µg/g). Although the consumption of such high levels of calcium through the use of fortified foods and supplements is possible, the effects of more moderate amounts of calcium can be observed if dietary phosphorus levels, especially dietary phytate levels, are high (Greger, 1982). For example, Fordyce et al. (1987) recently estimated that 81.6% of the variation in bone zinc levels in a series of studies with rats could be explained on the basis of dietary phytate X calcium/zinc ratios.

## SUMMARY

A number of physiological interactions can occur when dietary levels and forms of protein, phosphorus, electrolytes, and calcium are altered. Moreover, those nutrients occur in a complex milieu in foods. Thus, it is often difficult to predict the total effect of apparently simple suggested changes in the intake of real foods. More research needs to be done not only on the mechanisms of interactions but also on the practical effects of dietary changes.

## ACKNOWLEDGMENTS

This research was supported by the College of Agricultural and Life Sciences, University of Wisconsin project 2623, and the University of Wisconsin Graduate School.

## REFERENCES

Allen, L. H. 1982. Calcium bioavailability and absorption: a review. *Am. J. Clin. Nutr..* 35: 783.

Allen, L. H., Bartlett, R. S., and Block, A. D. 1979. Reduction of renal calcium reabsorption in man by consumption of dietary protein. *J. Nutr.* 109: 1345.

Blaustein, M. P. 1977. Sodium ions, calcium ions, blood pressure regulation and hypertension: a reassessment and a hypothesis. *Am. J. Cell Physiol.* 1: C165.

Bour, N. J. S., Soullier, B. A., and Zemel M. B. 1984. Effect of level and form of phosphorus and level of calcium intake on zinc, iron and copper bioavailability in man. *Nutr. Res.* 4: 371.

Burdette, D. C., Thomas, W. C., and Finlayson, B. 1976. Urinary supersaturation with calcium oxalate before and during orthophosphate therapy. *J. Urology* 115: 418.

Bushinsky, D. A., Fauvas, M. J., Schneider, A. B., Sen, P. K., Sherwood, L.

M., and Coe, F. L. 1982. Effect of metabolic acidosis on PTH and 1,25 $(OH)_2D_3$ response to low calcium diet. *Am. J. Physiol.* 243: F570.

Calvo, M. S., Bell, R. R., and Forbes, R. M. 1982. Effect of protein-induced calciuria on calcium metabolism and bone status in adult rats. *J. Nutr.* 112: 1401.

Castenmiller, J. J. M., Mensink, R. P., Vander Heijden, L., Kouwenhoven, T., Hautvast, G. A. J., de Leeuw, P. W., and Schaafsma, G. 1985. The effect of dietary sodium on urinary calcium and potassium excretion in normotensive men with different calcium intakes. *Am. J. Clin. Nutr.* 41: 52.

Cole, D. E. C. and Zlotkin, S. H. 1983. Increased sulfate as an etiological factor in the hypercalciuria associated with total parenteral nutrition. *Am. J. Clin. Nutr.* 37: 108.

Colin, M. A., Taper, L. J., and Ritchey, S. J. 1983. Effect of dietary zinc and protein levels on the utilization of zinc and copper by adult males. *J. Nutr.* 113: 1480.

Committee on Nutritional Misinformation. 1975. Hazards of overuse of vitamin D. *J. Am. Dietet. Assoc.* 66: 453.

DeLuca, H. F. and Schnoes, H. L. 1983. Vitamin D: recent advances. *Ann. Rev. Biochem.* 52: 411.

Doris, P. A. 1985. Sodium and hypertension: effect of dietary calcium supplementation on blood pressure. *Clin. and Exper. - Theory and Practice* A7: 1441.

Draper, H. H. and Scythes C. A. 1981. Calcium, phosphorus and osteoporosis. *Federation Proc.* 40: 2434.

Dudley, F. J. and Blackburn, C. R. B. 1970. Extraskeletal calcification complicating oral neutral-phosphate therapy. *Lancet* II: 628.

Erdman, J. W., Weingartner, K. E., Mustakas, G. C., Schmutz, R. D., Parker, H. M., and Forbes, R. M. 1980. Zinc and magnesium bioavailability from acid-precipitated and neutralized soybean protein products. *J. Food Sci.* 45: 1193.

Farquharson, R. F., Salter, W. T., Tribbetts, D. M., and Aub, J. C. 1931a. Studies of calcium and phosphorus metabolism. 2. The effect of the ingestion of acid-producing substances. *J. Clin. Invest.* 10: 221.

Farquharson, R. F., Salter, W. T., and Aub, J. C. 1931b. Studies of calcium and phosphorus metabolism. 13. The effect of ingestion of phosphates on the excretion of calcium. *J. Clin. Invest.* 10: 251.

Filteau, S. M. and Woodward, B. 1982. The effect of severe protein deficiency on serum zinc concentration of mice fed a requirement level or a very high level of dietary zinc. *J. Nutr.* 112: 1974.

Fordyce, E. J., Forbes, R. M., Robbins, K. R., and Erdman, J. W. Jr. 1987. Phytate X calcium/zinc molar ratios: are they predictive of zinc bioavailability? *J. Food Sci.* 53: 440.

Gafter, U., Kraut, J. A., Lee, D. B. N., Silis, V., Walling, M. W., Kurokawa, K., Hassler, M. R., and Coburn, J. W. 1980. Effect of metabolic acidosis on intestinal absorption of calcium and phosphorus. *Am. J. Physiol.* 239: G480.

Goldman, J. A., Ahn, Y. H., and Wheeler, M. F. 1985. Vitamin D and hypercalcemia. *J. Am. Med. Assoc.* 254: 1719.

Goldsmith, R. S., Jowsey, J., Dube, W. J., Riggs, B. L., Arnaud, C. D., and Kelly, P. J. 1976. Effect of phosphorus supplementation on serum parathyroid hormone and bone morphology in osteoporosis. *J. Clin. Endocrinol. Metab.* 43: 523.

Goldsmith, R. S., Killian, P., Ingbar, S. H., and Bass, D. E. 1969. Effect of phosphate supplementation during immobilization of normal men. *Metabolism* 18: 349.

Goldsmith, R. S., Woodhouse, C. F., Ingbar, S. H., and Segal D. 1967. Effect of phosphate supplements in patients with fractures. *Lancet* I: 687.

Gollaher, C. J., Wood, R. J., Holl, M., and Allen L.H. 1984. A comparison of amino acid-induced hypercalciuria in sham-operated and parathyroidectomized rats. *J. Nutr.* 114: 622.

Goulding, A., McIntosh, J., and Campbell, D. 1984. Effects of sodium bicarbonate and 1,25 dihydroxycholecalciferol on calcium and phosphorus balances in the rat. *J. Nutr.* 114: 653.

Graves, K. L. and Wolinsky, I. 1980. Calcium and phosphorus metabolism in pregnant rats ingesting a high protein diet. *J. Nutr.* 110: 2420.

Greger, J. L. 1982. Effects of phosphorus-containing compounds on iron and zinc utilization. In *Nutritional Bioavailability of Iron,* Kies, C. (Ed.), p. 107. American Chemical Society, Washington DC.

Greger, J. L., Krashoc, C.L., and Krzykowski, C. E. 1987a. Calcium, sodium and chloride interactions in rats. *Nutr. Res.* 7: 401.

Greger, J. L., Krzykowski, C. E., Khazen, R. R., and Krashoc, C. L. 1987b. Mineral utilization by rats fed various commercially available calcium supplements or milk. *J. Nutr.* 117: 717.

Greger, J. L. and Mulvaney, J. 1985. Absorption and tissue distribution of zinc, iron and copper by rats fed diets containing lactalbumin, soy and supplemental sulfur-containing amino acids. *J. Nutr.* 115: 200.

Greger, J. L., Smith, S. A., and Snedeker, S. M. 1981. Effect of dietary calcium and phosphorus levels on the utilization of calcium, phosphorus, manganese and selenium by adult males. *Nutr. Res.* 1: 315.

Greger, J. L. and Snedeker, S.M. 1980. Effect of dietary protein and phosphorus levels on the utilization of zinc, copper and manganese by adult males. *J. Nutr.* 110: 2243.

Hegsted, M. and Linkswiler, H. M. 1981. Long-term effects of level of protein intake on calcium metabolism in young adult women. *J. Nutr.* 111: 244.

Hegsted, M., Schuette, S. A., Zemel, M. B., and Linkswiler, H. M. 1981. Urinary calcium and calcium balance in young men as affected by level of protein and phosphorus intake. *J. Nutr.* 111: 553.

Henkin, R. I., Patten, B. M., Re, P. K., and Bronzert, D. A. 1975. A syndrome of acute zinc loss. *Arch. Neurol.* 32: 745.

Howe, J. C. and Beecher, G. R. 1981. Effect of dietary protein and phosphorus levels on calcium and phosphorus metabolism of the young fast growing rat. *J. Nutr.* 111: 708.

Howe, J. C. and Beecher, G. R. 1983. Dietary protein and phosphorus: effect on calcium and phosphorus metabolism in bone, blood and muscle of the rat. *J. Nutr.* 113: 2085.

Hsu, J. M. and Smith, J. C. Jr. 1983. Cysteine feeding affects urinary zinc excretion in normal and ethanol-treated rats. *J. Nutr.* 113: 2171.

Hulley, S. B., Vogel, J. M., Donaldson, C. L., Bayers, J. H., Friedman, R. J., and Rosen, S. N. 1971. The effect of supplemental oral phosphate on the bone mineral changes during prolonged bed rest. *J. Clin. Invest.* 50: 2506.

Johnson, W. T. and Canfield, W. K. 1985. Intestinal absorption and excretion of zinc in streptozotocin-diabetic rats affected by dietary zinc and protein. *J. Nutr.* 115: 1217.

Johnson, W. T. and Evans, G. W. 1984. Effects of the interrelationship between dietary protein and minerals on tissue content of trace metals in streptozotocin-diabetic rats. *J. Nutr.* 114: 180.

Kunkel, M. E., Roughead, Z. K., and Navia, J. M. 1986. The effect of dietary acid stress on bone metabolizm in young ovariectomized and intact rats. *Br. J. Nutr.* 55: 79.

Lau, K., Wolf, G., Nussbaum, P., Weiner, B.L, DeOreo, P., Slatopolsky, E., Agus, S., and Goldfarb, S. 1979. Differing effects of acid versus neutral phosphate therapy of hypercalciuria. *Kidney Intl.* 16: 736.

Lemann, J. Jr., Litzow, J. R., and Lennon, E. J. 1966. The effect of chronic acid loads in normal man: further evidence for the participation of bone mineral in the defense against chronic metabolic acidoses. *J. Clin. Invest.* 45: 1608.

Lennon, E. J. and Piering, W. F. 1970. A comparison of the effects of glucose ingestion and $NH_4Cl$ acidosis on urinary calcium and magnesium excretion in man. *J. Clin. Invest.* 49: 1458.

Linkswiler, H. M., Zemel, M. B., Hegsted, M., and Schuette, S. 1981. Protein-in duced hypercalciuria. *Federation Proc.* 40: 2429.

Lönnerdal, B. 1987. Protein-mineral interaction. In Nutrition '87, Levander, O. (Ed.), p. 32, American Institute of Nutrition, Bethesda, MD.

Lutz, J. 1984. Calcium balance and acid-base status of women as affected by increased protein intake and by sodium bicarbonate ingestion. *Am. J. Clin. Nutr.* 39: 281.

Lykken, G. I., Mahalko, J., Johnson, P. E., Milne, D., Sandstead, H. H., Garcia, W. J., Dintzis, F. R., and Inglett, G. E. 1986. Effect of browned and unbrowned corn products intrinsically labeled with $^{65}Zn$ on absorption of $^{65}Zn$ in humans. *J. Nutr.* 116: 795.

Mahalko, J. R., Sandstead, H. H., Johnson, L. K., and Milne D. B. 1983. Effect of a moderate increase in dietary protein on the retention and excretion of Ca, Cu, Fe, Mg, P and Zn by adult males. *Am. J. Clin. Nutr.* 37: 8.

Matkovíc, V., Kostial, K., Simonovíc, I., Buzina, R., Brodarec, A., and Nordin, B. E. C. 1979. Bone status and fracture rates in two regions of Yugoslavia. *Am. J. Clin. Nutr.* 32: 540.

McCarron, D. A., Lucas, P. A., Schneidman, R. J., LaCour, B., and Drueke, T. 1985. Blood pressure development of the spontaneously hypertensive rat after concurrent manipulations of dietary $Ca^{+2}$ and $Na^+$. *J. Clin. Invest.* 76: 1147.

McKenna, M. J., Freaney, R., Meade, A., and Moldowney, F. P. 1985. Hypervitaminosis D and elevated serum alkaline phosphatase in elderly Irish people. *Am. J. Clin. Nutr.* 41: 101.

Muldowney, F. P., Freaney, R., and Moloney, M. F. 1982. Importance of dietary sodium in the hypercalciuria syndrome. *Kidney Int.* 22: 292.

Nordin, B. E. C., Baker, M. R., Horsman, A., and Peacock, M. 1985. A prospective trial of the effect of vitamin D supplementation on metacarpal bone loss in elderly women. *Am. J. Clin. Nutr.* 42: 470.

Omdahl, J. L., Garry, P. J., Hansaker, L. A., Hunt, W. C., and Goodwin, J. S. 1982. Nutritional status in a healthy elderly population: vitamin D. *Am. J. Clin. Nutr.* 36: 1125.

Petito, S. L. and Evans, J. L. 1984. Calcium status of the growing rat as affected by diet acidity from ammonium chloride, phosphate and protein. *J. Nutr.* 114: 1049.

Price, N. O. and Bunce, G. E. 1972. Effect of nitrogen and calcium on balance of copper, manganese and zinc in preadolescent girls. *Nutr. Rep. Int.* 5: 275.

Price, N. O., Bunce, G. E., and Engel, R. W. 1970. Copper, manganese and zinc balance in preadolescent girls. *Nutr. Rep. Int.* 23: 258.

Recker, R. R. and Heaney R. P. 1985. The effect of milk supplements on calcium metabolism: bone metabolism and calcium balance. *Am. J. Clin. Nutr.* 41: 254.

Sandstead, H. A., Klevay, L. M., Jacob, R. A., Munoz, M., Logan, G. M., and Reck, S. J. 1979. Effect of dietary fiber and protein level on mineral element metabolism. In *Dietary Fibers: Chemistry and Nutrition,* Inglett, G. E. and Falkehag, S. I. (Ed.), p. 147. Academic Press, New York.

Sandstrom, B., Kivisto, B., and Cederblad, A. 1987. Absorption of zinc from soy protein meals in humans. *J. Nutr.* 117: 321.

Schuette, S. A., Hegsted, M., Zemel, M. B., and Linkswiler, H. M. 1981. Renal acid, urinary cyclic AMP, and hydroxyproline excretion as affected by level of protein, sulfur amino acid and phosphorus intake. *J. Nutr.* 111: 2106.

Schuette, S. A. and Linkswiler, H. M. 1982. Effects of Ca and P metabolism in humans by adding meat, meat plus milk, or purified proteins plus Ca and P to a low protein diet. *J. Nutr.* 112: 338.

Science and Education Administration. 1980. *Food and Nutrient Intakes of Individuals in 1 Day in the United States, Spring 1977.* USDA, Washington, DC.

Sherman, A.R., Helyar, L., and Wolinsky I. 1985. Effects of dietary protein concentration on trace minerals in rat tissues at different ages. *J. Nutr.* 115: 607.

Silver, J., Friedlaender, M. M., Rubinger, D., and Popovtzer, M. M. 1983. Sodium-dependent idiopathic hypercalciuria in renal-stone formers. *Lancet* II: 484.

Smith, K. J., Cousins, R. J., Silbon, B. L., and Failla M. L. 1978. Zinc absorption and metabolism by isolated, vascularly perfused rat intestine. *J. Nutr.* 108: 1849.

Snedeker, S. M. and Greger, J. L. 1981. Effect of dietary protein, sulfur amino acids and phosphorus on human trace element metabolism. *Nutr. Rep. Intl.* 23: 853.

Snedeker, S. M. and Greger, J. L. 1983. Metabolism of zinc, copper and iron as affected by dietary protein, cysteine and histidine. *J. Nutr.* 113: 644.

Snedeker, S. M., Smith, S. A., and Greger, J. L. 1982. Effect of dietary calcium and phosphorus levels on the utilization of iron, copper and zinc by adult males. *J. Nutr.* 112: 136.

Sowers, M. F. R., Wallace, R. B., Hollis, B. W., and Lemke, J. H. 1986. Parameters related to 25-OH-D levels in a population-based study of women. *Am. J. Clin. Nutr.* 43: 621.

Spencer, H., Asmussen, C. R., Holtzmann, R. B., and Kramer, L. 1979. Metabolic balances of cadmium, copper, manganese and zinc in man. *Am. J. Clin. Nutr.* 32: 1867.

Spencer, H., Kramer,L., and Osis, D. 1982. Zinc balances in humans. In *Clinical, Biochemical, and Nutritional Aspects of Trace Elements,* Prasad, A.S. (Ed.), p. 103. Alan R. Liss, Inc., New York.

Spencer, H., Kramer, L., Osis, D., and Norris, C. 1978a. Effect of phosphorus on the absorption of calcium and on calcium balance in man. *J. Nutr.* 108: 447.

Spencer, H., Kramer, L., Osis, D., and Norris, C. 1978b. Effect of a high protein (meat) intake on calcium metabolism in man. *Am. J. Clin. Nutr.* 31: 2167.

Suki, W. N., Eknoyan, G., Rector, F. C., and Seldin, D. W. 1969. The renal diluting and concentrating mechanism in hypercalciuria. *Nephron* 6: 50.

VanCampen D. and House, W. A. 1974. Effect of a low protein diet on retention of an oral doses of $^{65}$Zn and on tissue concentrations of zinc, iron and copper in rats. *J. Nutr.* 104: 84.

Wallwork, J. C., Johnson, L. K., Milne, D. B., and Sandstead, H. H. 1983. The effect of interaction between dietary egg white protein and zinc on body weight,bone growth and tissue trace metals in the 30-day-old rat. *J. Nutr.* 113: 1307.

Wapnir, R. A., Khani, D. E., Bayne, M. A., and Lifshitz, F. 1983. Absorption of zinc by the rat ileum: effects of histidine and other low-molecular weight ligands. *J. Nutr.* 113: 1346.

Wapnir, R. A. and Stiel, L. 1986. Zinc intestinal absorption in rats: specificity of amino acids as ligands. *J. Nutr.* 116: 2171.

Weaver, C. M. and Evans, G. H. 1986. Nutrient interactions and hypertension. *Food Technol.* 40(12): 99.

Whiting, S. J. and Cole, D. E. C. 1986. Effect of dietary anion composition on acid induced hypercalciuria in the adult rat. *J. Nutr.* 116: 388.

Whiting, S. J. and Draper, H. H. 1980. The role of sulfate in the calciuria of high protein diets in adult rats. *J. Nutr.* 110: 212.

Wood, R. J. and Allen, L. H. 1983. Evidence for insulin involvement in arginine- and glucose- induced hypercalciuria in the rat. *J. Nutr.* 113: 1561.

Woodard, J. C. and Jee, W. S. S. 1984. Effects of dietary calcium, phosphorus and magnesium on intranephronic calculosis in rats. *J. Nutr.* 114: 2331.

Yuen, D. E. and Draper, H. H. 1983. Long-term effects of excess protein and phosphorus on bone homeostatis in adult mice. *J. Nutr.* 113: 1374.

Zemel, M. B. and Bidari, M. T. 1983. Zinc, iron and copper availability as affected by orthophosphates, polyphosphates and calcium. *J. Food Sci.* 48: 567.

Zemel, M. B. and Linkswiler, H. M. 1981. Calcium metabolism in the young adult male as affected by level and form of phosphorus intake and level of calcium intake. *J. Nutr.* 111: 315.

Zemel, M. B. and Schuette, S. A., Hegsted, M. and Linkswiler, H. M. 1981. Role of the sulfur-containing amino acids in protein-induced hyper-calciuria in men. *J. Nutr.* 111: 545.

# 9

# Interactions Between Utilizable Dietary Carbohydrates and Minerals

**Sheldon Reiser**

Carbohydrate Nutrition Laboratory
Beltsville Human Nutrition Research Center
United States Department of Agriculture, Agricultural Research
  Station
Beltsville, Maryland

## INTRODUCTION

Most of the attention given to carbohydrate–mineral interactions has focused on the effects of the fiber component of the carbohydrate diet on the intestinal absorption and therefore the bioavailability of minerals. However, it is now apparent that interactions occur between utilizable dietary carbohydrates and a number of minerals that can influence the biological function of these minerals and thereby affect metabolic and physiological processes dependent upon them. In this discussion, studies will be described in which the feeding of carbohydrates, such as sucrose, lactose, fructose, glucose, glucose polymers, and starch, either alone or as part of diets to experimental animals or humans, has been reported to differentially affect the status of minerals, such as iron, copper, selenium, chromium, calcium, zinc, and magnesium. Although many of these effects appear to be mediated through modulations of intestinal absorption, a direct action of certain carbohydrates on the postabsorptive metabolic utilization of some of these minerals also appears to occur. The implications of some of these findings as they impact on human health will be evaluated and possible mechanisms of action underlying these interactions will be discussed.

## IRON

The absorption of iron from the gastrointestinal tract of humans appears to be enhanced by the presence of fructose. Brodan et al. (1967) measured the absorption of 132 mg of ferrous iron in the form of ferrous gluconate given alone or along with 50 g of fructose in seven healthy humans. The absorption of iron as determined by increases in plasma iron levels was significantly greater 4, 6, and 8 hours following the meal containing fructose. The absorption of iron given as [59]ferrous sulfate by normal humans as determined by whole body counting was increased by the presence of a 50:1 molar ratio of fructose to iron, but not by EDTA or diethylene-triaminepentaacetic acid (Davis and Dellar, 1967).

The enhancement of iron absorption by fructose appears to be relatively specific since it is not demonstrated by a structurally analogous sugar such as glucose. In a study conducted by Krause and Jenner (1970), the comparative effects of varying amounts of fructose or glucose on the absorption of 30 mg of radioactive ([55]Fe and [59]Fe) ferrous sulfate by healthy men as measured by levels of blood iron 14 days after the administration of the isotopes was determined. The results are shown in Table 9.1. The addition of 300, 600, and 2600 mg of fructose, as compared to glucose, increased iron absorption by 8, 28, and 97%, respectively. The 28 and 97% increases in iron absorption due to fructose were statistically significant. The enhancement of iron absorption by fructose as compared to glucose could not be attributed to a greater decrease in luminal pH producing a greater solubility of the ferrous salt (Daniel and Rehner, 1986). An increased absorption of iron due to fructose as compared to either glucose or galactose was also reported in rats by Pollack et al. (1964).

Of the common dietary mono- and disaccharides, fructose was reported to have the greater capacity to form complexes with iron when both were present in aqueous solutions (Charley et al., 1963). A ferric–fructose complex has been isolated and purified, and elemental analysis indicated that the complex contained two atoms of iron, two molecules of fructose, and one atom of sodium. The ability of fructose to form chelates with iron appears to be dependent on the dihydroxyacetone moiety found in the open chain form of fructose but not in any aldohexose or fructose-containing

**TABLE 9.1** Relative Absorption of 30 mg of Ferrous Sulfate in the Presence of Equal Amounts of Fructose or Glucose in Men

| Fructose or Glucose (mg) | Iron absorption ratio Fructose/Glucose[a] |
|---|---|
| 300 | 1.08 ± 0.15 |
| 600 | 1.28 ± 0.007[b] |
| 2600 | 1.97 ± 0.53[b] |

[a]Each value is the mean ± 95% confidence limit for 5 men.
[b]Significantly greater than 1 ($p < 0.05$) as determined by a Student's t-test.
*Adapted from Krause and Jenner (1970).*

disaccharide (Davis and Deller, 1966). The findings that sorbose and tagatose are as effective as fructose in forming chelates with iron and that glucose, galactose, mannose, sucrose, lactose, and maltose do not form chelates with iron are consistent with this structural prerequisite.

It appears that the iron–fructose complex facilitates the absorption and retention of iron to a greater extent than do other iron compounds. Table 9.2 presents the results of a study in which the whole body retention of either 28 μg or 280 μg of [59]ferric iron was determined in fasted guinea pigs

**TABLE 9.2** Retention of 28 μg or 280 μg of [59]Iron from the Indicated Chelates and Compounds of Iron by Guinea Pigs

| Iron compound | Percent retention[a] | |
|---|---|---|
| | 28 μg Iron | 280 μg Iron |
| Ferric fructose | 21 ± 5[b] | 9 ± 2[b] |
| Ferric citrate | 11 ± 3 | 6 ± 1 |
| Ferric nitrilotriacetate | 8 ± 6 | 6 ± 1 |
| Ferric EDTA | 7 ± 2 | 4 ± 1 |
| Ferric gluconate | — | 6 ± 2 |
| Ferrous sulfate | 7 ± 3 | 6 ± 2 |

[a]Each value is the mean ± standard deviation from either six animals (28 μg iron) or five animals (280 μg iron).
[b]Iron retention after ferric fructose was significantly greater ($p < 0.05$) than that after any other iron compound tested.
*Adapted from Bates et al. (1972).*

10 days after the oral administration of the iron in the form of various chelates and complexes (Bates et al., 1972). The retention of iron from $^{59}FeSO_4$ was used as a basis for comparison. The retention of iron from the ferric-fructose complex containing 28 µg of iron was two to three times higher than from the ferric complexes of either citrate, nitrilotriacetic acid, EDTA or from ferric gluconate and ferrous sulfate. The results using 280 µg of iron showed the same general pattern of retention but with the increase due to fructose being of lesser magnitude. The greater retention of iron from the fructose complex could not be explained on the basis of an increased energy supply provided by the metabolism of fructose. These results are consistent with previous studies that have shown that the ferric-fructose complex is effectively absorbed by humans, rats, rabbits, and swine (Bates et al., 1972).

The relevance of these findings as they apply to iron nutriture in humans remains questionable since formation of the iron-fructose complex would have to occur under normal dietary conditions. Serum iron was not found to be significantly different six weeks after rats were fed diets adequate or high in iron and containing 75% by weight of either fructose, glucose, sucrose, or cornstarch (Landes, 1975). Similarly, the feeding of diets containing 62% fructose as compared to either 62% glucose (Fields et al., 1984a) or cornstarch (Fields et al., 1984a; Fields et al., 1986a) did not produce an increase in the hepatic content of iron in rats. However, the addition of an iron-fructose complex to certain foods or its use as an iron supplement may be a useful and pleasurable way by which to effectively meet the requirements for iron.

The effect of carbohydrates other than fructose on iron status in rats has been the subject of some other studies. The retention of $^{59}$iron by iron-depleted rats was determined when the isotope was fed in diets containing either 60% lactose, glucose, sucrose, or starch for 4 days (Amine and Hegsted, 1971). Lactose produced the greatest retention of iron followed by sucrose and glucose. Starch produced the least retention. The stimulatory effect of lactose on iron retention persisted even when lactose replaced only 20% of the starch. The increase in iron retention due to lactose is usually attributed to the promotion of acidic conditions in the intestine by this disaccharide. In contrast to these results, Miller and Landes (1976a) reported that diets containing 72% glucose as compared to starch decreased iron stores in the liver and spleen when dietary iron was adequate and reduced serum iron hemoglobin concentrations when dietary iron was low. In another study by these investigators (Miller and Landes, 1976b), more hemoglobin was regenerated by anemic rats per mg of iron consumed when fed a diet containing 73% starch as compared to either

73% sucrose or 73% glucose for 11 days. The differences in the effects of dietary carbohydrates in these studies and that of Amine and Hegsted (1971) were attributed to dissimilarities in experimental design (e.g., time during which the anemic rats were fed the test diet).

A significant increase in the apparent absorption of iron expressed as percent of intake during a 5-day balance period was recently reported in rats consuming a diet containing 62% sucrose as to compared to 62% cornstarch (Johnson, 1986).

## COPPER

In most of the studies in which the metabolic effects of copper deficiency in experimental animals has been reported, sucrose has been used as the source of dietary carbohydrate. The rationale for the use of sucrose is that it is an extremely pure type of dietary carbohydrate and would not be expected to provide any impurities containing copper. However, it appears that the nature of the dietary carbohydrate can influence copper status in rats. In studies utilizing a copper-free salt mix (USP II), it was found that the feeding of diets containing sucrose (Cohen et al., 1974) or fructose (Cohen et al., 1977) produced signs of diabetes in genetically selected rats, while the feeding of diets containing starch protected against these signs of diabetes. These findings prompted the initiation of a series of experiments at the Beltsville Human Nutrition Research Center to determine the effects of different types of dietary carbohydrates on metabolic and physiological processes in male rats fed diets deficient or adequate in copper. The results of these studies conclusively demonstrate that the feeding of diets deficient in copper and containing either sucrose or fructose as compared to starch and, to a lesser extent, glucose dramatically exacerbate the severity of the resultant copper deficiency (Fields et al., 1983; Reiser et al., 1983; Fields et al., 1984a–e; Fields et al., 1985; Fields et al., 1986b; Babu et al., 1987; Lewis et al., 1987). The existence of this interaction has been confimed by work in other laboratories (Cornatzer and Klevay, 1986; Johnson, 1986; Johnson and Hove, 1986).

Signs of copper deficiency observed in experimental animals fed a copper-deficient diet include anemia, hypercholesterolemia, hypertriglyceridemia, hyperuricemia, abnormal electrocardiograms, cardiac hypertrophy, and often sudden death (Klevay, 1984). The most striking effect produced by differences in the kind of dietary carbohydrate fed to rats receiving diets deficient in copper was on the incidence of their sudden

**TABLE 9.3** Mortality of Rats Fed Diets Deficient in Copper and Containing Either 62% Cornstarch, Sucrose, Fructose, or Glucose as the Carbohydrate Source for 7–9 Weeks

| Carbohydrate | Fields et al. (1983) | | Reiser et al. (1983) | | Fields et al. (1984a) | |
|---|---|---|---|---|---|---|
| | Dietary copper (μg/g) | Mortality | Dietary copper (μg/g) | Mortality | Dietary copper (μg/g) | Mortality |
| Cornstarch | 0.97 | 0/10 | 0.97 | 1/10 | 0.86 | 3/12 |
| Sucrose | 0.97 | 6/10 | 0.99 | 6/20 | — | — |
| Fructose | — | — | 1.01 | 7/20 | 0.92 | 10/15 |
| Glucose | — | — | — | — | 0.82 | 4/15 |

death. Table 9.3 presents the mortality of rats during three studies in which copper-deficient diets containing by weight either 62% cornstarch, sucrose, fructose, or glucose was fed. It is apparent that the feeding of fructose, whether as the monosaccharide or as part of sucrose, to copper-deficient rats promotes mortality. In contrast, rats fed diets as deficient in copper but with cornstarch as the dietary carbohydrate generally survive. Dietary glucose appeared to produce the same rate of survival as did cornstarch. The sudden death of the rats was due to heart histopathology. Postmortem examination of the rats revealed a large quantity of clotted blood in the thoracic cavity. These results may reconcile apparently conflicting findings as to the effects of copper deficiency on mortality in rats. In studies where sucrose was the dietary carbohydrate, numerous deaths were reported after 7 weeks (Allen and Klevay, 1978; Viestenz and Klevay, 1982). When starch was the dietary carbohydrate, no deaths were reported (Cohen et al., 1982).

Table 9.4 summarizes some of the metabolic effects observed in rats fed diets deficient in or supplemented with copper and containing either cornstarch or fructose. Hepatic copper level and superoxide dismutase (SOD) activity, direct indices of copper status, were significantly decreased in rats fed a copper-deficient diet as compared to the corresponding values from rats fed a copper-supplemented diet. However, in copper-deficient rats fed fructose as compared to starch, hepatic copper and SOD were significantly reduced by an additional 45 and 37%, respectively. Hearts from copper-deficient rats fed fructose, but not starch, were significantly hypertrophied. Copper-deficient rats fed fructose as compared to starch developed a severe anemia. Copper-deficient rats fed starch had hemoglobin and hematocrit levels not significantly different from copper-supplemented rats. Total blood cholesterol was significantly increased by copper deficiency in both the starch-fed and fructose-fed rats. However, the magnitude of increase was 79% greater ($p < 0.05$) in rats fed fructose than in those fed starch. Hepatic ATP levels were significantly reduced only in those copper-deficient rats fed fructose. It is clear from these results that the inclusion of fructose as compared to cornstarch as part of a copper-deficient diet fed to rats produces a copper deficiency of greater magnitude and impacts negatively on a wide variety of metabolic functions.

One possible mechanism by which this interaction can occur would involve an inhibition of the intestinal absorption of copper by fructose or sucrose as compared to glucose-containing carbohydrates. A number of studies have investigated the effect of dietary carbohydrates on indicators of copper absorption in rats. Landes (1975) compared the effects of diets

**TABLE 9.4** Some Metabolic Parameters of Rats Fed Diets Deficient in or Supplemented with Copper and Containing Either 62% Cornstarch or Fructose[a]

| Dietary status | Hepatic Cu (µg/g) | Hepatic SOD[b] (U/g wet wt.) | Heart wt. (g/100 g B.W.) | Hemoglobin (g/100 ml) | Hematocrit (%) | Blood cholestrol (mg/100 ml) | Hepatic ATP (µmol/g wet wt.) |
|---|---|---|---|---|---|---|---|
| Cu-deficient, starch | 1.8 ± 0.1[w] | 1038 ± 60[w] | 0.45 ± 0.01[w] | 13.9 ± 0.2[w] | 47 ± 1.3[w] | 113 ± 8[w] | 0.90 ± 0.12[w] |
| Cu-deficient, fructose | 1.0 ± 0.1[x] | 660 ± 40[x] | 0.63 ± 0.04[x] | 9.6 ± 0.9[x] | 31.7 ± 2.9[x] | 202 ± 30[x] | 0.55 ± 0.12[x] |
| Cu-supplemented, starch | 4.2 ± 0.1[y] | 1384 ± 100[y] | 0.40 ± 0.01[w] | 14.5 ± 0.3[w] | 46.5 ± 0.9[w] | 70 ± 5[y] | 1.04 ± 0.06[w] |
| Cu-supplemented, fructose | 3.8 ± 0.1[z] | 1330 ± 85[y] | 0.38 ± 0.03[w] | 13.7 ± 0.3[w] | 46.1 ± 0.5[w] | 71 ± 3[y] | 1.22 ± 0.10[w] |

[a]Each value represents the mean ± SEM from at least five rats.
[b]Superoxide dismutase.
[w-z]Mean in a column with different superscript letters are significantly different from each other ($p < 0.05$) as determined by Duncan's multiple range test.
*Adapted from Reiser et al. (1983) and Fields et al. (1984e).*

containing 73% cornstarch, sucrose, glucose, or fructose on copper absorption. The feeding of glucose produced the lowest level of copper absorption. Using an isotope dilution method to measure $^{67}$Cu absorption, Johnson and Bowman (1986) reported that there was no difference in the absorption of 0.4 ppm copper when rats were fed diets containing either fructose, glucose, sucrose, or cornstarch. Based on a 5-day balance period, the feeding of a diet containing 63% sucrose as compared to 63% cornstarch resulted in a small (12.5%) but statistically significant decrease in the apparent absorption of copper when expressed as percent of intake when fed at levels of 1.11 ppm (Johnson, 1986). This effect was reversed when the copper content of the diet was raised to 9 ppm. In a study by Fields et al. (1986a), the retention of copper in the gastrointestinal tract was determined following the intragastric intubation of $^{67}$Cu-containing meals into rats previously fed 62% fructose or cornstarch diets deficient in (0.6 ppm) or supplemented with (6 ppm) copper for 5 weeks (Table 9.5). Although significant only at 48 and 96 hours after the load, copper-deficient rats fed fructose retained more radioactivity in the gastrointestinal contents than did copper-deficient rats fed cornstarch throughout the study. No significant differences in copper retention were observed in copper-supplemented rats fed fructose as compared to cornstarch.

These studies do not provide conclusive evidence for an interaction between fructose feeding and copper status at the absorption level. In all of these studies no consistent effect on copper absorption due to fructose or sucrose as compared to cornstarch or glucose was observed when the rats were not copper-deficient. The decrease in copper absorption observed in studies where copper-deficient rats were used (Fields et al., 1986a; John-

**TABLE 9.5** Gastrointestinal Retention of $^{67}$Cu at Various Times after Intragastric Administration to Copper-deficient or -supplemented Rats Fed Fructose or Cornstarch

| Dietary Status | Percent of radioactivity administered[a] | | | |
|---|---|---|---|---|
| | 8 hr | 24 hr | 48 hr | 96 hr |
| Cu-deficient, fructose | 57 | 60 | 29[b] | 21[b] |
| Cu-deficient, cornstarch | 34 | 43 | 18 | 6 |
| Cu-supplemented, fructose | 60 | 48 | 22 | 13 |
| Cu-supplemented, cornstarch | 57 | 37 | 22 | 12 |

[a]Each value represents the mean from five rats.
[b]Significantly greater ($p < 0.05$) than corresponding starch-fed value as determined by orthogonal contrasts.
*Adapted from Fields et al. (1986a).*

son, 1986) could be attributed to the generalized decrease in tissue function, such as that observed in the exocrine pancreas (Lewis et al., 1987) or thymus (Babu et al., 1987) in these rats.

Further evidence against an interaction between fructose and copper at the intestinal level is provided by studies using female rats. Strause et al. (1986) recently reported that female rats fed copper-deficient diets containing fructose for 1 year did not die. A study was therefore initiated to determine whether the same type of interaction between copper status and dietary fructose found in male rats also occurs in female rats (Fields et al., 1986c). Table 9.6 compares indices of copper status and signs of copper deficiency in male and female rats fed fructose or cornstarch diets deficient in copper (0.6 ppm) for 8 weeks. Although direct indicators of copper status such as blood ceruloplasmin, hepatic SOD activities, and hepatic copper levels were not different between the male and female rats fed fructose, only the male rats exhibited a relative heart hypertrophy, anemia, hypercholesterolemia, and sudden death. While the reason for the protection against the lethal manifestations of the fructose–copper interaction shown by the females is not clear, the apparently identical levels of indicators of copper status argues against an effect on intestinal copper absorption. It has been reported that estrogen secretion associated with the estrous cycle as well as exogenous estrogen alters the subcellular distribution of copper in the liver (Russanov et al., 1981; Saylor and Downer, 1986). Since fructose metabolism occurs primarily in the liver, a difference in the postabsorptive utilization of copper in female as compared to male rats which protects against the effects of fructose is suggested. An effect of fructose on copper utilization is also indicated from the results of a study by Fields et al. (1986d) in which the uptake of $^{67}$Cu by the liver of male rats was always greater when the isotope was administered with fructose as compared to glucose.

An interaction between dietary fructose and copper similar to that demonstrated in rats has been observed in studies with pigs and suggested by one study with humans. The pig appears to be a useful animal model since it has a cardiac architecture similar to that of humans. The feeding of copper-deficient diets (1.5 ppm) containing 20% of the calories as fructose as compared to glucose for 10 weeks resulted in a marked heart hypertrophy (Steele et al., 1986; Steele et al., 1987), decreased levels of aortic lysyl oxidase activity (Ono et al., 1986), and deterioration of cardiac morphology (Steele et al., 1987). Diets consumed by people living in the United States contain relatively high levels of fructose-containing carbohydrates and fructose (Harvey et al., 1987). The intake of copper in the United States appears to be well below the 2 mg per day considered to be adequate

TABLE 9.6 Indices of Copper Status and Signs of Copper Deficiency in Male and Female Rats Fed Copper-deficient Diets Containing Either Fructose or Cornstarch for 8 Weeks[a]

| Parameter | Fructose-fed | | Cornstarch-fed | |
|---|---|---|---|---|
| | Male | Female | Male | Female |
| Mortality | 16/40 | 0 | 0 | 0 |
| Ceruloplasmin (U/l) | ND[b] | ND | ND | ND |
| Hepatic SOD (U/g) | $623 \pm 102^x$ | $609 \pm 238^x$ | $1078 \pm 57^{xy}$ | $1518 \pm 204^y$ |
| Hepatic copper (μg/g wet wt) | $1.83 \pm 0.19^x$ | $1.58 \pm 0.30^x$ | $2.11 \pm 0.22^x$ | $2.23 \pm 0.26^x$ |
| Heart weight (g/100 B.W.) | $0.74 \pm 0.03^x$ | $0.45 \pm 0.03^y$ | $0.44 \pm 0.02^y$ | $0.38 \pm 0.01^y$ |
| Hematocrit (%) | $25 \pm 1.5^x$ | $38 \pm 1.6^y$ | $45 \pm 1.0^z$ | $44 \pm 1.0^z$ |
| Cholesterol (mg/100 ml) | $204 \pm 30^x$ | $138 \pm 10^y$ | $71 \pm 8^z$ | $82 \pm 10^z$ |

[a] Each value represents the mean ± SEM from six rats.
[b] ND = Not detectable.
x–z Means within a row not sharing the same superscript letter are significantly different from each other ($p < 0.05$) as determined by Duncan's multiple range test.

*Adapted from Fields et al. (1986c).*

by the Food and Nutrition Board of the National Academy of Sciences (Holden et al., 1979; Klevay et al., 1979). If the same type of interaction between fructose and copper status found in rats and pigs also occurs in humans, then the high intake of fructose and/or sucrose may produce signs of decreased copper nutriture. A study was therefore carried out to determine direct indicators of copper status in 20 men consuming a typical American diet providing an average of 1 mg of copper per day and containing either 20% of the calories as fructose or cornstarch in a crossover design (Reiser et al., 1985). Fructose ingestion as compared to starch had no effect on serum copper or serum ceruloplasmin activity but did significantly reduce the activity of SOD in erythrocytes. Repletion of the subjects with 3 mg of copper per day significantly increased SOD activity in subjects previously fed fructose but not starch. These results suggest that the type of dietary carbohydrate fed humans can differentially affect indices of their copper status.

## SELENIUM

There has recently been compelling evidence that adequate selenium nutriture is essential for human health and well-being. The incidence of Keshan disease, a cardiomyopathy endemic in certain regions of the People's Republic of China, was dramatically reduced after an intervention trial with sodium selenite (Keshan Disease Research Group, 1979). Epidemiological studies also suggest a relationship between the low intake of selenium and the increased incidence of both cardiovascular disease (Salonen et al., 1982) and colon and mammary cancer (Griffin, 1979). From these reports it is evident that any nutritional interaction that produces a decrease in the bioavailability or metabolic utilization of selenium may have serious consequences on human health.

The activity of the selenoenzyme glutathione peroxidase in blood platelets has been reported to be a good indicator of selenium status in rats (Levander et al., 1983). Rats made copper-deficient by the feeding of diets based on evaporated milk have been shown to have a decreased activity of glutathione peroxidase as compared to rats supplemented with copper (Balevska et al., 1981; Jenkinson et al., 1982). These results suggest that an interaction exists between dietary copper and selenium status. Both of these trace minerals are involved in reactions that protect against tissue peroxidation and membrane damage (Fridovich, 1976). Since the signs of copper deficiency are much more severe in rats fed diets containing fructose or sucrose as compared to starch or glucose, an interaction between the type of dietary carbohydrate and selenium status is suggested. The existence of such an interaction is indicated by the results of a study summarized in Table 9.7. The activity of hepatic glutathione peroxidase was significantly reduced by 50% in copper-deficient rats fed fructose. In contrast, copper-deficient rats fed cornstarch showed no significant difference in enzyme activity as compared to copper-supplemented rats. An increase in hepatic mitochondrial peroxidation was associated with the decreased activity of glutathione peroxidase in copper-deficient rats fed fructose. These results are consistent with a relationship between copper and selenium status and indicate that the feeding of fructose to copper-deficient rats can negatively affect both of these minerals.

Table 9.7 shows that the feeding of fructose as compared to cornstarch also significantly decreased glutathione peroxidase activity by about 25% in copper-supplemented rats. These results indicate a direct effect of fructose feeding on selenium status that is independent of copper status. Two recent studies using rats supports the existence of an interaction between fructose-containing carbohydrates and selenium. It was reported that a gradual decrease of glutathione peroxidase activity in blood and liver of both selenium-deficient and supplemented rats occurred with rats fed sucrose or fructose as compared to starch (Wong et al., 1985). Thornber and Eckhert (1984) reported that the retinopathy that has been found in rats after the long-term consumption of diets high in sucrose and containing

**TABLE 9.7**  Effect of Dietary Carbohydrate and Copper Status on Liver Glutathione Peroxidase Activity and Mitochondrial Peroxidation in Rats[a]

| Dietary status | Glutathione peroxidase (μmol NADPH oxidized/mg protein/min) | Mitochondrial peroxidation (nmol malonaldehyde/mg protein/hr) |
|---|---|---|
| Starch, Cu-deficient | $1.00 \pm 0.15^x$ | $4.1 \pm 0.4^x$ |
| Fructose, Cu-deficient | $0.50 \pm 0.10^y$ | $6.3 \pm 0.2^y$ |
| Starch, Cu-supplemented | $1.09 \pm 0.15^x$ | $3.6 \pm 0.3^x$ |
| Fructose, Cu-supplemented | $0.75 \pm 0.02^z$ | $3.3 \pm 0.3^x$ |

Each value represents the mean ± SEM from five rats.
[x-z]Values within a column with different superscript letters are significantly different from each other ($p < 0.05$) according to Duncan's multiple range test.
*Adapted from Fields et al. (1986).*

no added selenium (Cohen et al., 1972; Papachristodoulou and Heath, 1977) was prevented when selenium was incorporated into diets containing 65% sucrose and no added selenium.

## CHROMIUM

Chromium is an essential trace mineral required by both animals and humans. The intake of chromium in the United States appears to be marginal (Anderson, 1981). A relationship exists between chromium and carbohydrate metabolism. It is believed that chromium is involved in carbohydrate metabolism by its ability to potentiate the action of insulin (Anderson and Mertz, 1977). Chromium is mobilized from body stores when glucose metabolism is increased, producing increased levels in both plasma and urine (Liu and Morris, 1978; Anderson et al., 1982).

The effect of the type and amount of dietary carbohydrate on urinary chromium losses has been determined in two human studies. Kozlovsky et al. (1986) reported an increased urinary loss of chromium ranging from 10–300% in 27 of 37 subjects when they consumed a diet containing 35% of calories from sugars and 15% of calories from complex carbohydrates as compared to a diet containing 15% of calories from sugars and 35% from complex carbohydrates. Insulin responses to a glucose tolerance test were also greater after the subjects consumed a high-sugar diet as compared to a low-sugar diet (Reiser et al., 1986). The results suggested that the consumption of diets high in simple sugars stimulate the loss of chromium from body stores and, coupled with suboptimal chromium intake, may lead to marginal chromium deficiency.

A more definitive association between the type and amount of dietary carbohydrate and urinary chromium excretion was reported by Anderson et al. (1987). Twenty subjects were given drinks containing either (per kg body weight) 1 g glucose, 1 g glucose followed 20 min later by 1.75 g fructose, 0.9 g cornstarch, 0.9 g cornstarch followed 20 min later by 1.75 g fructose, or water followed 20 min later by 1.75 g fructose on each of five mornings separated by at least 2 weeks. Table 9.8 presents the summed insulin responses 30–90 minutes after the drinks were consumed (Reiser et al., 1987) and the urinary chromium levels 90 minutes after the drinks were given. The greatest insulin response was observed after the drink containing glucose followed by fructose. Insulin responses to the other drinks decreased in the following order: glucose alone, starch followed by fructose, starch alone, and water followed by fructose. The urinary excretion of chromium followed a similar pattern. These results indicate that the exis-

**TABLE 9.8** Relationship Between the Insulin Response to Carbohydrates and the Urinary Excretion of Chromium in Humans[a]

| Carbohydrate drink (g/kg body wt.) | Summation plasma insulin ($\mu$Units/ml)(30–90 min[b]) | Urinary chromium excretion (ng)(90 min[c]) |
|---|---|---|
| 1 g Glucose followed 20 min later by 1.75 g fructose | $380 \pm 42^{x}$ | $22.4 \pm 3.5^{x}$ |
| 1 g Glucose | $282 \pm 31^{y}$ | $17.3 \pm 1.8^{xy}$ |
| 0.9 g Uncooked cornstarch followed 20 min later by 1.75 g fructose | $255 \pm 19^{y}$ | $13.8 \pm 1.9^{y}$ |
| 0.9 g Uncooked cornstarch | $139 \pm 12^{z}$ | $12.1 \pm 1.3^{y}$ |
| Water followed 20 min later by 1.75 g fructose | $136 \pm 8^{z}$ | $11.0 \pm 1.2^{y}$ |

[a]Each value is the mean ± SEM from 20 subjects.
[b]Time after first drink consumed.
[c]Urine collected 90 min after subjects voided and consumed first drink.
x-zMeans in a column with different superscript letters are significantly different from each other ($p < 0.05$) as determined by Duncan's multiple range test.
*Adapted from Anderson et al. (1987) and Reiser et al. (1987).*

tence of an interaction between dietary carbohydrate and chromium is based on the ability of the carbohydrate to stimulate insulin secretion. It would therefore be expected that individual carbohydrates or carbohydrate combinations producing the highest insulin response would be the least capable of maintaining desirable chromium status.

## CALCIUM, MAGNESIUM, ZINC

Studies examining the possible interactions between dietary carbohydrate and calcium, magnesium, and zinc in both experimental animals and humans have focused on effects on intestinal absorption. Since several of these studies have compared the effects of carbohydrates on the absorption of two or more of these minerals under the same experimental conditions, the results of these studies will be discussed as a unit.

Bei et al. (1986) investigated the effect of glucose polymer (see below) on the intestinal absorption of calcium, magnesium, and zinc in eight normal subjects. The perfusion of the jejunum with a solution containing 4 mM glucose polymer markedly and significantly increased the net absorption of each of these minerals as compared to absorption in the absence of glucose polymer (Table 9.9). A similar enhancement of calcium absorption by glucose polymer has been reported in both normal subjects and patients with calcium malabsorption (Kelly et al., 1984) and in rats (Zheng et al., 1985). These results suggest that the inclusion of glucose polymer in the diet may be of clinical relevance for subjects in which the malabsorption of these minerals is present (e.g., those with Crohn's disease or ileal resections).

The mechanism by which glucose polymer enhances the absorption of these minerals is not apparent. Glucose polymers are polysaccharides derived from the partial hydrolysis of cornstarch and are readily hydrolyzed to glucose by intestinal enzymes. Since glucose has been reported to enhance the intestinal absorption of calcium in humans (Norman et al., 1980) and rats (Vaughan and Filer, 1960) and zinc in humans (Steinhardt and Adibi, 1984), it is possible that the enhancement of mineral absorption produced by glucose polymer is mediated by the action of free glucose. However, when the level of glucose polymer was raised from 4 mM to 8 mM, no additional increase in the absorption of calcium, magnesium, and zinc was evident despite a 30% increase in glucose absorption (Bei et al., 1986). In addition, other studies using the rat have shown no effect of glucose on calcium absorption (Wasserman and Comar, 1959; Urban and Pena, 1977) and even an inhibitory effect of

**TABLE 9.9** Effect of Glucose Polymer on the Absorption of Calcium, Magnesium, and Zinc from Perfused Human Jejunum

| Glucose polymer in perfusate (mM) | Absorption ($\mu$mol/30 cm/hr)[a] | | |
|---|---|---|---|
| | Calcium | Magnesium | Zinc |
| 0 | 95 ± 52 | −14 ± 48[b] | 13 ± 5 |
| 4 | 488 ± 112[c] | 393 ± 113[c] | 29 ± 4[c] |

[a]Each value is the mean ± SEM for four subjects.
[b]Negative number denotes secretion.
[c]Significantly greater than corresponding absorption value in the absence of glucose polymer.
*Adapted from Bei et al. (1986).*

glucose on zinc absorption (Chikosi et al., 1985). It therefore appears that the beneficial effects of glucose polymer on mineral absorption cannot be adequately explained by an action of the liberated glucose.

The presence of lactose in the small intestine of the rat facilitates the absorption of many minerals including calcium (Lengemann, 1959; Au and Raisz, 1967; Leichter and Tolensky, 1975; Armbrecht and Wasserman, 1976; Ghishan et al., 1982; Zheng et al., 1985) magnesium (Fournier and Digaud, 1971), and zinc (Lengemann, 1959; Ghishan et al., 1982). Lactose has also been shown to increase calcium absorption (Cochet et al., 1983) and calcium retention (Condon et al., 1970) in humans. A direct stimulation of calcium transport by lactose in rats at the intestinal epithelium cell level has been demonstrated both in situ using ligated ileal segments (Lengemann and Comar, 1961) and in vitro using everted sacs from the ileum (Chang and Hegsted, 1964; Armbrecht and Wasserman, 1976). Based on results obtained by measuring unidirectional steady-state calcium flux in vitro under short-circuited conditions in segments of rat ileum, Favus and Angeid-Backman (1984) suggested that lactose may increase the mucosal transmembrane potential (interior more negative with respect to the luminal surface), thus enhancing calcium influx across the brush border and thereby increasing calcium absorption. The results of these studies indicate that the consumption of lactose-containing foods (e. g., dairy products) as part of meals should be beneficial in optimizing the intestinal absorption of minerals.

## CONCLUSIONS

In this chapter interactions between utilizable dietary carbohydrates and minerals such as iron, copper, selenium, chromium, calcium, magnesium, and zinc in both experimental animals and humans have been described. It is apparent from these results that there exists a multitude of possibilities for such interactions to occur and that these interactions can profoundly influence the bioavailability of minerals and thus their recommended dietary intakes that have been established as adequate and safe. An understanding of the nature of these interactions and the resultant effects they produce on metabolic and physiological processes is an area of nutrition research that merits increased emphasis.

## REFERENCES

Allen, K. G. D. and Klevay, L. M. 1978. Cholesterolemic and cardiovascular abnormalities in rats caused by copper deficiency. *Atherosclerosis* 29: 81.

Amine, E. K. and Hegsted, D. M. 1971. Effect of diet on iron absorption in iron-deficient rats. *J. Nutr.* 101: 927.

Anderson, R. A. 1981. Nutritional role of chromium. *Sci. Total Environ.* 17: 13.

Anderson, R. A. and Mertz, W. 1977. Glucose Tolerance Factor: an essential dietary agent. *Trends Biochem. Sci.* 2: 277.

Anderson, R., Polansky, M. M., Bryden, N. A., Roginski, E. E., Mertz, W., and Glinsmann, W. 1982. Urinary chromium excretion of human subjects: effects of chromium supplementation and glucose loading. *Am. J. Clin. Nutr.* 36: 1184.

Anderson, R. A., Bryden, N. A., Polansky, M. M., Powell, A. S., and Reiser, S. 1987. Urinary chromium excretion and insulinogenic properties of carbohydrates. *Fed. Proc.* 46: 1007.

Armbrecht, H. J. and Wasserman, R. H. 1976. Enhancement of $Ca^{++}$ uptake by lactose in the rat small intestine. *J. Nutr.* 106: 1265.

Au, W. Y. W. and Raisz, L. G. 1967. Restoration of parathyroid responsiveness in vitamin D-deficient rats by parenteral calcium or dietary lactose. *J. Clin. Invest.* 46: 1572.

Babu, U., Failla, M. L., and Caperna, T. J. 1987. Fructose attenuates humoral immunity in rats with moderate copper deficiency. *Fed. Proc.* 46: 568.

Balevska, P. S., Russanov, E. M., and Kassabova, T. A. 1981. Studies on lipid peroxidation in rat liver by copper deficiency. *Int. J. Biochem.* 13: 489.

Bates, G. W., Boyer, J., Hegenauer, J. C., and Saltman, P. 1972. Facilitation of iron absorption of ferric-fructose. *Am. J. Clin. Nutr.* 25: 983.

Bei, L., Wood, R. J. and Rosenberg, I. H. 1986. Glucose polymer increases jejunal calcium, magnesium, and zinc absorption in humans. *Am. J. Clin. Nutr.* 44: 244.

Brodan, V., Brodanova, M., Kuhn, E., Kordac, V., and Valek, J. 1967. Influence of fructose on iron absorption from the digestive system of healthy subjects. *Nutr. Dieta.* 9: 263.

Chang, Y. -O. and Hegsted, D. M. 1964. Lactose and calcium transport in gut sacs. *J. Nutr.* 82: 297.

Charley, P. J., Bibudhendra, S., Stitt, C. F., and Saltman, P. 1963. Chelation of iron by sugars. *Biochim. Biophys. Acta* 69: 313.

Chikosi, S. F., McMaster, D., and Love, A. H. G. 1985. Absorption and transport of zinc by the mammalian gut: influence of the presence of some sugars in the lumen. *Nutr. Res. Suppl.* I: 259.

Cochet, B., Jung, A., Griessen, M., Bartholdi, P., Schaller, P., and Donath, A. 1983. Effects of lactose on intestinal calcium absorption in normal and lactose-deficient subjects. *Gastroenterology* 84: 935.

Cohen, A. M., Michaelson, I. C., and Yanko, L. 1972. Retinopathy in rats with disturbed carbohydrate metabolism following a high sucrose diet. *Am. J. Opthalomol.* 73: 863.

Cohen, A. M., Teitelbaum, A., Briller, S., Yanko, L., Rosenmann, E., and Shafrir, E. 1974. Experimental models of diabetes. Ch. 29. In *Sugars in Nutrition,* Sipple, H. L. and McNutt, K. W. (Ed.), p. 485. Academic Press, New York.

Cohen, A. M., Teitelbaum, A., and Rosenmann, E. 1977. Diabetes induced by a high fructose diet. *Metabolism* 26: 17.

Cohen, A. M., Teitelbaum, A., Miller, E., Ben-Tor, V., Hirt, R., and Fields, M. 1982. The effect of copper on carbohydrate metabolism in rats. *Isr. J. Med. Sci.* 18: 840.

Condon, J. R., Nassim, J. R., Millard, F. J. C., Hilbe, A., and Stainthorpe, E. M. 1970. Calcium and phosphorus metabolism in relation to lactose tolerance. *Lancet* 1: 1027.

Cornatzer, W. E. and Klevay, L. M. 1986. Fructose lowers cardiac copper in rats. *Fed. Proc.* 45: 357.

Daniel, H. and Rehner, G. 1986. Effect of metabolizable sugars on the mucosal surface of pH of rat intestine. *J. Nutr.* 116: 768.

Davis, P. S. and Deller, D. J. 1966. Prediction and demonstration of iron chelating ability of sugars. *Nature* 212: 404.

Davis, P. S. and Deller, D. J. 1967. Effect of orally administered chelating agents EDTA, DTPA, and fructose on radioiron absorption in man. *Aust. Ann. Med.* 16: 70.

Favus, M. J. and Angeid-Backman, E. 1984. Effects of lactose on calcium absorption and secretion by rat ileum. *Am. J. Physiol.* 246: G281.

Fields, M., Ferretti, R. J., Smith, Jr., J. C., and Reiser, S. 1983. Effect of copper deficiency on metabolism and mortality in rats fed sucrose or starch diets. *J. Nutr.* 113: 1335.

Fields, M., Ferretti, R. J., Smith, Jr., J. C., and Reiser, S. 1984a. The interaction of type of dietary carbohydrates with copper deficiency. *Am. J. Clin. Nutr.* 39: 289.

Fields, M., Ferretti, R. J., Smith, J.C. Jr., and Reiser, S. 1984b. Impairment of glucose tolerance in copper-deficient rats: dependency on the type of dietary carbohydrate. *J. Nutr.* 114: 393.

Fields, M., Ferretti, R. J., Smith, J. C. Jr., and Reiser, S. 1984c. Effect of dietary carbohydrates and copper status on blood pressure of rats. *Life Sci.* 34: 763.

Fields, M., Ferretti, R. J., Reiser, S. and Smith, J. C. Jr., 1984d. The severity of copper deficiency in rats is determined by the type of dietary carbohydrate. *Proc. Soc. Exp. Biol. Med.* 175: 530.

Fields, M., Ferretti, R. J., Smith, J.C. Jr., and Reiser, S. 1984e. Interactions between dietary carbohydrate and copper nutriture on lipid peroxidation in rat tissues. *Biol. Trace Elem. Res.* 6: 379.

Fields, M., Ferretti, R. J., Judge, J. M., Smith, J. C., and Reiser, S. 1985. Effects of dietary carbohydrates on hepatic enzymes of copper-deficient rats. *Proc. Soc. Exp. Biol. Med.* 178: 362.

Fields, M., Holbrook, J., Scholfield, D., Smith, J.C. Jr., Reiser, S., and Los Alamos Medical Research Group. 1986a. Effect of fructose or starch on copper-67 absorption and excretion by the rat. *J. Nutr.* 116: 625.

Fields, M., Holbrook, J., Scholfield, D., Rose, A., Smith, J. C., and Reiser, S. 1986b. Development of copper deficiency in rats fed fructose or starch: weekly measurements of copper indices in blood. *Proc. Soc. Exp. Biol. Med.* 181: 120.

Fields, M., Lewis, C., Scholfield, D. J., Powell, A. S., Rose, A. J., Reiser, S., and Smith, J. C. 1986c. Female rats are protected against the fructose induced mortality of copper deficiency. *Proc. Soc. Exp. Biol. Med.* 183: 145.

Fields, M., Lewis, C. G., Rose, A., Smith, J. C., and Reiser, S. 1986d. Uptake of radiolabeled copper from portal blood containing fructose or glucose. *Biol. Trace Elem. Res.* 10: 335.

Fournier, P. and Digaud, A. 1971. Influence de l'ingestion de lactose sur l'absorption et la retention du zinc. C. R. Acad. Sci. Ser. D. 272: 3061.

Fridovich, I. 1976. Oxygen radicals, hydrogen peroxide and oxygen toxicity. Ch. 6. In Free Radicals in Biology, Vol. 1, Pryer, W. A. (Ed.), p. 239. Academic Press, New York.

Ghishan, F. K., Stroop, S., and Meneely, R. 1982. The effect of lactose on the intestinal absorption of calcium and zinc in the rat during maturation. Pediatr. Res. 16: 566.

Griffin, A. C. 1979. Role of Se in the chemoprevention of cancer. Adv. Cancer Res. 29: 419.

Harvey, D., Barry, R. D., and Gray, F. 1987. Sugar and sweetner situation and outlook report. U. S. Department of Agriculture, Economic Research Service, SSRV12N1: 35, Washington, D.C.

Holden, J. M., Wolf, W. F., and Mertz W. 1979. Zinc and copper in self-selected diets. J. Am. Diet. Assoc. 75: 23.

Jenkinson, S. G., Lawrence, R. A., Burk, R. F., and Williams, D. M. 1982. Effects of copper deficiency on the activity of the selenoenzyme glutathione peroxidase and on excretion and tissue retention of $^{75}SeO_3$. J. Nutr. 112: 197.

Johnson, M. A. 1986. Interaction of dietary carbohydrate, ascorbic acid and copper with the development of copper deficiency in rats. J. Nutr. 116: 802.

Johnson, M. A. and Hove, S. S. 1986. Development of anemia in copper-deficient rats fed high levels of dietary iron and sucrose. J. Nutr. 116: 1225.

Johnson, P. E. and Bowman, T. D. 1986. Isotope dilution measurement of copper absorption and excretion in rats fed different carbohydrates. Fed. Proc. 45: 234.

Kelly, S. E., Chawla-Singh, K., Sellin, J. H., Yasillo, N. J., and Rosenberg, I. H. 1984. Effect of meal composition on calcium absorption: enhancing effect of carbohydrate polymers. Gastroenterology 87: 596.

Keshan Disease Research Group. 1979. Observation on effect of sodium selenite in prevention of Keshan Disease. Chin. Med. J. 92: 471.

Klevay, L. M. 1984. The role of copper, zinc and other chemical elements in ischemic heart disease. Ch. 7. In Metabolism of Trace Metals in Man, Rennert, O. W. and Chan, W. Y. (Ed.), p. 129. CRC Press, Boca Raton, FL.

Klevay, L. M., Reck, S., and Barcome, D. F. 1979. Evidence of dietary copper and zinc deficiencies. J. Am. Med. Assoc. 241: 1916.

Kozlovsky, A. S., Moser, P. B., Reiser, S., and Anderson, R. A. 1986. Effects of diets high in simple sugars on urinary chromium losses. *Metabolism* 35: 515.

Krause, U. and Jenner, H. 1970. The effect of fructose on the absorption of iron. *Acta Soc. Med. Upsal.* 75: 266.

Landes, D. R. 1975. Influence of dietary carbohydrate on copper, iron and zinc status of the rat. *Proc. Soc. Exp. Biol. Med.* 150: 686.

Leichter, J. and Tolensky, A. F. 1975. Effect of dietary lactose on the absorption of protein, fat and calcium in the postweaning rat. *Am. J. Clin. Nutr.* 28: 238.

Lengemann, F. W. 1959. The site of action of lactose in the enhancement of calcium utilization. *J. Nutr.* 69: 23.

Lengemann, F. W. and Comar, C. L. 1961. Distribution of absorbed strontium-85 and calcium-45 as influenced by lactose. *Am. J. Physiol.* 200: 1051.

Levander, O. A., DeLoach, D. P., Morris, V. C., and Moser, P. B. 1983. Platelet glutathione peroxidase activity as an index of selenium status in rats. *J. Nutr.* 113: 55.

Lewis, C. G., Fields, M., Craft, N., Yang, C. -Y., and Reiser, S. 1987. Alteration of pancreatic enzyme activities in small intestine of rats fed a high- fructose, low-copper diet. *J. Nutr.* 117: 1447.

Liu, V. J. and Morris, J. S. 1978. Relative chromium response as an indicator of chromium status. *Am. J. Clin Nutr.* 31: 972.

Miller, J. and Landes, D. R. 1976a. Modification of iron and copper metabolism by dietary starch and glucose in rats. *Nutr. Rep. Int.* 13: 187.

Miller, J. and Landes, D. R. 1976b. Effects of starch, sucrose, and glucose on iron absorption by anemic rats. *Nutr. Rep. Int.* 14: 7.

Norman, D. A., Morawski, S. G., and Fordtran, J. S. 1980. Influence of glucose, fructose, and water movement on calcium absorption in the jejunum. *Gastroenterology* 78: 22.

Ono, K., Steele, N., Richards, M., Darcey, S., Fields, M., Scholfield, D., Smith, J., and Reiser, S. 1986. Copper-carbohydrate interaction in the pig: effects on lysyl oxidase activities (LOA). *Fed. Proc.* 45: 357.

Papachristodoulou, D. and Heath, H. 1977. Ultrastructural alterations during the development of retinopathy in sucrose-fed and streptozotocin-diabetic rats. *Exp. Eye Res.* 25: 371.

Pollack, S., Kaufmanm, R. M., and Crosby, W. H. 1964. Iron absorption:

effect of sugars and reducing agents. *Blood* 24: 577.

Reiser, S., Ferretti, R. J., Fields, M., and Smith, J. C. Jr., 1983. Role of dietary fructose in the enhancement of mortality and biochemical changes associated with copper deficiency in rats. *Am. J. Clin. Nutr.* 38: 214.

Reiser, S., Smith, J. C. Jr., Mertz, W., Holbrook, J. T., Scholfield, D. J., Powell, A. S., Canfield, W. K., and Canary, J. J. 1985. Indices of copper status in humans consuming a typical American diet containing either fructose or starch. *Am. J. Clin. Nutr.* 42: 242.

Reiser, S., Hallfrisch, J., Fields, M., Powell, A., Mertz, W., Prather, E. S., and Canary, J. J. 1986. Effects of sugars on indices of glucose tolerance in humans. *Am. J. Clin. Nutr.* 43: 151.

Reiser, S., Powell, A. S., Yang, C. -Y., and Canary, J. J. 1987. An insulinogenic effect of oral fructose in humans during postprandial hyperglycemia. *Am. J. Clin. Nutr.* 45: 580.

Russanov, E., Banskalieva, V., and Ljutakova, S. 1981. Influence of sex hormones on the subcellular distribution of copper in sheep liver. *Res. Vet. Sci.* 30: 233.

Salonen, J. T., Alfthan, G., Huttunen, J. K., Pikkarainen, J., and Puska, P. 1982. Association between cardiovascular death and myocardial infarction and serum selenium in a matched-pair longitudinal study. *Lancet* 2: 1975.

Saylor, W. W. and Downer, J. V. 1986. Copper and zinc distribution in the liver and oviduct of estrogen and testrosterone treated hens. *Nutr. Res.* 6: 181.

Steele, N., Richards, M., Darcey, S., Fields, M., Smith, J., and Reiser, S. 1986. Copper-carbohydrate interaction in the growing pig. *Fed. Proc.* 45: 357.

Steele, N., Richards, M., Virmani, R., Fields, M., Smith, J., and Reiser, S. 1987. Copper-carbohydrate interaction in the growing pig: cardiac histology. *Fed. Proc.* 46: 910.

Steinhardt, H. J. and Adibi, S. A. 1984. Interaction between transport of zinc and other solutes in human intestine. *Am. J. Physiol.* 247: G176.

Strause, L. G., Hegenauer, J., Saltman, P., Cone, R., and Resnick, D. 1986. Effect of long-term dietary manganese and copper deficiency on rat skeleton. *J. Nutr.* 116: 135.

Thornber, J. M. and Eckhert, C. D. 1984. Protection against sucrose-induced retinal capillary damage in the Wistar Rat. *J. Nutr.* 114: 1070.

Urban, E. and Pena, M. 1977. Failure of lactose and glucose to influence in vivo intestinal calcium transport in normal rats. *Digestion* 15: 18.

Vaughan, O. W. and Filer, L. J. Jr. 1960. The enhancing action of certain carbohydrates on the intestinal absorption of calcium in the rat. *J. Nutr.* 71: 10.

Viestenz, K. E. and Klevay, L. M. 1982. A randomized trial of copper therapy in rats with electrocardiographic abnormalities due to copper deficiency. *Am. J. Clin. Nutr.* 35: 258.

Wasserman, R. H. and Comar, C. L. 1959. Carbohydrates and gastrointestinal absorption of radiostrontium and radiocalcium in the rat. *Proc. Soc. Exp. Biol. Med.* 101: 314.

Wong, L., Deagen, J. T., and Whanger, P. D. 1985. Effects of dietary carbohydrates on selenium metabolism and glutathione peroxidase activity in rats. *Fed. Proc.* 44: 1509.

Zheng, J. -J., Wood, R. J., and Rosenberg, I. H. 1985. Enhancement of calcium absorption in rats by coadministration of glucose polymer. *Am. J. Clin. Nutr.* 41: 243.

# 10

# Fiber–Mineral and Fiber–Vitamin Interactions

**Steven L. Ink**

The Quaker Oats Company
Barrington, Illinois

## INTRODUCTION

Dietary fiber is considered to be an important component of a nutritionally balanced diet. Epidemiologic, clinical, and animal study results suggest that fiber may have a significant role to play in an overall dietary plan designed to reduce the risk for certain diseases, such as heart disease and colon cancer. The mechanisms by which fiber may impact chronic disease have not been well defined; as a result, specific recommendations regarding appropriate levels in the diet would seem to be premature at this point.

Unfortunately, many have advocated increased consumption of dietary fiber without consideration of fiber's impact on other nutrients, such as vitamins and minerals. Is there sufficient data at this point to suggest that the commonly recommended level of fiber for prophylactic use by the public will negatively affect mineral and vitamin nutriture? This question will be reviewed based on recent studies, which have examined the influence of dietary fiber on mineral and vitamin bioavailability.

Dietary fiber is typically defined as those components of plants that cannot be digested by the endogenous secretions of the human digestive tract. Therefore, any degradation that does occur is a result of metabolic activities of lower intestinal microflora. Dietary fiber is predominantly a mixture of the carbohydrate polymers cellulose, hemicellulose, gums, pectins, and mucilages, but also includes noncarbohydrate materials, such as lignin, waxes, and cutin.

Dietary fiber is often conceptually divided into a water-soluble fraction and an insoluble fraction due to relative solubilities and other chemical properties. Pectins, gums, and some hemicellulose compounds are considered soluble dietary fiber, since they can be solubilized in hot water following removal of lipids and most starches. Cellulose, lignin, and the remaining hemicelluloses constitute most of the insoluble fiber isolated from foodstuffs.

Although this stratification of dietary fiber into soluble and insoluble fractions is based on methods of extraction and quantification, it also lends itself to apparent division according to physiological function. The soluble fibers, especially pectins and gums, generally have the potential for increasing the viscosity of intestinal contents, slowing of glucose and lipid absorption in the small intestine, impacting cholesterol metabolism, and promoting short-chain fatty acid production in the colon (Hurt and Ink, 1987). The insoluble fibers have much less impact on the above-mentioned gastrointestinal events and tend to influence large bowel functions, such as stool weight, defecation frequency, and gastrointestinal transit rates.

Dietary fiber could theoretically influence mineral and vitamin bioavailability via several mechanisms. The solubility of minerals may be decreased as a result of ionic or coordination/chelation interactions with dietary fiber. Vitamin solubilities could be reduced due to fiber disruption of micellar formation. Dietary fiber in some cases may increase lumenal viscosities and reduce rates of migration of nutrients from the bulk phase of the lumen to the mucosal surface (Blackburn and Johnson, 1981). This possibility in combination with reduced transit times through the gut, entrapment of minerals and vitamins in a fiber matrix, and possible changes in gut surface morphology provide further potential mechanisms for reduced mineral and vitamin bioavailabilities (Cummings et al., 1978; Kriek, et al., 1982).

## BIOAVAILABILITY OF MINERALS

In vitro binding studies have confirmed that dietary fibers, such as wheat bran, soy bran, corn bran, rice bran, cellulose, and lignin, bind various minerals thus reducing solubilities and possibly bioavailabilities (Thompson, and Weber, 1979, 1981; Camire and Clydesdale, 1981; Reinhold et al., 1981; Fernandez and Phillips, 1982; Rendleman, 1982; Lyon, 1984; Platt and Clydesdale, 1985; Platt and Clydesdale, 1986). In addition to providing information on the relative mineral-binding capacities of various isolated fibers, these in vitro studies have elucidated some of the important factors that may determine in vivo binding of minerals by dietary fiber in a complex food matrix. The factors that must be considered include the following: (1) the presence of chelators such as ascorbate, citrate, oxalate, phytate, and some amino acids (Reinhold et al., 1981; Fernandez and Phillips, 1982); (2) the pH and heat treatment effects (Camire and Clydesdale, 1981); (3) concentrations of other minerals that may compete for a given binding site; and (4) fermentability of the fiber source in the colon and potential for colonic uptake of previously bound minerals (Cummings and Englyst, 1987).

As a result of the complexity of interactions that may take place between minerals, dietary fiber, and other components of a mixed diet, it becomes very difficult to extrapolate results from in vitro studies to expected mineral balances in the human. Therefore, this review will focus on in vivo research, with primary emphasis on human metabolic balance studies.

Animal and human studies of the effect of dietary fiber on mineral bioavailability have been recently reviewed (Munoz, 1986). These studies have examined the question of bioavailability via radiolabeled absorption tests, effect on mineral levels in the blood, and metabolic balance techniques.

In the case of iron, zinc, and calcium, combined animal and human data show that in about two-thirds of the studies, utilization of these minerals was compromised by dietary fiber. However, few of the approximately 30 studies are of the type that can be considered carefully controlled metabolic balance studies. As a result, it may be inferred that mineral uptake is reduced by dietary fiber, but it should not be concluded that mineral pool sizes in the body have been decreased.

Metabolic balance studies seem to be the technique of choice for answering the question, "At what level does dietary fiber significantly compromise the mineral status of an individual?" The balance study should have carefully controlled dietary intakes and adequate collection periods for measurement of fecal and urinary mineral losses. The length of the study is also very critical, since the body must be allowed sufficient time to adapt to the new dietary regimen. In addition, some measure of physiological utilization should be included for the nutrient in question. This could be enzyme activities for which the mineral serves as a cofactor or an appropriate marker of tissue levels. The effect of dietary fiber on mineral balances will now be assessed based on some recent metabolic balance study data.

Table 10.1 shows the results of several metabolic balance studies examining the influence of various dietary fibers on calcium balance (Cummings et al, 1979a,b; Slavin and Marlett, 1980; VanDokkum et al., 1982; Kelsay and Prather, 1983; Spencer et al., 1987). Insoluble dietary fiber sources, such as wheat bran and cellulose, have significant negative effects

**TABLE 10.1**  Calcium

| | Fiber | Calcium | |
| Source | Intake (g/day) | Intake (mg/day) | Balance (mg/day) |
|---|---|---|---|
| Wheat bran[a] | 35 | 1087 | −30 |
| Wholemeal bread | 22 | 1022 | −10 |
| Control | 9 | 956 | +14 |
| Wheat bran + wholemeal bread[b] | 53 | 1302 | −77 |
| Control | 23 | 995 | +27 |
| Cellulose[c] | 23.5 | 600 | −199 |
| Control | 9.5 | 585 | −16 |
| Fruit and vegetables[d] | 24.0 | 1037 | −11 |
| Fruit and vegetables + spinach | 26.5 | 1064 | −73 |
| Control + spinach | 4.9 | 953 | +44 |
| Oat bran[e] | 43.2 | 800 | −32 |
| Control | 22.6 | 800 | −31 |
| Pectin[f] | 46 | 1357 | −37 |
| Control | 15 | 1280 | −35 |

[a]VanDokkum et al. (1982).
[b]Cummings et al. (1979a).
[c]Slavin and Marlett (1980).
[d]Kelsay and Prather (1983).
[e]Spencer et al. (1987).
[f]Cummings et al. (1979b).

**TABLE 10.2** Magnesium

| | Fiber | Magnesium | | |
| Source | Intake[a] (g/day) | Intake[a] (mg/day) | Balance (mg/day) |
|---|---|---|---|
| Cellulose[b] | 23.5 | 300 | −15 |
| Control | 9.5 | 276 | +14 |
| Fruit and vegetables + spinach[c] | 26.5 | 350 | −10 |
| Fruit and vegetables | 24.0 | 326 | +18 |
| Control + spinach | 4.9 | 308 | +20 |
| Wheat bran[d] | 35 | 537 | −3 |
| Control | 9 | 213 | −8 |

[a]Approximate.
[b]*Slavin and Marlett (1980).*
[c]*Kelsay and Prather (1983).*
[d]*VanDokkum et al. (1982).*

on calcium balance when present at levels approximating 20–25 g/day. On the other hand, fiber sources containing considerable soluble fiber, such as oat bran, fruits, and vegetables, seem to have little impact on calcium balance. Of interest is the apparent negative effect of a combination of fruit and vegetable fiber with oxalate from spinach (Kelsay and Prather, 1983). However, when this study was repeated with a 6-week period of feeding instead of 4 weeks, calcium balances were no longer negative, indicating that the body may adapt to this dietary regimen with time (Kelsay et al., 1987).

The results of several metabolic balance studies that examined the effect of dietary fiber on magnesium nutriture are found in Table 10.2 (Slavin and Marlett, 1980; VanDokkum et al., 1982; Kelsay and Prather, 1983). The data are suggestive of a trend toward negative magnesium balance with the higher fiber diets. However, the cellulose-induced negative magnesium balance was not significantly different from the control, the negative balance found after 4 weeks of fruit, vegetable, and spinach consumption was not apparent after 6 weeks on the diet (Kelsay et al., 1987a,b), and although the wheat bran diet reduced the percentage of magnesium absorbed, overall magnesium balance was no different from the control. These results illustrate three inherent difficulties in designing metabolic balance studies to assess mineral nutriture. The number of subjects is usually limited, appropriate length of the study is either not known or not feasible, and control of all dietary components is impossible.

Iron bioavailability was studied as early as 1942, when McCance and Widdowson reported that high extraction wholemeal bread decreased iron absorption relative to white bread. Unfortunately, this well-conceived study was conducted before the dietary fiber hypothesis had been proposed and was not designed to answer questions relative to fiber and mineral bioavailability. However, recent studies have focused on the affect of dietary fiber on iron nutriture (Table 10.3). Dietary fiber derived from soy, cellulose, gums, fruits, vegetables, and wheat bran had no negative impact on iron balance in subjects consuming approximately 20–25 g of dietary fiber/day (Widdowson and McCance, 1942; Kelsay et al., 1979 a,b; VanDokkum et al., 1982; Behall et al., 1983; Tsai et al., 1983). However, when the level of dietary fiber was increased to 35 g/day with addition of wheat bran, a negative iron balance resulted. This result is surprising in light of apparent increased iron absorption associated with wheat bran consumption (16 g) found in another study (Sandberg et al., 1982). These studies together suggest a threshold level of dietary fiber approximating 30 g (equivalent to about 70 g of wheat bran) beyond which iron balance may be adversely affected.

Table 10.4 contains the results of several studies that have examined the effect of fiber on zinc bioavailability. Wheat bran and corn bran were found to have little impact on overall zinc nutriture (Sandstead et al., 1978; VanDokkum et al., 1982), while fruit and vegetables had a significant adverse affect on zinc balance after 4 weeks (Kelsay et al., 1979 a,b; Slavin and Marlett, 1980) but no apparent affect after 6 weeks (Prather et al., 1987).

**TABLE 10.3**   Iron

|  | Fiber | Iron | |
| --- | --- | --- | --- |
| Source | Intake (g/day) | Intake (mg/day) | Balance (mg/day) |
| Wheat bran[a] | 35 | 12.2 | −2.9 |
| Wheat bran | 22 | 11.8 | +1.2 |
| Control | 9 | 8.3 | +0.8 |
| Fruits and vegetables[b] | 24 | 26.4 | +4.6 |
| Control | 4.6 | 21.8 | +3.8 |
| Soy[c] | (11.3)[d] | | NS[d] |
| Cellulose | (15–20) | | NS |
| Carboxymethylcellulose | (15–20) | | NS |
| Locus bean gum | (15–20) | | NS |

[a]*VanDokkum et al. (1982).*
[b]*Kelsay et al. (1979a).*
[c]*Tsai et al. (1983); Behall et al. (1983).*
[d]Number in parentheses indicates approximate amount of fiber added to basal diet. NS indicates no significant change in iron balance defined as intake–fecal in this case.

**TABLE 10.4**   Zinc

| | Fiber | Zinc | |
|---|---|---|---|
| Source | Intake (g/day) | Intake (mg/day) | Balance (mg/day) |
| Wheat bran[a] | 35 | 13.5 | −0.6 |
| Wheat bran | 22 | 11.2 | −0.1 |
| Control | 9 | 9 | −0.6 |
| Fruits & vegetables[b] | 23.8 | 12.6 | −0.9 |
| Control | 46 | 13.2 | +3.5 |
| Corn bran[c] | (+23.9) | 10–16 | +2 |
| Wheat bran | (+11.4) | 10–16 | +2 |
| Control[d] | Not Determined | 10–16 | +1 |

[a]*VanDokkum et al. (1982).*
[b]*Kelsay et al. (1979b).*
[c]*Sandstead et al. (1978).*
[d]Control dietary fiber intake was not measured but would be expected to be about 10–15 g. Values in parentheses indicate amount of fiber added to control diet.

The effect of dietary fiber on the bioavailability of other minerals, such as copper, selenium, and phosphorus, has been the subject of several studies but will not be reviewed here since few conclusions can be drawn based on previous results.

When examining the influence of dietary fiber from food sources on mineral bioavailabilities, one must consider other food components that could also interact with minerals. Among the components most often suggested to have potential for adversely affecting mineral balances are chelators, such as phytate and oxalates. Phytate has been found to bind minerals in vitro and to reduce mineral absorption in humans (Rendleman, 1982; Lyon, 1984; Turnlund et al., 1984). However, studies by Morris and co-workers compared whole bran muffins to dephytinized bran muffins and found no differences in mineral balances (Harland and Morris, 1985). Interestingly, subjects consuming the phytate-containing whole bran muffins showed reduced Zn, Fe, Mn, Cu, and Ca absorption during the initial 5 days of the study, but for the subsequent 10 days had greater apparent absorption of these minerals than the dephytinized bran muffin group. This study along with the above-mentioned studies examining mineral balances during consumption of foods containing fiber and oxalates (Kelsay et al., 1987; Prather et al., 1987) suggest that the body may adapt positively with time to factors which bind minerals in vitro. Although it is not the intent of this review to thoroughly examine the role of phytate in mineral bioavailability, it is apparent that absorption studies involving phytate should be interpreted with appropriate regard for fac-

tors, such as length of the study and source of phytate, since food sources
of phytate may have different binding characteristics than purified phytate (Rendleman, 1982).

## BIOAVAILABILITY OF VITAMINS

The effects of dietary fiber on vitamin bioavailability have not been extensively studied; as a result, little is known about the ways in which dietary
fiber affect vitamin absorption in the gastrointestinal tract.

Since some high fiber diets increase fecal excretion of lipids, such as
cholesterol, possibly by impairment of micelle formation, fat-soluble
vitamins may be influenced by dietary fiber in a similar way. This
mechanism was suggested as a potential explanation for the reduced β-
carotene utilization found in chicks consuming hemicellulose, lignin, or
pectin at a 7% supplemental level (Erdman et al., 1986). However, Mongeau et al. (1986) found that wheat bran at graded doses of 4–20% in the
diet increased fecal fat excretion in rats without significantly reducing
vitamin E status. Additional studies involving fat-soluble vitamins and
fiber have been previously reviewed (Kasper, 1986) with no clear-cut conclusions apparent.

Bioavailability studies of water-soluble vitamins in the presence of
dietary fiber have provided mixed results (Kasper, 1986; Lewis et al., 1986).
Wheat bran and cellulose enhance the absorption of riboflavin, nicotinic
acid, and ascorbic acid, have little effect on vitamin $B_6$, adversely influence vitamin $B_{12}$ status, and have variable impact on folic acid. The
mechanisms involved in fiber–vitamin interactions are poorly understood. Binding of the vitamin to the fiber source, impairment of
vitamin interaction with an intestinal receptor, such as the ileal receptor
for vitamin $B_{12}$, and decreased activity of brush border enzymes, required
for absorption in the case of riboflavin and folacin, are all potential
modes of interaction between dietary fiber and vitamins. In the case of
folacin, dietary fiber could even theoretically reduce the bioavailability of
this vitamin by depleting tissue zinc since brush border folate conjugase is
a zinc-dependent enzyme.

Unfortunately, at this time the breadth and scope of studies examining
the effects of dietary fiber on vitamin bioavailabilities are not sufficient
for general recommendations to be made about a given vitamin and/or
fiber type.

## SUMMARY

Although it is virtually impossible to determine with certainty what the optimal dietary fiber intake of a population should be for overall health, available evidence tends to suggest that current daily intakes of 10–15 g in the U.S. population are somewhat below the levels associated with potential health benefits. The question that must be addressed relative to mineral and vitamin nutriture is "Will increases in typical intakes of dietary fiber on the order of 50–100% have any adverse effect on these nutrients?"

The answer is probably not, provided that the increase in dietary fiber is associated with an overall dietary plan that will provide adequate intakes of all nutrients (Hallfrisch et al., 1987). Addition of dietary fiber to the diet should be accomplished by choosing foods not only rich in dietary fiber but also providing a well-rounded complement of other nutrients. Fresh fruits and vegetables, wholegrains, and food sources of fiber in general should be emphasized over purified fiber sources and fiber supplements. Populations that may be somewhat susceptible to the impact of mineral–fiber interactions are those consuming diets that are marginally adequate, such as the elderly, pregnant or lactating women, and the adolescent.

## REFERENCES

Behall, K. M., Lee, K. Wilson, A., and Prather, E. S. 1983. Effect of purified fibers added to a basic diet on apparent mineral balance of male subjects. *Fed. Proceedings* 42: 1063.

Blackburn, N. A. and Johnson, I. T. 1981. The effect of guar gum on the viscosity of the gastrointestinal contents and on glucose uptake from the perfused jejunum in the rat. *Brit. J. Nutr.* 46: 239.

Camire, A. L. and Clydesdale, F. M. 1981. Effect of pH and heat treatment on the binding of calcium, magnesium, zinc, and iron to wheat bran and fractions of dietary fiber. *J. Food Sci.* 46: 548.

Cummings, J. H., Branch, W., Jenkins, D. J. A., Southgate, D. A. T., Houston, H., and James, W. P. T. 1978. Colonic response to dietary fibre from carrot, cabbage, apple, bran, and guar gum. *Lancet* 1(8054): 5.

Cummings, J. H. and Englyst, H. N. 1987. Fermentation in the human large intestine and the available substrates. *Am. J. Clin. Nutr.* 45: 1243.

Cummings, J. H., Hill, M. J., Jivraj, T., Houston, H., Branch, W. J., and Jenkins, D. J. A. 1979a. The effect of meat protein and dietary fiber on colonic function and metabolism. *Am. J. Clin. Nutr.* 32: 2086.

Cummings, J. H., Southgate, D. A. T., Branch, W. J., and Wiggins, H. S. 1979b. The digestion of pectin in the human gut and its effect on calcium absorption and large bowel function. *Br. J. Nutr.* 41: 477.

Erdman, J. W., Fahey, G. C., and White, C. B. 1986. Effects of purified dietary fiber sources on β-carotene utilization by the chick. *J. Nutr.* 116: 2415.

Fernandez, R. and Phillips, S. F. 1982. Components of fiber bind iron in vitro. *Am. J. Clin. Nutr.* 35: 100.

Hallfrisch, J., Powell, A., Carafelli, C., Reiser, S., and Prather, E. S. 1987. Mineral balances of men and women consuming high fiber diets with complex or simple carbohydrate. *J. Nutr.* 117: 48.

Harland, B. F. and Morris, E. R. 1985. Fibre and mineral absorption. In *Dietary Fibre Perspectives Reviews and Bibliography.* Leeds A. R. (Ed.). John Libbey, London.

Hurt, H. D. and Ink, S. L. 1987. Nutritional implications of gums. *Food Technol.* 41(1): 77.

Kasper, H. 1986. Effects of dietary fiber on vitamin metabolism. In *CRC Handbook of Dietary Fiber in Human Nutrition.* Spiller, G. A. (Ed.). CRC Press, Inc., Boca Raton, FL.

Kelsay, J. L., Behall, K. M., and Prather, E. S. 1979a. Effect of fiber from fruits and vegetables on metabolic responses of human subjects. II. Calcium, magnesium, iron, and silicon balances. *Am. J. Clin. Nutr.* 32: 1876.

Kelsay, J. L., Jacob, R. A., and Prather, E. S. 1979b. Effect of fiber from fruits and vegetables on metabolic responses of human subjects: III. Zinc, copper, and phosphorus balances. *Am. J. Clin. Nutr.* 32: 2307.

Kelsay, J. L. and Prather, E. S. 1983. Mineral balances of human subjects consuming spinach in a low-fiber diet and in a diet containing fruits and vegetables. *Am. J. Clin. Nutr.* 38: 12.

Kelsay, J. L., Prather, E. S., and Canary, J. J. 1987. Calcium and magnesium balances of men fed a diet containing fiber in fruits and vegetables and oxalic acid in spinach for six weeks. In *Abstracts of Papers.* Federation of American Societies for Experimental Biology.

Kriek, N. P. J., Sly, M. R., DeBruyn, D. B., De Klerk, W. A., Benan, M. J., VanSchalkwyk, D. J., and Van Rensburg, S. J. 1982. Dietary wheaten bran in baboons: Long-term effect on the morphology of the digestive tract and aorta, and on tissue mineral concentrations. *Br. J. Exp. Path.* 63: 251.

Lewis, N. M., Kies, C., and Fox, H. M. 1986. Vitamin $B_{12}$ status of humans as affected by wheat bran supplements. *Nutrition Reports International* 34(4): 495.

Lyon, D. B. 1984. Studies on the solubility of Ca, Mg, Zn, and Cu in cereal products. *Am. J. Clin. Nutr.* 39: 190.

Mongeau, R., Behrens, W. A., Madere, R., and Brassard, R. 1986. Effects of dietary fiber on vitamin E status in rats: dose-response to wheat bran. *Nutrition Research* 6: 215.

Munoz, J. M. 1986. Overview of the effects on dietary fiber on the utilization of minerals and trace elements. In *CRC Handbook of Dietary Fiber in Human Nutrition.* Spiller, G. A. (Ed.). CRC Press, Inc., Boca Raton, FL.

Platt, S. R. and Clydesdale, F. M. 1985. Binding of iron by lignin in the presence of various concentrations of calcium, magnesium, and zinc. *J. Food Sci.* 50: 322.

Platt, S. R. and Clydesdale, F. M. 1986. Effects of iron alone and in combination with calcium, zinc and copper on the mineral-binding capacity of wheat bran. *J. Food Protection* 49: 37.

Prather, E. S., Kelsay, J. L., and Clark, W. M. 1987. Zinc and copper balances of men fed a diet containing fiber in fruits and vegetables and oxalic acid in spinach for six weeks. *Fed. Proceedings* 46: 1167.

Reinhold, J. G., Garcia, J. S., and Garzon, P. 1981. Binding of iron by fiber of wheat and maize. *Am. J. Clin. Nutr.* 34: 1384.

Rendleman, J. A. 1982. Cereal complexes: Binding of calcium by bran and components of bran. *Cereal Chemistry* 59(4): 302.

Sandberg, A. S., Hasselblad, C., and Hasselblad, K. 1982. The effect of wheat bran on the absorption of minerals in the small intestine. *Br. J. Nutr.* 48: 185.

Sandstead, H. H., Munoz, J. M., Jacob, R. A., Klevay, L. M., Reck, S. J., Logan, G. M., Dintzis, F. R., Inglett, G. E., and Shuey, W. C. 1978. Influence of dietary fiber on trace element balance. *Am. J. Clin. Nutr.* 31: S180.

Slavin, J. L. and Marlett, J. A. 1980. Influence of refined cellulose on human bowel function and calcium and magnesium balance. *Am. J. Clin. Nutr.* 33: 1932.

Spencer, H., Derler, J., and Osis, D. 1987. Calcium requirement, bioavailability, and loss. *Fed. Proceedings* 46: 631.

Thompson, S. A. and Weber, C. W. 1979. Influence of pH on the binding of copper, zinc, and iron in six fiber sources. *J. Food Sci.* 44: 752.

Thompson, S. A. and Weber, C. W. 1981. Copper and zinc binding to dietary fiber sources: An ion exchange column method. *J. Food Sci.* 47: 125.

Tsai, A. C., Mott, E. L., Owen, G. M., Bennick, M. R., Lo, G. S., and Steinke, F. H. 1983. Effects of soy polysaccharide on gastrointestinal functions, nutrient balance, steroid excretions, glucose tolerance, serum lipids, and other parameters in humans. *Am. J. Clin. Nutr.* 38: 504.

Turnlund, J. R., King, J. C., Keyes, W. R., Gong, B., and Michel, M. C. 1984. A stable isotope study of zinc absorption in young men: effects of phytate and α-cellulose. *Am. J. Clin. Nutr.* 40: 1071.

VanDokkum, W., Wesstra, A., and Schippers, F. A. 1982. Physiological effects of fibre-rich types of bread. *Br. J. Nutr.* 47: 451.

Widdowson, E. M. and McCance, R. A. 1942. Iron exchanges of adults on white and brown bread diet. *Lancet* 1: 588.

# 11

# Interacting Effects of Carbohydrate and Lipid on Metabolism

**Carolyn D. Berdanier**

University of Georgia
Athens, Georgia

## INTRODUCTION

Students of intermediary metabolism have long recognized that fat and carbohydrate share some pathways of metabolism within the cell. Complex carbohydrates are hydrolyzed and converted to glucose, which is then either oxidized or converted to fatty acids for energy storage. Lipids, likewise, are hydrolyzed to their simple components of glycerol, fatty acids, and cholesterol and either stored or oxidized for energy. Fatty acids, glycerol, and cholesterol can be synthesized from glucose and glucose can be synthesized from glycerol as well as from lactate, pyruvate, and other metabolites. It stands to reason, therefore, that the pathways of lipogenesis, lipolysis, glycolysis, and gluconeogenesis can be influenced by dietary carbohydrate and lipid. That these pathways can also be influenced by an interaction of these dietary components is also reasonable, albeit poorly understood.

Why would there be interest in this dietary interaction effect on metabolism? The answer lies, in part, with the observations of epidemiologists who suggest that diseases such as cardiovascular disease (CVD) and diabetes occur as a result of an interaction of environmental and genetic factors. One of the environmental factors is diet. Epidemiologists have pointed out that the incidence of CVD is well correlated to the percentage of the diet that is fat (Fig. 11.1). Other epidemiologists have shown that just as good a correlation can be drawn between CVD and sugar intake. High sugar and/or high fat intakes seem to be associated with high incidences of CVD and diabetes. There are a few notable exceptions, however. One which has caught the attention of nutritionists and nonnutritionists alike is the fat intake (high) of Greenland Eskimos and their incidence (low) of CVD (Bang and Dyerberg, 1972; Dyerberg et al., 1975, 1978; Bang et al. 1976, 1980; Dyerberg and Bang, 1979). The fat intake of these people arises mainly from the marine foods they consume: seals, fish, walrus, whale, and other sea creatures. The fat in these foods contain a significant amount of omega-3 fatty acids. These fatty acids are long-

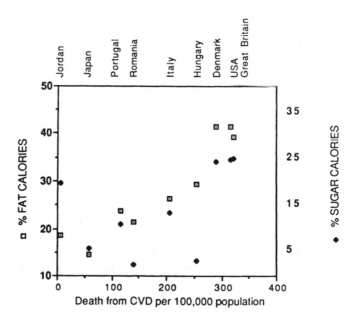

**FIG. 11.1**  Epidemiological evidence which suggests that dietary fat and sugar intake may be related to the incidence of cardiovascular disease (CVD).

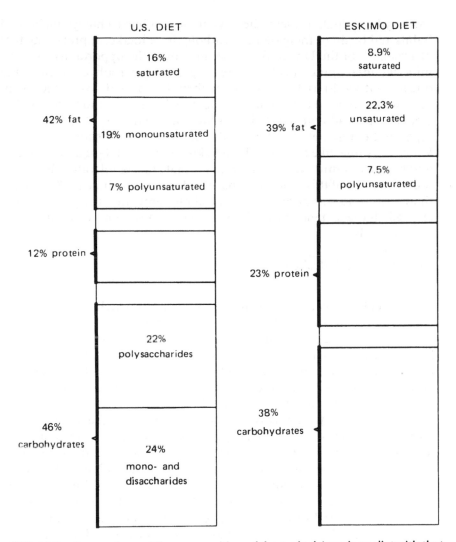

**FIG. 11.2** Comparison of the composition of the typical American diet with that of the Greenland Eskimos. Figures given are % of calories.

chain fatty acids of the linolenic acid family and consist largely of eicosapentaenoic (20:5ω3) and docosahexaenoic (22:6ω3) acids. The metabolism of these fatty acids is being studied extensively, and there is considerable interest in the possible protective effect the omega-3 fatty acids might have on CVD development and progression. Compared to the average American diet (Fig. 11.2), the Greenland Eskimo diet is a low carbohydrate–high protein diet. As these Eskimos change their food habits and begin to increase their carbohydrate intake, approaching that of Denmark or the United States, we may have the opportunity to determine the interacting effect of increasing dietary carbohydrate (with the omega-3 fatty acid–rich dietary fat) on their disease patterns. In the meantime, we will have to rely on the data obtained from rats and mice to tell us what to expect 40 years hence in this human population. Of interest is the report of Feldman et al. (1975) on the Alaskan Arctic Eskimo. These Eskimos consume, like the Greenland Eskimos, a high fat–high protein diet and have low serum triglycerides with normal glucose tolerance. However, those Eskimos of this group who have added carbohydrate (50% sugar) to their diets had elevations in blood cholesterol and triglycerides and showed diabetic tendencies despite their continuing consumption of marine foods.

## STUDIES IN ANIMALS

Much data have been obtained from rats and mice, which have the advantage over humans due to their size, speed of metabolic response, and cost. Traditionally, rodent diets are low in fat ($\sim$5–6%) because low fat diets are less expensive to make and spoil less easily than diets replicating that of the human. Whereas the human in the United States (Fig. 11.2) consumes about 40% of his or her calories as fat, the rat consuming a 5% fat diet consumes only 11.7% of its calories as fat. The primary energy source in the rodent diet is carbohydrate. This poses some problems in the design and interpretation of results of animal experiments. The diets are not analogous to those of the human. However, the rat response to the extreme diet with respect to high/low carbohydrate and fat helps us understand how the human responds to variation in dietary energy sources. Yet, the investigator must decide whether to examine the effect of change in type (only!) of carbohydrate and/or fat or to compound the issue and examine the effect of change in both type and amount of dietary carbohydrate and/or fat on the rat's metabolism. Most investigators elect the first paradigm rather than the second. These investigators use the proximate composition of the rodent stock diet for the formulation of their experimental diets. Thus, the reports on the effects of starch versus sucrose or glucose utilize a 65% carbohydrate diet in which one of these is the sole carbohydrate. Similarly, effects of type of dietary fat may be studied using a single fat source as 5 or 6% of the diet. Studies using the rat on the interacting effects of carbohydrate and fat type are few, but those published and reviewed here use 64–65% carbohydrate and 5–6% fat.

## APPROPRIATE ANIMAL MODELS FOR THE HUMAN

Regardless of the feeding paradigm and the proximate composition of the diet used as a reference, the investigator must decide whether to examine the dietary interaction effects on a particular metabolic pathway *only* or to ask the question as to the long-term effect of these diet manipulations on CVD and diabetes. The first question can be answered simply by using one of the many different strains of rats and mice available for nutrition studies. The second question is more difficult. Just as humans vary in their susceptibility to CVD and diabetes, so too do rodents. The investigator must choose a rodent that is genetically susceptible to the disease in question if the nutrition studies are to have application to the human population with comparable genetic programming. Thus, the choice of the animal is critical if one wishes to determine whether diet, in particular the type of carbohydrate and lipid, can affect the time course of disease development and intensity and therefore be analogous to the human. Table 11.1 lists a few of the many rodent strains available for these kinds of studies.

Long-term studies, that is, studies of the effects of a lifetime of consuming diets varying in amount and kind of fat and carbohydrate on longevity and disease, have not been conducted except in BHE and Wistar rats. The results of studies on the influence of dietary factors on longevity and histopathology of the BHE and Wistar rat were published in a USDA report by Adams in 1964. Over 1000 rats were studied, and the diets contained a variety of fats as well as carbohydrates. Sucrose, glucose, and starch, peanut butter, whole egg, beef tallow, safflower oil, and corn oil were used.

**TABLE 11.1** Some Rodent Strains Available for Dietary Studies

| Rodent | Key features | Human disease analogy |
|---|---|---|
| Zucker fatty rat | Early onset obesity, lipemia | Early onset obesity, lipemia |
| ob/ob mouse | Early onset obesity | Early onset obesity |
| SHR rat | Hypertension | Hypertension |
| db/db mouse | Early onset diabetes | Early onset diabetes |
| BB Wistar rat | Early onset diabetes | Early onset diabetes |
| LA/N-corpulent rat | Late onset diabetes, obesity | Late onset diabetes, obesity |
| BHE rat | Late onset diabetes, lipemia | Late onset diabetes, lipemia |

The most striking findings were that the BHE rats were more lipemic, had fattier livers and more renal disease than Wistar rats, and that these differences were more pronounced if the dietary carbohydrate was sucrose and the source of dietary fat was whole egg or beef tallow (Adams, 1964; Durand et al., 1964, 1968). Rats consuming these diets gained weight faster and died sooner than rats consuming a starch–corn oil diet. The biochemical and genetic reasons for the shortened lifespans were not addressed in these reports. Longevity as influenced by an interaction of diet ingredients in other genotypes for obesity or diabetes have not been conducted. Michaelis et al. (1986) have reported that feeding sucrose rather than starch affects the longevity and pathology of LA/cp rats. Substituting menhaden oil for corn oil was without effect on these rats (Schoene et al., 1986).

## CARBOHYDRATE–FAT INTERACTION EFFECTS IN HUMANS

Except for the reports by epidemiologists (e.g., Kannel, 1985) on the incidence of CVD and diabetes and the usual food intakes of the population, there are no long-term studies of the influence of diet on disease progression in humans. The reasons for this deficit are obvious—it would be far too expensive and totally unethical to manipulate the diet in order to determine its influence on the time course of degenerative disease development. Some short-term studies have, however, been conducted. Among the earliest are those of Little et al. (1970), Antar et al. (1970), and Birchwood et al. (1970). These investigators studied volunteers consuming defined diets in which the available carbohydrate portion was either sucrose or starch and the fat portion was either corn oil or a mixture of fats. In one study (Little et al., 1970) the fat mixture was derived from turkey, vegetable oil, and corn oil margarine, while in another study (Antar et al., 1970) the fat mixture consisted of the fats in butter, lard, egg yolk powder, and cocoa. Low (10% of calories), moderate (35% of calories), and high (65% of calories) fat levels were studied at a constant protein intake and corresponding (to the fat) decrements of carbohydrates. Serum triglycerides, cholesterol, and phospholipid levels were determined.

The volunteers were all lipemic. Some were responsive to dietary fat, some to dietary carbohydrate, while others were responsive in terms of elevated blood lipids to both dietary fat and carbohydrate. Several of the carbohydrate-induced lipemics were also diabetic. When the subjects were given the diet relatively high in saturated fat, and which contained sucrose instead of starch, there was a rise in serum triglycerides. However, the magnitude of this rise depended on the type of lipemia. All subjects had elevated serum lipids initially, and these lipid levels rose or fell in response to manipulation of the dietary fat level and type *and* carbohydrate type (Fig. 11.3). Other investigators (Kuo and Bassett, 1965; Kaufmann et al., 1966; Kuo et al., 1967) as well as those cited above have shown that lipemic subjects respond to these types of dietary manipulations but that their lipemia does not totally subside. Phillipson et al. (1985) reported that lipemic (Type IIb and Type V) subjects consuming fish oil rather than vegetable oil or a mixture of dietary fats evidenced a significant fall in serum cholesterol and triglyceride levels. These workers did not study the interacting effect of dietary carbohydrate on the response to fish oil in the diet. However, they concluded that the inclusion of fish oil in the diets of lipemic subjects would be useful in lowering their serum lipids.

Studies with normal subjects likewise showed some response to changes in the type of dietary fat or carbohydrate. However, not all subjects were affected significantly nor were the diets varied both in type and amount of fat *and* type and amount of carbohydrate. Some of the early

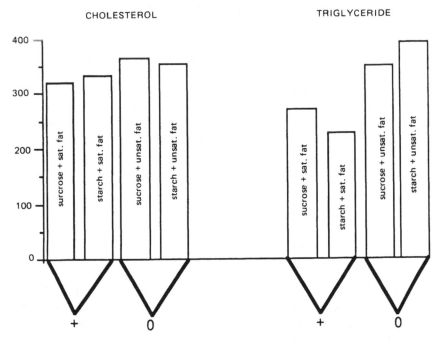

**FIG. 11.3** Serum lipid levels of lipemic volunteers consuming diets which varied in fat and carbohydrate source. Summary of findings of Antar et al. (1970) (+) and Little et al. (1970) (0).

workers, e.g., Keys et al. (1960), MacDonald and Braithwaite (1964), Antar and Ohlson (1965), McGandy et al. (1966), and Akinyanju et al. (1968), reported that substituting sucrose for starch increased serum triglyceride and cholesterol levels. In contrast, Irwin et al. (1964) and Grande et al. (1965) could not show a significant sucrose effect on serum lipids. The lack of agreement between the studies might be related to the fat source used by the respective investigators. As shown by Antar et al. (1970), fat type as well as carbohydrate type influences serum lipids. Harris et al. (1984) examined the influence of omega-3 fatty acid–rich menhaden oil on carbohydrate-induced hyperlipemia. Seven mildly lipemic subjects were fed, as % of calories, a 75% carbohydrate–15% fat diet or a 45% carbohydrate–45% fat diet. The carbohydrate was primarily starch, whereas the fat was either a mixture of naturally occurring plant and animal fats or primarily fish oil. Fish oil was found to reverse the lipemia that developed when the high carbohydrate diet was fed. Unfortunately, these investigators did not examine the subjects fed a low carbohydrate diet nor did they vary the type of carbohydrate.

All of these studies used diets consisting of foods normally present in the marketplace and consumed by the subjects. Few studies using human subjects have been able to control the composition of the diet (as well as the environment) as closely as is possible for the experiments using rodents. In addition, in studies with humans it is rarely possible to examine individual tissue metabolism in the detail necessary for conclusions to be made about the effects of diet on the different cell types. Thus, the remaining portion of this paper will review the literature on the effects of dietary carbohydrate and lipid on lipogenesis, lipolysis, glycolysis, gluconeogenesis, and pentose shunt activity in rats.

## CARBOHYDRATE EFFECTS ON LIPOGENESIS IN RATS

Numerous studies have shown that the substitution of sucrose for starch in the diets of normal rats results in an increase in hepatic fatty acid synthesis and in a transient rise in serum triglycerides. More than 500 papers have reported a differential response to carbohydrate type in the diet. Studies of a variety of hepatic enzymes in the pathways used for the conversion of glucose to fatty acids (Fig. 11.4) revealed that sucrose feeding resulted in an increase in the activities of a number of the rate-limiting enzymes. Those reactions whose enzymes were increased by sucrose feeding are shown by the heavy arrows. Glucokinase, glucose 6-phosphate dehydrogenase (G6PD), 6-phosphogluconate dehydrogenase, aldolase, dihy-

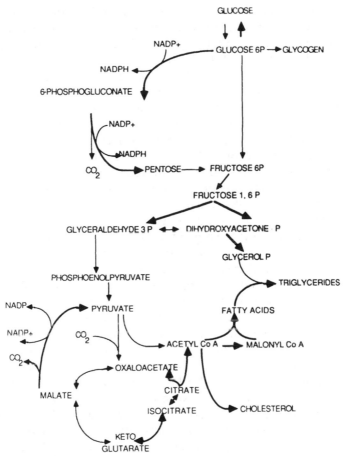

**FIG. 11.4** Schematic representation of metabolic pathways for glycolysis, pentose shunt, citric acid cycle, lipogenesis, and cholesterogenesis.

droxyacetone phosphate dehydrogenase, citrate cleavage enzyme, malic enzyme, isocitrate dehydrogenase, acetyl CoA carboxylase, fatty acid synthetase, HMG CoA carboxylase, and pyruvate kinase were all more active in sucrose-fed rats than in starch-fed rats. Long-term sucrose feeding in normal rats usually resulted in an adaptation to the high sucrose intake, and the initial increases in lipogenesis and serum lipids partially disappeared. However, when genetically obese and/or lipemic rats were fed these diets, adaptation did not occur to a significant degree, and degenerative changes in a variety of tissues were observed. In the SHR/N-corpulent rat (Michaelis et al., 1986) long-term sucrose feeding resulted in lipemia, a deterioration of glucose tolerance, renal sclerosis, and thickened glomerular tufts. These changes were not seen in the starch-fed rats. Feeding sucrose or glucose to obese Zucker rats also resulted in lipemia and an increase in hepatic lipid and lipogenesis (Zucker, 1965; Berka and Kaplan, 1983; Basilico et al., 1984). Similar findings have been reported for genetically lipemic/glycemic BHE rats (Taylor et al., 1967; Berdanier, 1974a, b; Berdanier et al., 1979a, b) and, in addition, resulted in an increase in hepatic gluconeogenesis (Park et al., 1983, 1986), decreased mitochondrial coupling of oxygen consumption to ATP synthesis (Bouillon and Berdanier, 1983; McCusker et al., 1983), increased mitochondrial membrane phospholipid fatty acid saturation (Wander and Berdanier, 1985), increased cytosolic redox and phosphorylation states (Berdanier et al., 1979b), shortened lifespans (Adams, 1964), and renal disease (Durand et al., 1964, 1968).

Most of the response or adaptation to dietary carbohydrate occurs in the liver because this is the organ that first "sees" this nutrient. Starch and sucrose are absorbed as their component monosaccharides and transported via the portal system to the liver. There, the glucose or fructose moiety is metabolized to glycogen, fatty acids, cholesterol, a variety of metabolic intermediates or oxidized. Not all of the glucose stops in the liver for further processing., The glucose moiety is also metabolized by other cells as it is circulated through the body. However, the same is not true for fructose. More than 90% of the fructose is initially metabolized by the hepatocyte. This is because the initial first step, fructose phosphorylation, requires the presence of a kinase unique to the liver, fructokinase. Other cell types have a less specific kinase, hexokinase, which will phosphorylate fructose. However, this kinase is about 1/10 as active as the hepatic fructokinase. Thus, the majority of the fructose consumed, whether as the free sugar or in sucrose, is metabolized in the liver. This, then, increases the activity of the various pathways needed to dispose of this fructose and results in an accumulation of its end products. The hepatic response to a load dose of starch, sucrose, glucose, or fructose is also affected by how much of the carbohydrate is presented to it and the solubility of that carbohydrate. Thus, the amount of carbohydrate consumed *and* its solubility are factors in determining the magnitude of the hepatic response with respect to lipogenesis.

Another factor in determining the hepatic response to different carbohydrates is the frequency of consumption. Rats that were starved for 48 hours and refed for 48 hours had a much larger lipogenic response to simple versus complex carbohydrates than rats always having the same food available (Williams and Berdanier, 1982). Sucrose- and glucose-refed rats had higher enzyme activities and higher de novo fatty acid synthetic rates than starch-refed rats. Rats starved and refed a high fat diet did not increase their lipogenic capacity as significantly as rats refed the sugar-rich diets.

## CARBOHYDRATE–LIPID INTERACTIONS

If the fat in the diet is also varied then the responses of rats to starvation-refeeding are further modified. The substitution of coconut oil for corn oil in the starch diet was not nearly as potent a stimulator of lipogenesis as this same substitution in a sucrose diet (Baltzell and Berdanier, 1985). Shown in Fig. 11.5, the G6PD response in the starch–corn oil, starved-refed rats is less than the starch–coconut oil, starved–refed rats. In turn, both of these groups had less of a G6PD response than rats fed a sucrose-coconut oil diet. The same response pattern was observed when the parameter measured was the incorporation of tritium (from water) into

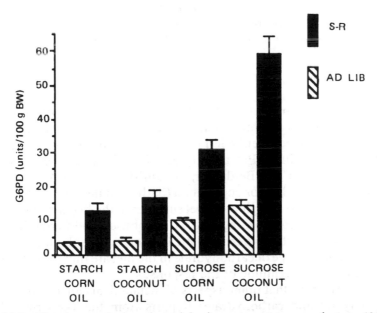

**FIG. 11.5**   The glucose-6-phosphate dehydrogenase response of rats to 48 hours starvation–48 hours refeeding or fed ad libitum. The diet refed the rats contained 65% carbohydrate as either cornstarch or sucrose and 5% fat as either corn oil or hydrogenated coconut oil. *(From Baltzell and Berdanier, 1985.)*

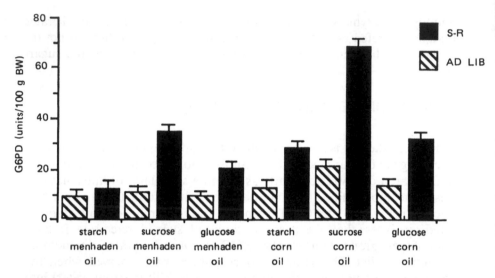

**FIG. 11.6** The glucose-6-phosphate dehydrogenase response of rats to 48 hours of starvation followed by 48 hours of refeeding or fed ad libitum. The diet refed the rats contained 65% carbohydrate as either cornstarch, sucrose, or glucose and 5% fat as either corn oil (CO) or menhaden oil (MO). *(From Johnson and Berdanier, 1987.)*

fatty acids or percent liver lipid. When menhaden oil was substituted for corn oil in a diet containing starch, sucrose, or glucose (Fig. 11.6) the reverse was observed: menhaden oil suppressed the sucrose- or glucose-induced increase in G6PD activity and liver lipid due to starvation–refeeding (Johnson and Berdanier, unpublished results). Menhaden oil compared to beef tallow or coconut oil (both saturated fats) or to corn oil (an unsaturated fat) in a glucose diet suppressed the hepatic lipogenic response to starvation–refeeding (Fig. 11.7) (Johnson and Berdanier, 1987). Thus, in a dietary state where hepatic lipogenesis is enhanced, both the type of fat and the type of carbohydrate in the diet are important factors in determining the magnitude of the response.

Is there evidence that other aspects of metabolism can also be affected by dietary fat and carbohydrate? Reports from the Teppermans' laboratory (Tepperman et al., 1978; Sun et al., 1979) indicated that the gluconeogenic and glycogenolytic response to insulin and glucagon was greater in glucose-fed rats than in lard-fed rats. These workers interpreted these results as an indication that the diet fed the rats prior to the isolation of the liver cells affected the composition of the plasma membranes, which in turn determined the responsiveness of the liver cells to insulin and glucagon. These two hormones must bind to the plasma membrane prior to, or concurrent with, their effects on cellular metabolism. If the diet affected the membrane lipids such that the activity of the receptor sites for these proteins were affected, then the differential responses of the cells could be explained. Unfortunately, these workers used only two diets: a fat-free–glucose diet and a lard diet. They did not vary the types of carbohydrate or fat. Yet, their findings of a diet effect on a membrane-associated function indicated a need for further investigation of the role of diet on membrane function and the role of the membrane in metabolic control.

De Schrijver and Privett (1983) studied the interacting effects of coconut oil and safflower oil with glucose or sucrose on hepatic desaturase activity. They found that desaturase activity was high in glucose- and sucrose-fed rats. This activity remained high when coconut oil was fed but dropped when safflower oil was substituted for the coconut oil. The levels of microsomal fatty acids (20:4n6 and 20:3n9) were not affected by the type of carbohydrate fed. Both carbohydrates serve as insulin secretogogues, and insulin is a potent stimulator of the desaturases, particularly the delta-9 desaturase. These enzymes were probably maximally stimulated by the

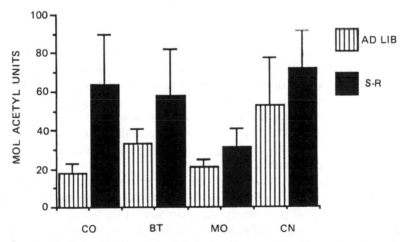

**FIG. 11.7** Tritium (from water) incorporation into hepatic fatty acids by rats either ad libitum fed or starved for 48 hours and refed for 48 hours. The diet contained 65% glucose and 5% fat as either corn oil (CO), beef tallow (RT), menhaden oil (MO), or hydrogenated coconut oil (CN). One mCi tritium was injected and rats were killed 30 minutes later. Fatty acid synthesis is expressed as moles of acetyl units formed per hour per gram tissue. (*From Johnson and Berdanier, 1987.*)

simple sugars so as to compensate for the relative lack of dietary unsaturated fatty acids. This compensation was necessary so that the animal
could maintain optimal membrane fluidity. This is particularly important
to the intracellular membrane since the membrane plays an important
role in metabolic control. Other researchers (Mahfouz et al., 1984; Shimp
et al., 1982) have also indicated that the type of dietary fat can affect the
delta-9 desaturase, which, in turn, may affect the needs for dietary
arachidonic acid since this desaturase is important in the conversion of
linoleic to arachidonic acid. These diet variables affect not only the
desaturation of long-chain fatty acids but also prostaglandin synthesis
since arachidonic acid is a precursor for many of these compounds
(Lands and Kulmacz, 1986). Although no reports of a dietary carbohydrate–lipid interaction effect on prostaglandin synthesis were found,
such an effect is likely in view of the foregoing discussion.

As mentioned earlier, the most likely effect of a carbohydrate and lipid
interaction is on the membranes within and around the cells and is likely
manifested as a change in membrane function. The plasma membrane
contains receptors and translocators for a number of hormones, substrates, products, and metabolites. Alterations in the plasma membrane
thus might be expected to result in alterations in receptor activity, as
shown by the work from the Teppermans' laboratory. Triglyceride secretion by the hepatocyte would also be affected. If secretion is impaired,
triglyceride will accumulate in the liver. Such has been shown by Bird and
Williams (1982). Rats fed a diet containing sucrose, fructose, or glucose
with either corn or coconut oil were studied. Triglyceride secretion was
decreased in rats fed the coconut oil irrespective of the type of dietary carbohydrate. However, the coconut oil–fructose-fed rats secreted 50% more
triglyceride than did the rats fed coconut oil with either glucose or sucrose.
No such carbohydrate effect was noted in the rats fed corn oil. As reported
by others, Bird and Williams (1982) confirmed that fructose-fed rats had
higher liver lipid levels than glucose-fed rats and that this difference was
potentiated by the feeding of coconut oil.

Not only is fatty acid synthesis affected by the interaction of carbohydrate and lipid, so too is glucose oxidation to $CO_2$ and water. Hepatic and
adipose tissue from genetically diabetic BHE rats fed diets containing
either sucrose or starch and either corn oil or coconut oil oxidized glucose
differently depending on the diet fed (Fig. 11.8) (Baltzell, 1984; Berdanier
and Baltzell, 1986). Coconut oil–fed rats oxidized more glucose in the
presence of insulin and synthesized fewer fatty acids in both liver and
adipose tissue than corn oil–fed rats when the diet contained starch.
When the diet contained sucrose the suppression of glucose oxidation by

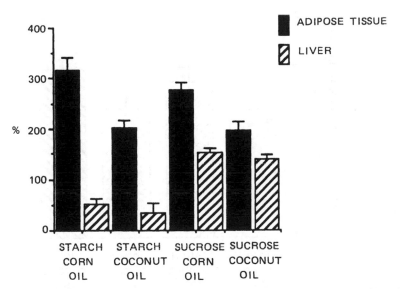

**FIG. 11.8** Percent stimulation of glucose oxidation by 100 μ units of insulin. Adipose and hepatic tissue slices were incubated for 2 hours in the presence or absence of insulin and $^{14}C$-U-glucose. $^{14}CO_2$ released was measured. The rats were fed a diet containing 65% carbohydrate as cornstarch or sucrose and 5% fat as corn oil or coconut oil. *(From Baltzell, 1984; Berdanier and Baltzell, 1986.)*

coconut oil was not seen, and the adipose synthesis of fatty acids was un-changed. However, in the liver, fatty acid synthesis was increased. Feeding sucrose to BHE rats results in an increase in membrane phospholipid saturated fatty acids and fatty acid synthesis (Berdanier and Burrell, 1980; Wander and Berdanier, 1985). Substituting coconut oil for corn oil in either the sucrose or starch diet (Deaver et al., 1986 and unpublished data) not only increased the saturation of the phospholipid fatty acids but also perturbed the coupling of site I oxygen consumption to ATP synthesis.

Because ATP synthesis as well as mitochondrial membrane transport can affect gluconeogenesis, studies of glucose synthesis were conducted (Wander and Berdanier, 1986). Hepatocytes from rats fed coconut oil with either starch or sucrose synthesized significantly more glucose from lac-tate than hepatocytes from rats fed corn oil as 6% of the diet (Fig. 11.9). With corn oil as 6% of the diet, glucose synthesis did not differ with car-bohydrate source. When the diet contained only 5% fat as corn oil, however, sucrose feeding resulted in a significant increase in glucose syn-thesis (Fig. 11.10) (Park et al., 1986). Thus, the interacting effects of dietary fat and carbohydrate on gluconeogenesis depends on the level of fat used.

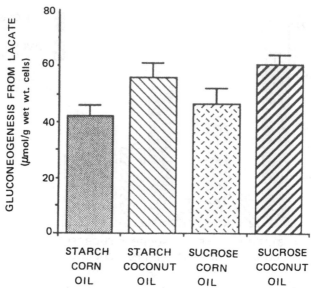

**FIG. 11.9** Glucose synthesis from lactate by hepatocytes isolated from 48-hour starved BHE rats. Prior to starvation rats were fed a diet containing 64% carbohydrate as either cornstarch or sucrose and 6% fat as corn oil or hydrogenated coconut oil. (*From Park et al., 1986.*)

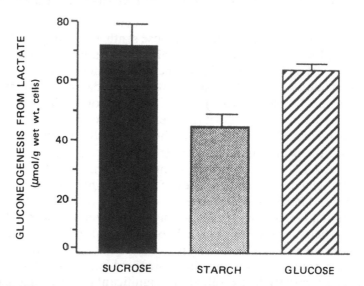

**FIG. 11.10** Glucose synthesis from lactate by hepatocytes isolated from 48-hour starved BHE rats. Prior to starvation the rats were fed a diet containing 65% carbohydrate as sucrose, starch, or glucose and 5% fat as corn oil. (*From Wander and Berdanier, 1986.*)

At 5% fat, the type of carbohydrate was important; at 6% it was not and the influence of type of fat was predominant.

## SUMMARY

An understanding of the tissue differences in metabolism as well as genetic differences in response to carbohydrate–lipid manipulations is also needed. Although it may appear, at present, that the inclusion of marine oil or omega-3 fatty acids is desirable in terms of CVD and diabetes, consideration of its interaction with the dietary carbohydrates of importance on processes other than the regulation of serum lipids is desirable so as to avoid untoward and unexpected results. From the many studies on the responses of animals and humans to fluctuations in either dietary carbohydrate or fat, it would appear that more studies on their interactions are needed before dietary intake recommendations can be made. The foregoing review has shown that carbohydrate–lipid interactions do occur. The reasons for these interactions are multiple. They include type and level of carbohydrate and type and level of fat as well as the genotype of the consumer.

## REFERENCES

Adams, M. 1964. Diet as a factor in length of life and in structure and composition of tissues of the rat with aging. Home Economics Research Report No. 24, USDA. U.S. Government Printing Office, Washington, DC.

Akinyanju, P. A., Qureshi, R. U., Salter, A. J., and Yudkin, J. 1968. Effect atherogenic diet containing starch or sucrose on the blood lipids of young men. *Nature* (Lond.) 218: 975.

Antar, M. A., Little, J. A., Lucas, C., Buckley, G. C., and Csima, A. 1970. Interrelationship between kinds of dietary carbohydrate and fat in hyperlipoproteinemic patients. *Atherosclerosis* 11: 191.

Baltzell, J. 1984. Studies on the influence of diet on lipogenesis and insulin status in rats. M. S. thesis, University of Georgia.

Baltzell, J. K. and Berdanier, C. D. 1985. Effect of the interaction of dietary carbohydrate and fat on the responses of rats to starvation-refeeding. *J. Nutr.* 115: 104.

Bang, H. O., Dyerberg, J., and Sinclair, H. M. 1980. The composition of the

Eskimo food in Northwestern Greenland. *Am. J. Clin. Nutr.* 33: 2657.

Bang, H. O., Dyerberg, J., and Hjorne, N. 1976. The composition of food consumed by Greenland Eskimos. *Acta Med. Scand.* 200: 69.

Bang, H. O. and Dyerberg, J. 1972. Plasma lipids and lipoproteins in Greenlandic West Coast Eskimos. *Acta Med. Scand.* 192: 85.

Basilico, M. A., Chanussot, F., Villaume, C., Lombardo, Y. B., and Debry. 1984. Effect of carbohydrate type upon obesity and hyperlipemia in the Zucker fa/fa rat. *Ann. Nutr. Metab.* 28: 253.

Berdanier, C. D. 1974a. Metabolic abnormalities in BHE rats. *Diabetologia* 10: 691.

Berdanier, C. D. 1974b. Metabolic characteristics of the carbohydrate sensitive BHE strain of rats. *J. Nutr.* 104: 1246.

Berdanier, C. D., Tobin, R. B., and DeVore, V. 1979a. Studies on the control of lipogenesis: Strain differences in hepatic metabolism. *J. Nutr.* 109: 247.

Berdanier, C. D., Tobin, R. B., and DeVore, V. 1979b. Effect of age strain and dietary carbohydrate on the hepatic metabolism of male rats. *J. Nutr.* 109: 261.

Berdanier, C. D. and Baltzell, J. K. 1986. Comparative studies of the responses of two strains of rats to an essential fatty acid deficient diet. *Comp. Biochem. Physiol.* 85A: 725.

Berke, B. M. and Kaplan, M. L. 1983. Effects of high fat and high carbohydrate diets on development of hepatic and adipose lipogenesis in fa/fa and non fa/fa rats. *J. Nutr.* 113: 820.

Birchwood, B. L., Little, J. A., Antar, M. A., Lucas, C., Buckley, G. C., Csima, A., and Kallos, A. 1970. Interrelationship between the kinds of dietary carbohydrate and fat in hyperlipoproteinemic patients. Part 2. *Atherosclerosis* 11: 183.

Bird, M. I. and Williams, M. A. 1982. Triacylglycerol secretion in rats: Effects of essential fatty acids and influence of dietary sucrose glucose or fructose. *J. Nutr.* 112: 2267.

Bouillon, D. J. and Berdanier, C. D. 1983. Effect of maternal carbohydrate intake on mitochondrial activity and on lipogenesis by the young and mature progeny. *J. Nutr.* 113: 2205.

Deaver, O. E., Wander, R. C., McCusker, R. H., and Berdanier, C. D. 1986. Diet effects on membrane phospholipid fatty acids and membrane function in BHE rats. *J. Nutr.* 116: 1148.

DeSchrijver, R. and Privett, O. S. 1983. Hepatic fatty acids and acyl desaturases in rats: Effects of dietary carbohydrate and essential fatty acids. *J. Nutr.* 113: 2217.

Durand, A. M. A., Fisher, M., and Adams, M. 1968. The influence of type of dietary carbohydrate: Effect on histological findings in two strains of rats. *Arch. Path.* 85: 318.

Durand, A. M. A., Fisher, M., and Adams, M. 1968. The influence of type of dietary carbohydrate: Effect on histological findings in two strains of rats. *Arch. Path.* 85: 318.

Dyerberg, J. , Bang, H. O., Stoffersen, E., Moncada, S., and Vane, J. R. 1978. Eicosapentaenoic acid and prevention of thrombosis and atherosclerosis. *Lancet* 11: 117.

Dyerberg, J. and Bang, H. O. 1979. Haemostatic function and platelet polyunsaturated fatty acids in Eskimos. *Lancet* I: 433.

Dyerberg, J. , Bang, H. O., and Hjorne, N. 1975. Fatty acid composition of the plasma lipids in Greenland Eskimos. *Am. J. Clin. Nutr.* 28: 958.

Feldman, S. A., Rubenstein, A. H., Ho, K-J., Taylor, C. B., Lewis, L. A., and Mikkelson, B. 1975. Carbohydrate and lipid metabolism in the Alaskan Artic Eskimos. *Am. J. Clin. Nutr.* 28: 588.

Grande, F., Anderson, J. T., and Keys, A. 1965. Effect of carbohydrates of leguminous seeds, wheat and potatoes on serum cholesterol concentration in man. *J. Nutr.* 86: 313.

Harris, W. S., Conner, W. E., Inkeles, S. B., and Illingworth, D. R. 1984. Dietary omega-3 fatty acids prevent carbohydrate-induced hypertriglyceridemia. *Metabolism* 33: 1016.

Irwin, M. I., Taylor, D. D., and Feeley, R. M. 1964. Serum lipid levels, fat, nitrogen and mineral metabolism of young men associated with kind of dietary carbohydrate. *J. Nutr.* 82: 338.

Johnson, B. J. and Berdanier, C. D. 1987. Effect of menhaden oil on the responses of rats to starvation-refeeding. *Nutr. Rep. Internat.* 36: 809.

Kannel, W. B. 1985. Lipids, diabetes and coronary heart disease: Insights from the Framingham study. *Am. Heart J.* 110: 1100.

Kaufmann, N. A., Poznanski, R., Blondheim, S. H., and Stein, Y. 1966. Changes in serum lipid levels of hyperlipemic patients following the feeding of starch, sucrose and glucose. *Am. J. Clin. Nutr.* 18: 261.

Keys, A., Anderson, J. T., and Grande, F. 1960. Diet type and blood lipid in man. *J. Nutr.* 70: 257.

Kuo, P. T., Feng, L., Cohen, N. N., Fitts, W. T., and Miller, L. D. 1967. Dietary carbohydrates in hyperlipemia, hepatic and adipose tissue lipogenic activities. *Am. J. Clin. Nutr.* 20: 116.

Kuo, P. T. and Bassett, D. R. 1965. Dietary sugar in the production of hyperglyceridemia. *Ann. Internal. Med.* 62: 1199.

Lands, W. E. M. and Kulmacz, R. J. 1986. The regulation of the biosynthesis of prostaglandins and leukotrienes. *Prog. Lipid Res.* 25: 105.

Little, J. A., Birchwood, B. L., Simmons, D. A., Antar, M. A., Kallos, A., Buckley, G. C., and Cisma, A. 1970. Interrelationship between the kinds of dietary carbohydrate and fat in hyperlipoproteinemic patients. *Atherosclerosis* 11: 173.

MacDonald, I. and Braithwaite, D. M. 1964. The influence of dietary carbohydrates on lipid pattern in serum and in adipose tissue. *Clin. Sci.* 27: 23.

Mahfouz, M. M., Smith, T. L., and Kummerow, F. A. 1984. Effect of dietary fats on desaturase activities and the biosynthesis of fatty acids in rat liver microsomes. *Lipids* 19: 214.

McCusker, R. H., Deaver, O. E., and Berdanier, C.D. 1983. Effect of sucrose or starch feeding on the hepatic mitochondrial activity of BHE and Wistar rats. *J. Nutr.* 113: 1327.

McGandy, R. B., Hegsted, D. M., Myers, M. L., and Stare, F. J. 1966. Dietary carbohydrate and serum cholesterol levels in man. *Am. J. Clin. Nutr.* 18: 237.

Michaelis, O. E., Ellwood, K. C., Judge, J. M., Schoene, N. W., and Hansen, C. T. 1984. Effect of dietary sucrose on the SHR/N-corpulent rat: a new model for insulin-independent diabetes. *Am. J. Clin. Nutr.* 39: 612.

Michaelis, O. E., Ellwood, K. C., Tulp, O. L., and Greenwood, M. R. C. 1986. Effect of feeding sucrose or starch diets on parameters of glucose tolerance in the LA/N corpulent rat. *Nutr. Res.* 6: 95.

Park, J. H. Y., Berdanier, C. D., Deaver, O. E., and Szepesi, B. 1986. Effects of dietary carbohydrate on hepatic gluconeogenesis in BHE rats. *J. Nutr.* 116: 1193.

Park, J. H. Y., Berdanier, C. D., and Szepesi, B. 1983. Effects of dietary sucrose on the gluconeogenic capacity of isolated hepatocytes from BHE rats. *Nutr. Rep. Internat.* 28: 287.

Phillipson, B. E., Rothrock, D. W., Conner, W. E., Harris, W. S., and Illingworth, D. R. 1985. Reduction of plasma lipid lipoproteins and apoproteins by dietary fish oils in patients with hypertriglyceridemia. *N. Eng. J. Med.* 312: 1210.

Schoene, N. W., Church, J. P., Michaelis, O. E., Carswell, N., and Hansen, C. T. 1986. Effect of dietary menhaden oil on the development of type II diabetes mellitus characteristics in the corpulent spontaneously hypertensive rat. (Abstract) Washington Spring Symposium, George Washington University School of Medicine.

Shimp, J. L., Bruckner, G., and Kinsella, J. E. 1982. Effects of dietary trilinoelaidin on fatty acid and acyl desaturases in rat liver. *J. Nutr.* 112: 722.

Sun, J. V., Tepperman, H. M., and Tepperman, J. 1979. Lipid composition of liver plasma membranes isolated from rats fed a high glucose or a high fat diet. *J. Nutr.* 109: 193.

Taylor, D. D., Conway E. S., Schuster, E. M., and Adams, M. 1967. Influence of dietary carbohydrate on liver content and on serum lipids in relation to age and strain of rat. *J. Nutr.* 91: 275.

Tepperman, H. M., DeWitt, J. and Tepperman, J. 1978. Hormone effects on glycogenolysis, gluconeogenesis, and cAMP production by liver cells from rats fed diets high in glucose or lard. *J. Nutr.* 108: 1924.

Wander, R. C. and Berdanier, C. D. 1985. Effects of dietary carbohydrate on mitochondrial composition and function in two strains of rats. *J. Nutr.* 115: 190.

Wander, R. C. and Berdanier, C. D. 1986. Effects of type of dietary fat and carbohydrate on gluconeogenesis in isolated hepatocytes from BHE rats. *J. Nutr.* 116: 1156.

Williams, B. H. and Berdanier, C. D. 1982. Effects of diet composition and adrenalectomy on the lipogenic responses of rats to starvation-refeeding. *J. Nutr.* 112: 534.

Zucker, L. 1965. Hereditary obesity in the rat associated with hyperlipemia. *Ann. NY Acad. Sci.* 131: 447.

# 12

# Vitamin–Vitamin Interactions

**Lawrence J. Machlin**

Hoffmann-La Roche Inc.
Nutley, New Jersey


**Lillian Langseth**

Nutrition Research Newsletter
Palisades, New York

## INTRODUCTION

Much of the experimental work conducted with vitamin deficiency disease has focused on the deficiency of a single vitamin, with all other vitamins and nutrients held constant. While this experimental approach is essential to our understanding of the function of individual vitamins, it does not accurately reflect nutrient deficiency profiles in the real world, where vitamin deficiencies generally do *not* occur as single entities, but rather as multiple deficiencies. The classic example is that of pellagra, where deficiencies of niacin and tryptophan are usually accompanied by deficiencies of vitamin $B_6$ and riboflavin (Sauberlich, 1980).

While acute vitamin deficiency diseases such as pellagra are no longer common in the United States, there is some concern that marginal deficiencies of vitamins and minerals may occur in a significant proportion of Americans. On the other hand, excessive intakes of vitamins have also been of concern to many nutritionists.

The fact that there are individuals who have multiple marginal deficiencies while others are at risk for excesses suggests that the whole question of vitamin–vitamin interactions merits exploration in greater depth.

Data are available about the effect of one vitamin on the absorption, metabolism, catabolism, and excretion of another. This review presents some examples of vitamin–vitamin interactions that have been reasonably well defined and are illustrative of the kinds of interactions that may occur.

## TYPES OF INTERACTIONS

The interactions between vitamins may involve the processes of absorption, metabolism, catabolism and excretion. Some vitamins may be able to spare the requirements of others, while others may have potentially adverse effects. Some examples are presented in Table 12.1.

One vitamin may be required for optimal absorption of another. Two such examples are $B_6/B_{12}$ and folate/thiamin. In other instances a vitamin (particularly at high levels) may interfere with the absorption of another, as in the cases of E/K, $B_6$/niacin, and thiamin/riboflavin.

**TABLE 12.1** Examples of Types of Vitamin–Vitamin Interactions

One vitamin needed for optimum absorption of another:
  Vitamin $B_6$/Vitamin $B_{12}$
  Folate/Thiamin
A high level of one vitamin may interfere with absorption or metabolism of another:
  Vitamin E/Vitamin K
  Vitamin $B_6$/Niacin
  Thiamin/Riboflavin
One vitamin needed for metabolism of another:
  Riboflavin/Vitamin $B_6$ and Niacin
  Vitamin $B_6$/Niacin
One vitamin can protect against excess catabolism or urinary losses of another:
  Vitamin C/Vitamin $B_6$
One vitamin can protect against oxidative destruction of another:
  Vitamin E/Vitamin A
  Vitamin C/Vitamin E
A high level of one vitamin can obscure the diagnosis of deficiency of another:
  Folate/Vitamin $B_{12}$

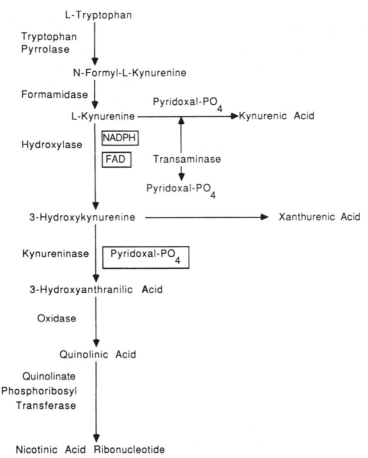

**FIG. 12.1** Involvement of vitamin B₆ (pyridoxal phosphate), riboflavin (FHD), and niacin (NADPH) in the conversion of tryptophan to nicotinic acid. (*From Sauberlich, 1980.*)

A vitamin may be required for the metabolism of another vitamin. Classic examples are the need for riboflavin for the metabolism of $B_6$ and niacin, and of $B_6$ for the metabolism of niacin. These classic dependent interactions are shown in Fig. 12.1.

One vitamin can protect against excessive catabolism or urinary losses of another. For example, vitamin C reduces urinary losses of vitamin $B_6$. One vitamin can protect against oxidative destruction of another. For example, vitamin C protects against destruction of vitamin E, whereas vitamin E protects against destruction of vitamin A. A high level of one vitamin can obscure the diagnosis of deficiency of another, for example, folate and vitamin $B_{12}$.

The vitamin interactions that are of most concern are those that could create or potentiate acute or marginal deficiencies. These include the examples noted above and, in addition, the less frequently observed phenomenon whereby massive dosing with one vitamin increases the need for another vitamin, for example, when excess vitamin A intake increases the requirement for vitamin E. Another example is that of the interaction between anticoagulant therapy and vitamins E and K. These will all be discussed in greater detail later in this chapter.

## VITAMIN A

Vitamin A has been shown to interact with the other fat-soluble vitamins D and E, as well as with the water-soluble vitamin C. These interactions are summarized in Table 12.2.

**TABLE 12.2**  Interactions Involving Vitamin A

| Vitamin acted upon | Interactions | References |
|---|---|---|
| Vitamin C | Under conditions of hypervitaminosis A in *humans,* tissue levels of vitamin C may be reduced and urinary excretion of vitamin C may increase. | Bauernfeind (1980) |
| Vitamin D | In animals, vitamin A, in large doses, protects against some signs of vitamin D toxicity. | Morgan et al. (1937); Clark and Smith (1964); Belanger and Clark (1967); Taylor et al. (1968); Metz et al. (1985). |
| Vitamin E | In the chick, high levels of vitamin A increase the vitamin E requirement. | Sklan and Donoghue (1982); Frigg and Broz (1984) |
| Vitamin K | Under conditions of hypervitaminosis A in *humans,* hypoprothrombinemia may occur. It can be corrected by administering supplementary vitamin K. | Bauerfeind (1980); Suttie (1984) |

## Vitamin A and Vitamin D

When both are consumed in excessive amounts, vitamin A appears to mitigate the toxicity of vitamin D. In rats, for example, large doses of vitamin A have been shown to reduce the hypercalcemia and deposition of calcium in soft tissues caused by high doses of vitamin D (hypervitaminosis D) (Morgan et al., 1937; Belanger and Clark, 1967). In studies in chicks, it has been shown that vitamins A and D fed at excessive levels have an antagonistic effect on plasma calcium, phosphorus, and acid phosphatases (Taylor et al., 1968). Studies in turkey poults have shown that the antagonism between vitamins A and D ameliorates the decreased growth rates and reduced bone mineral content observed when either of these vitamins are given alone at high doses (Metz et al., 1985).

Although the exact mechanism of action is not well known, it has been suggested that vitamin A protects against vitamin D toxicity through increased mucopolysaccharide and collagen turnover (Clark and Smith, 1964). This interaction may be of significance in humans, because many cases of vitamin toxicity involve both vitamins A and D. This is particularly true in children, since many pediatric vitamin supplements contain both vitamins.

## Vitamin A and Vitamin E

Most of the interactions between vitamins A and E are beneficial interactions. As noted earlier, vitamin E protects against oxidative destruction of Vitamin A. However, there are also interactions between these two vitamins that may have potentially negative effects.

Animal studies indicate that high levels of vitamin A increase the need for vitamin E (Sklan and Donoghue, 1982; Frigg and Broz, 1984). Pudelkiewicz et al. (1964) demonstrated this action in chicks: at dietary doses of 5000 mg vitamin A acetate/kg diet it was observed that tocopherol content of liver and plasma was markedly depleted in this species. Sklan and Donoghue (1982), also working with chicks, showed that the interaction of vitamin E and high doses of vitamin A (360,000 µg retinyl palmitate/kg diet) resulted in enhanced oxidation of dietary vitamin E prior to intestinal absorption, increased vitamin E turnover due to increased conjugation to glucuronides, and changes in enzymes such as superoxide dismutase and glutathione peroxidase, which protect cells from oxidative damage. Frigg and Broz (1984), again working with chicks, were able to show that a high single oral dose of vitamin A (100,000 IU) reduced the apparent absorption of vitamin E. However, when the vitamin A was administered parenterally to avoid interactions in the gastrointestinal tract, distribution of vitamin E was unaffected.

## Vitamin A and Vitamin K

High levels of vitamin A intake have been shown to have an adverse effect on the action of vitamin K. Hypoprothrombinemia has been observed in both animals and humans with chronic hypervitaminosis A. The mechanism of this interaction is not clear: it may be a general effect on nonpolar lipid absorption or a specific vitamin/vitamin antagonism (Suttie, 1984). Whether vitamin A influences intestinal synthesis of vitamin K, absorption of vitamin K, or other factors is uncertain (Bauernfeind, 1980). However, it is known that the effect is rapid and reversible: in rats it has been observed after only 5 days of excess vitamin A intake (Maddock et al., 1948), and it is easily corrected by administering supplementary vitamin K or discontinuing high-level vitamin A intake.

## Vitamin A and Vitamin C

Some of the toxic effects of vitamin A appear to be ameliorated by vitamin C. Toxic manifestations of hypervitaminosis A in guinea pigs and rainbow trout, both of which species require a dietary source of vitamin C, resemble those of scurvy. In guinea pigs, the toxic effect of vitamin A is greater when the animals are vitamin C-deficient and the effects can be lessened by administration of vitamin C (Bauernfeind, 1980). In humans, some evidence suggests that urinary excretion of vitamin C is increased in hypervitaminosis A, indicating a reduction in tissue levels of vitamin C (Bauernfeind, 1980).

## VITAMIN E

Vitamin E has been shown to interact beneficially with vitamin A, to be necessary for normal metabolism of vitamin $B_{12}$, and, at high levels, to affect coagulation in vitamin K-deficient subjects (see Table 12.3). High vitamin E intakes may induce deficiencies of the other fat-soluble vitamins if these are present at borderline levels in the diet, due to competition for absorption at the level of the mucosal cell in the intestinal tract.

**TABLE 12.3** Interactions Involving Vitamin E

| Vitamin acted upon | Interactions | References |
|---|---|---|
| Vitamin A | Vitamin E is necessary for normal vitamin A metabolism, spares vitamin A, and protects against some signs of vitamin A toxicity. In *humans*, vitamin E improved the therapeutic efficacy of vitamin A; toxic manifestations of vitamin A did not occur when *children* were given vitamin E along with intermittent massive doses of vitamin A. In rats, large doses of vitamin E reduced storage of vitamin A derived from beta-carotene. | McLaren (1959); Ames (1969); Bennett et al. (1965); Johnson and Baumann (1948); Arnrich and Arthur (1980) |
| Vitamin $B_{12}$ | Administration of vitamin E stopped urinary excretion of methylmalonic acid, an indicator of $B_{12}$ deficiency, in *humans* at risk of vitamin E deficiency. Vitamin E may be necessary for conversion of vitamin $B_{12}$ to its coenzyme form. | Barness (1967); Pappu et al. (1978) |
| Vitamin K | Large doses of vitamin E (1200 IU/day) increase the vitamin K requirement in *humans* taking anticoagulants. However, vitamin E has not been confirmed to have a detrimental effect on vitamin K nutrition in normal individuals not taking anti-coagulants or deficient in vitamin K. | Corrigan and Marcus (1974); Helson (1984) |

## Vitamin E and Vitamin A

Extensive animal evidence and a limited amount of human data show that there is a beneficial interaction between vitamins E and A at both extremes of vitamin A nutrition: vitamin E can help to ameliorate the symptoms of both hypovitaminosis and hypervitaminosis A.

The signs of vitamin A deficiency in animals are markedly accelerated if vitamin E reserves are low (McLaren, 1959), and utilization of supplementary vitamin A is markedly impaired in vitamin E-deficient animals (Ames, 1969).

Vitamin E apparently spares vitamin A in several ways, by (a) protecting vitamin A from oxidation in the gut; (b) increasing vitamin A absorption; (c) increasing vitamin A utilization; and (d) increasing vitamin A storage (Green and Bunyan, 1969).

Vitamin E can protect against vitamin A toxicity, apparently by preventing the disruption of membrane structure by high levels of vitamin A (Bauernfeind, 1980). In animals, vitamin E has prevented growth retardation and congenital anomalies caused by hypervitaminosis A (Bauernfeind, 1980) and a high level of vitamin E has removed the effects of excess vitamin A, reducing the relative weights of the liver, spleen and testes of rats (Jenkins and Mitchell, 1975).

Human evidence includes several single cases in which administration of vitamin E improved the therapeutic efficacy of vitamin A. Bennett et al. (1965) reported that the injection of vitamin E with vitamin A in children with cystic fibrosis increased serum vitamin A to normal and lowered urinary creatine:creatinine ratios. In several studies, children at risk of vitamin A deficiency have been given vitamin E along with intermittent massive doses of vitamin A, without toxic manifestations from the vitamin A. No simultaneous controls have been reported; however, in other studies transient symptoms of hypervitaminosis A have occurred when similar doses of vitamin A were administered alone (Reddy, 1969; Swaminathan et al., 1970).

Although most of the supporting evidence comes from animal rather than human studies, some authorities recommend that vitamin E be administered along with vitamin A when the latter is given in intermittent massive doses as an emergency measure (Bauernfeind et al., 1974; Bauernfeind, 1980). In the treatment of children, it has been recommended that 200,000 IU of vitamin A and 50–200 IU of vitamin E in an oily solution be administered orally every six months (Bauernfeind et al., 1974). There are no known risks from this practice and vitamin E could be valuable in three ways: (a) by assuring the proper utilization of the administered vitamin A; (b) by counteracting any inadequacy in vitamin E levels (such inadequacies are present in some individuals in populations where vitamin A deficiency is common); and (c) by lessening any transient symptoms of hypervitaminosis A that might follow massive dosing (Bauernfeind et al., 1974).

Animal studies have provided additional data on the interaction between vitamins E and A. When rats were fed 10 mg of vitamin E along with beta-carotene, storage of vitamin A derived from the beta-carotene was lessened (Johnson and Baumann, 1948). Vitamin E did not cause this effect if fed 8 hours after the carotene or if preformed vitamin A was fed. Later studies by Arnrich and Arthur (1980) confirmed that hepatic storage of vitamin A from carotene is lessened when rats are fed 10 mg vitamin E daily (equivalent to about 4000 IU in the human adult male) and almost completely inhibited at 50 mg. High levels of vitamin E evidently do not interfere with the activity of the enzyme involved in carotene conversion. Instead, they somehow interfere with the accessibility of carotene to the enzyme, possibly by an antioxidant action (Arnrich and Arthur, 1980).

## Vitamin E and Vitamin K

Interactions between vitamins E and K have been observed in animal studies since 1945 (Anonymous, 1983). Oral intake of high levels of vitamin E has not produced blood coagulation abnormalities in normal humans (Farrell and Bieri, 1975; Tsai et al., 1978; Corrigan, 1982). In vitamin K-deficient individuals, however, the administration of high levels of vitamin E can exacerbate coagulation defects. Vitamin K deficiency can be caused by malabsorption syndromes, diet, or interaction with the anticoagulation drug warfarin. Warfarin is a vitamin K antagonist and can induce changes in blood coagulation similar to those seen with other forms of vitamin K deficiency (Corrigan, 1982).

In one case study (Corrigan and Marcus, 1974), a patient receiving warfarin therapy and 1200 IU vitamin E per day had prolonged prothrombin times, ecchymosis, and depressed vitamin K-dependent clotting factors. Upon discontinuing the vitamin, clinical evidence of bleeding disappeared and clotting factors improved. In a more extensive study (Corrigan, 1982) with cardiology patients taking warfarin, 100 or 400 IU of vitamin E was administered to six patients. These patients showed no significant change in prothrombin time or factor II (prothrombin) coagulant activity and no evidence of bleeding. However, there was a significant reduction in the ratio of factor II coagulant activity to immunoreactive protein. These data and related animal studies (Corrigan, 1982) suggest that vitamin E acts at a step mediated by vitamin K and not in the synthesis of the factor II precursor in the liver. Vitamin E could be interfering with the oxidation of reduced vitamin K during the gamma carboxylation reaction. It has also been suggested that a metabolite of vitamin E could interfere with vitamin K activity. Specifically, it has been reported that alpha-tocopheryl quinone is a more potent antagonist of vitamin K than alpha-tocopherol (Bettger and Olson, 1982).

Zipursky et al. (1980), in a double-blind randomized trial, found that vitamin E did not affect the blood levels of vitamin K-dependent coagulation factors in premature infants. Detailed studies of plasma coagulation factors were performed after 1 to 6 weeks of therapy, and none of the factors was significantly changed between the treated and control groups.

Low levels of prothrombin and other vitamin K-dependent clotting factors were observed in neuroblastoma patients who were given intravenous injections of vitamin E therapeutically. The blood vitamin E levels were increased to 11–12 mg/dl, which is approximately 10 times normal. The clotting defects were overcome by the administration of vitamin K (Helson, 1984). In all other reports in humans where vitamin K status was intact, oral vitamin E supplementation had no effect on coagulation (Farrell and Bieri, 1975; Tsai et al., 1978) and furthermore had no effect on vitamin K concentration in plasma (Sadowski et al., 1987).

In cystic fibrosis, patients with a pre-existing vitamin K deficiency, administration of vitamin E exacerbated the deficiency; oral vitamin K supplementation prevented this (Corrigan, 1982).

Although intravenous vitamin E may need to be administered with caution, as Helson's observations indicate, the general consensus is that oral vitamin E does not have any significant effects on coagulation factors in humans unless they are vitamin K-deficient or are being treated with anticoagulants (Helson, 1984).

## Vitamin E and Vitamin $B_{12}$

Adequate amounts of vitamin E appear to be necessary for normal metabolism of vitamin $B_{12}$ (Machlin and Gabriel, 1980). Urinary excretion of methylmalonic acid, an indicator of vitamin $B_{12}$ deficiency, has been observed in a limited number of individuals at risk for vitamin E deficiency, such as premature infants and cystic fibrosis patients. Administration of vitamin E corrected the problem (Barness, 1967). Experiments in rats suggest that the mechanism of this interaction is a defect in the conversion of vitamin $B_{12}$ to its coenzyme form (Pappu et al., 1978).

# VITAMIN C

Vitamin C interacts with numerous other vitamins (see Table 12.4).

## Vitamin C and Vitamin B$_{12}$

It has been claimed but hotly disputed that large amounts of ascorbic acid might destroy vitamin B$_{12}$ in food and in the bloodstream. Herbert and Jacob (1974) reported that substantial losses of vitamin B$_{12}$ occurred when sample meals were incubated with added ascorbic acid (to simulate the consumption of an ascorbic acid tablet shortly after a meal). They advised that "daily ingestion of 0.5 g or more of ascorbic acid without regular evaluation of vitamin B$_{12}$ status is probably unwise."

Although more than a decade has passed since this warning appeared in a widely read medical journal and although supplementation with 0.5 g/day or more of ascorbic acid is common, ascorbic acid-induced cases of vitamin B$_{12}$ deficiency have not been reported. The only clinical data supporting Herbert and Jacob's contention are their own reports of low serum vitamin B$_{12}$ levels (but not clinical illness) in four paraplegic veterans taking vitamin C as a urine acidifier and Hines' (1975) description of low vitamin B$_{12}$ levels in three individuals who took vitamin C supplements. Hines later retracted his report (Marcus, 1981).

Experiments using official analytical procedures have not demonstrated destruction of vitamin B$_{12}$ by vitamin C in vitro (Newmark et al., 1976) or in vivo (Jaffe, 1984). On the other hand, Herbert et al. (1977) continued to report destruction of vitamin B$_{12}$ by vitamin C in food and suggested that this might occur in the bloodstream as well.

Experiments by Marcus et al. (1980) have shown the basis for this apparent discrepancy. The vitamin B$_{12}$ extraction method used by the Herbert group did not employ cyanide; standard methods do. In most analyses for B$_{12}$, cyanide is added and the temperature is raised in order to enhance the extraction of protein-bound B$_{12}$ and to convert unstable forms of B$_{12}$ to the stable cyanocobalamin. In the absence of cyanide, B$_{12}$ is lost from foods containing high levels of added ascorbic acid. In the presence of cyanide, this loss does not occur.

The destruction of B$_{12}$ by vitamin C does not occur while the foods are incubated at body temperature; it occurs during the boiling step of the extraction procedure (Marcus et al., 1980). Thus, in vivo destruction of B$_{12}$ by vitamin C seems unlikely.

Herbert et al. (1978) have shown that cyanide is needed to protect B$_{12}$ from destruction by vitamin C when it is extracted from blood serum as

**TABLE 12.4** Interactions Involving Vitamin C

| Vitamin acted upon | Interactions | References |
|---|---|---|
| Vitamin A | In rats, ascorbic acid in amounts less than 250 mg/kg body weight increased the conversion of beta-carotene to vitamin A. Larger amounts had no effect or may have decreased its utilization. | Mayfield and Roehm (1956) |
| Vitamin $B_6$ | Vitamin C has been reported to elevate the urinary excretion of a $B_6$ metabolite, but later studies have failed to confirm this at C doses as high as 1 g/day in *humans*. In *human volunteers* who were made deficient in vitamin C, urinary pyridoxine excretion increased. | Selivanova (1960); Shultz and Leklem (1982); Baker et al. (1971) |
| Vitamin $B_{12}$ | Excess vitamin C, in food or the *human* bloodstream, was claimed to destroy $B_{12}$. However, this was later found to be an artifact of a particular analytical procedure and it does not occur under physiological conditions. | Herbert and Jacob (1974); Marcus et al. (1980); Hogenkamp (1980) |
| Vitamin E | Vitamins C and E are synergistic in their antioxidant effects. Vitamin C can spare vitamin E by regenerating tocopherol from tocopheroxyl radicals. There is also some evidence that vitamin E can spare vitamin C. | Leung et al. (1981); Lambelet et al. (1985); Chen (1981) |

well as from food. They noted that serum $B_{12}$ determinations performed without added cyanide may be low due to the presence of ascorbate in the serum and that such low levels do not necessarily reflect tissue deficiency of the vitamin. They advised that in order to interpret a low serum $B_{12}$ level adequately, it is important to know if the patient has been taking large doses of vitamin C.

Hogenkamp (1980) has observed that of the several forms of $B_{12}$ only one, aquacobalamin, is likely to be destroyed by ascorbic acid at body temperature. Aquacobalamin is only a very minor component of $B_{12}$ in foods (Marcus, 1981) and is a minor form of $B_{12}$ in the body except in a few patients suffering from rare errors of $B_{12}$ metabolism (Hogenkamp, 1980). Thus, vitamin C consumed with meals or circulating in the bloodstream would not be expected to cause significant losses of $B_{12}$ in normal persons.

## Vitamin C and Vitamin E

Vitamins C and E are both antioxidants and, as such, help to protect other substances in the body, including other nutrients, from oxidative destruction. They may also help to inhibit cancer formation by blocking the production of carcinogenic nitrosamines. Both vitamins have been shown to prevent nitrosamine formation in vitro and in vivo (Newmark et al., 1976). Since vitamin C is water-soluble, it would exert its hypothesized protective effects in the aqueous phase while vitamin E, a fat-soluble substance, would do so in the lipid phase.

There is in vitro evidence that vitamins C and E are synergistic in their antioxidant properties. A mixture of the two vitamins was much more effective in suppressing the peroxidation of rat liver microsomes than the sum of both vitamins alone (Leung et al., 1981). Lambelet et al. (1985) showed that vitamins C and E act as free radical scavengers in vitro and act synergistically in lipid antioxidation. In one human study, supplementary vitamins C and E acted synergistically in reducing serum lipid peroxides (Kunert and Tappel, 1983).

Vitamin C has been shown to regenerate tocopherol from the tocopheroxyl radical, thus restoring vitamin E to its active antioxidant form. In this way, vitamin C can help to control lipid peroxidation (Bendich et al., 1986). It has long been known that vitamins C and E interact this way in vitro, but until recently the relevance of this interaction in vivo was questioned because vitamin E is membrane-bound, whereas vitamin C is located in aqueous fluids. However, studies using model membranes have recently shown that vitamins C and E can interact in biomembranes and thus are likely to do so in vivo (Bendich et al., 1986).

It has been shown in vivo that high levels of ascorbic acid increase the concentration of lung vitamin E in guinea pigs and that vitamin C-deficient diets decrease lung, liver, and adrenal vitamin E in the same species (Kanazawa et al., 1981; Hruba et al., 1982).

Chen (1981) reported that in rats, addition of 1.5 g ascorbic acid to a diet containing an adequate level of vitamin E resulted in increased red blood hemolysis and suggested that the vitamin E requirement may be increased by high levels of vitamin C supplementation. However, there are inconsistencies in the data, which suggest that the conclusions are not clear-cut. Specifically, the addition of vitamin C to the diet barely altered plasma levels of vitamin C, and furthermore there was no relationship between plasma vitamin E levels and red blood cell hemolysis.

## Vitamin C and Vitamin $B_6$

Selivanova (1960) reported that high doses of vitamin C increase the degradation of vitamin $B_6$ to its metabolite, 4-pyridoxic acid. However, Schultz and Leklem (1982) failed to confirm this but demonstrated normal $B_6$ metabolism with supplementation of vitamin C at the level of 1 g/day in humans.

In a study of scurvy in human volunteers, a statistically significant increase in urinary pyridoxine excretion occurred despite a constant intake of recommended levels of pyridoxine (Baker et al., 1971). Ascorbic acid evidently has some effect on the catabolism and/or availability of vitamin $B_6$.

## Vitamin C and Vitamin A

In rats, small amounts of ascorbic acid may increase the conversion of carotene to vitamin A, while large amounts (equivalent in humans to 20 g or more per day) may have no effect, or may decrease its utilization (Mayfield and Roehm, 1956). No evidence is available suggesting any inhibitory effect of vitamin C on conversion of carotene to vitamin A in humans.

## B VITAMINS

Many critical metabolic processes require the concerted action of several B-complex vitamins. Adequate amounts of all the B vitamins are needed for optimal functioning, and a deficiency of one B vitamin may lead to abnormalities in the metabolism of another. Two mechanisms predominate in B vitamin interactions. Because of the interlocking, interdependent nature of the biochemical pathways of energy and protein metabolism, one B vitamin may be needed for the metabolism of another. Also, some B vitamins have been shown to be necessary for the optimum absorption of others from the intestine (see Table 12.5).

Although these interactions have generally been demonstrated only in animals, they should be expected in humans as well, because the biochemical pathways involving B vitamins are similar in all species. In humans, the need for one vitamin for the absorption or metabolism of another may worsen the problem of multiple marginal vitamin deficiencies. Mild B vitamin deficiencies might lead to malabsorption or disrupted metabolism of other vitamins, thus contributing to malfunctions of others in a kind of nutritional domino effect. More research is needed to determine whether these interactions are of practical importance in human marginal deficiency states.

## Vitamin $B_6$

Vitamin $B_6$ is required for the formation of niacin from tryptophan (see Fig. 12.1). This metabolic route is impaired in $B_6$-deficient subjects (Sauberlich, 1980).

In a clinical trial of supplementary vitamin $B_6$ (25 mg/kg body weight), some young children with Down's syndrome developed biochemical evidence of niacin deficiency as measured by urinary metabolite levels (Coleman et al., 1985). Some patients receiving $B_6$ for more than 4 years also developed blisters after sun exposure, which could be prevented by a "small dose" of niacin.

At the same time, niacin and riboflavin are needed for the interconversions of the various forms of vitamin $B_6$ (Fig. 12.2). Therefore, niacin, riboflavin, and biotin may act synergistically with vitamin $B_6$ (Driskell, 1984).

There are two interactions between vitamin $B_6$ and thiamin: (a) $B_6$ has been reported to have a protective effect against thiamin excess in rats (Omaye, 1984); and (b) $B_6$ deficiency can be produced in rats by an overdose of thiamin (Driskell, 1984).

A deficiency of $B_6$ has been shown to have various effects on the status of other vitamins. $B_6$ deficiency has been reported to impair $B_{12}$ absorption in the rat (Sauberlich, 1980). A deficiency of $B_6$ (or $B_{12}$) can induce thiamin

**TABLE 12.5** Interactions Involving B Vitamins

| Vitamin causing the effect | Vitamin acted upon | Interactions | References |
|---|---|---|---|
| Vitamin $B_6$ Vitamin $B_{12}$ | Thiamin | $B_6$ and $B_{12}$ are necessary for thiamin absorption in rats. | Nishino and Itokawa (1977) |
| Vitamin $B_6$ | Vitamin C | In *human volunteers* made deficient in $B_6$, plasma vitamin C levels dropped. | Baker et al. (1964) |
| Vitamin $B_6$ | Thiamin | $B_6$ protects against the effects of excess thiamin in rats. | Omaye (1984) |
| Vitamin $B_6$ | Niacin | Vitamin $B_6$ (25 mg/kg body weight) results in biochemical evidence of niacin deficiency in *children* with Down's syndrome. Vitamin $B_6$ deficiency reduces the production of niacin from tryptophan. | Coleman et al. (1985); Snyderman et al. (1953) |
| Vitamin $B_6$ Vitamin $B_{12}$ | Vitamin $B_{12}$ Thiamin | $B_6$ is needed for $B_{12}$ absorption in rats. A mild $B_{12}$ deficiency increased urinary loss of thiamin and increased the severity of thiamin deficiency in rats. | Sauberlich (1980) Peifer and Cleland (1987) |
| Vitamin $B_{12}$ Biotin, Folic Acid | Pantothenic Acid Pantothenic Acid | $B_{12}$ can spare pantothenic acid in the chick. Biotin and folic acid are needed for the utilization of pantothenic acid in rats. | Fox (1984) Fox (1984) |
| Folic Acid | Thiamin | Folic acid is needed for absorption of low doses in thiamin in rats. | Howard et al. (1974) |
| Folic Acid | Vitamin $B_{12}$ | In *humans*, excess folic acid can obscure $B_{12}$ deficiency by curing its hematological but not neurological symptoms. This has been observed at an oral folate dose of 5 mg/day. | Herbert (1963a, b); Brody et al. (1984) |

| | | | |
|---|---|---|---|
| Pantothenic Acid | Biotin | Pantothenic acid is needed for the overall reactions involving biotin. | Sauberlich (1980) |
| Pantothenic Acid | Vitamin C | Pantothenic acid may be necessary for efficient utilization of vitamin C. Vitamin C may help to protect against pantothenic acid deficiency | Fox (1984) |
| Riboflavin/Niacin/Biotin | Vitamin $B_6$ | Niacin and riboflavin are needed for inter-conversion of the various forms of $B_6$; niacin, riboflavin, and biotin may act synergistically with $B_6$. | Driskell (1984) |
| Riboflavin | Folic Acid | Riboflavin is necessary for the conversion of folic acid to its coenzyme form. | Cooperman and Lopez (1984) |
| Riboflavin | Vitamin $B_{12}$ | Riboflavin can spare $B_{12}$ in the rat and chick. | Cooperman and Lopez (1984) |
| Thiamin | Riboflavin | In *humans*, high levels of thiamin increase riboflavin excretion. However, riboflavin deficiency has not been demonstrated except with parenteral administration of massive thiamin doses. | Klopp et al. (1943); Fujiwara (1954); Inoue et al. (1956) |
| Thiamin | Vitamin $B_6$ | In rats, thiamin overdose may produce $B_6$ deficiency. | Driskell (1984) |
| Thiamin/$B_6$/Riboflavin | Niacin | Vitamin $B_6$, thiamin, and riboflavin are necessary for the formation of niacin from tryptophan. | Sauberlich (1980) |

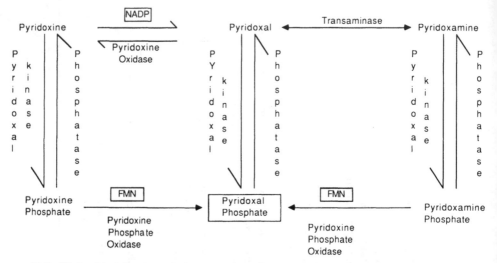

FIG. 12.2  Participation of riboflavin (FMN) and nicotinic acid (NADP) in vitamin B₆ (pyridoxal phosphate) formation. *(From Sauberlich, 1985.)*

deficiency in rats (Nishino and Itokawa, 1977). A study of $B_6$-deficient humans showed that plasma vitamin C levels dropped due to $B_6$ deficiency despite a constant level of vitamin C intake (Baker et al., 1964).

## Vitamin $B_{12}$

Some animal studies have shown interactions between $B_{12}$ and riboflavin and pantothenic acid. Cooperman and Lopez (1984) observed that riboflavin can partially replace the requirements of the rat and chick for vitamin $B_{12}$ because of a structural similarity between the two vitamin molecules—both contain the benzimidazole moiety. Meanwhile, vitamin $B_{12}$ can spare pantothenic acid in the chick (Fox, 1984).

## Folic Acid (Folates)

Deficiencies of either $B_{12}$ or folate produce hematologically indistinguishable macrocytic megaloblastic anemias. Vitamin $B_{12}$ deficiency also causes progressive neurological damage. Because folate and $B_{12}$ are metabolically interrelated, large doses of either vitamin can "cure" the anemia caused by a deficiency of the other (Herbert, 1963a, b). However, folate cannot prevent or reverse the neurologic lesions of $B_{12}$ deficiency. Hematologic responses in $B_{12}$-deficient patients have been observed when folate was administered intramuscularly at a dose of 400 μg/day or orally at 5 mg/day (Brody et al., 1984). Thus the indiscriminate use of large doses of folic acid should be avoided because it can prevent or delay a diagnosis of vitamin $B_{12}$ deficiency, possibly allowing irreversible neurological damage to take place (Brody et al., 1984).

In folate-deficient rats, thiamin is malabsorbed at low thiamin doses, but not at high doses (Howard et al., 1974). Riboflavin is necessary for the conversion of folic acid to its coenzyme form (Cooperman and Lopez, 1984).

## Pantothenic Acid

Pantothenic acid is necessary for the overall metabolic reactions involving biotin; thus a biotin deficiency may be aggravated by a simultaneous pantothenic acid deficiency (Sauberlich, 1980).

Pantothenic acid may also be necessary for the efficient utilization of vitamin C. Large amounts of ascorbic acid have beeen shown to delay the onset and to reduce symptoms of pantothenic acid deficiency in animals (Fox, 1984). Utilization of pantothenic acid requires biotin as well as folic acid, as shown in rat studies (Fox, 1984).

In summary, the review of interactions of B vitamins demonstrates the point made earlier—that many critical metabolic processes require the concerted action of several B-complex vitamins. Looking at the actions of each vitamin separately does not provide an accurate explanation, as it is essentially a dissection of parts of the whole picture.

The B-complex vitamins are just that, a complex, working together in many of the metabolic reactions of the cell. For example, one of the key reactions of metabolism is the conversion of pyruvate to acetyl-CoA. It is a key reaction because it is at an intersection between the breaking down of sugar for energy and the making of fat and protein. Since B vitamins work as coenzymes, four different vitamins (niacin, thiamin, pantothenic acid, and biotin) are required in this one enzymatic conversion. Thus, a deficiency of any one of these four vitamins compromises the efficiency of the other three. This model could explain why biotin and folic acid are needed for the utilization of pantothenic acid in rats and why pantothenic acid is needed for the overall reactions involving biotin.

## SUMMARY

Vitamins can interact with each other in many complex ways. One vitamin might be necessary for: (a) absorption or metabolism of another; (b) protection of another vitamin from oxidative destruction; or (c) protection of another vitamin from excessive catabolism or excretion.

The net effects of these kinds of interactions is that a deficiency of one vitamin—sometimes even a marginal deficiency—can exacerbate a deficiency or increase the requirement of another vitamin.

The concern that a high level of one vitamin might cause an "imbalance" or increased need for another vitamin has been demonstrated only under extraordinary circumstances. The two significant examples which nutritionists should be aware of are that: (1) high levels of folic acid might obscure the diagnosis of $B_{12}$ deficiency; and (2) very high levels of vitamin E (1200 IU) can provoke bleeding in patients receiving anticoagulant therapy.

Finally, there is a need for additional study of vitamin–vitamin interactions and their health implications in realistic situations of multiple marginal vitamin deficiency. Our current knowledge indicates that deficiencies of one vitamin often affect the metabolism of others, but these interactions have been studied almost exclusively in unrealistic conditions in which intake of one vitamin is seriously deficient while intake of all others are normal. It is difficult to extrapolate from this to the more realistic condition of multiple marginal deficiency. Epidemiological studies, rather than laboratory studies, may be needed to explore this relatively uncharted area.

## REFERENCES

Ames, S. R. 1969. Factors affecting absorption, transport, and storage of vitamin A. *Am. J. Clin. Nutr.* 22: 934.

Anonymous. 1983. Megavitamin E supplementation and vitamin K-dependent carboxylation. *Nutr. Rev.* 41: 268.

Arnrich, L. and Arthur, V. A. 1980. Interactions of fat-soluble vitamins in hypervitaminoses. *Proc. N.Y. Acad. Sci.* 355: 109.

Baker, E. M., Canham, J. E., Nunes, W. T., Sauberlich, H. E., and McDowell, M. E. 1964. Vitamin $B_6$ requirement for adult men. *Amer. J. Clin. Nutr.* 15: 59.

Baker, E. M., Hodges, R. E., Hood, J., Sauberlich, H. E., March, S. C., and

Canham, J. E. 1971. Metabolism of 14C- and 3H-labeled L-ascorbic acid in human scurvy. *Am. J. Clin. Nutr.* 24: 444.

Barness, L. A. 1967. Vitamin $B_{12}$ deficiency with emphasis on methylmalonic acid as a diagnostic aid. *Am. J. Clin. Nutr.* 20: 573.

Bauernfeind, J. C. 1980. *The Safe Use of Vitamin A.* A report of the International Vitamin A Consultative Group. Nutrition Foundation, Washington, D.C.

Bauernfeind, J. C., Newmark, H., and Brin, M. 1974. Vitamins A and E nutrition via intramuscular or oral route. *Am. J. Clin. Nutr.* 27: 234.

Belanger, L. F. and Clark, I. 1967. Alpharadiographic and histological observations on the skeletal effects of hypervitaminoses A and D in the rat. *Anat. Rec.* 158: 443.

Bendich, A., Machlin, L. J., and Scandurra, O. 1986. The antioxidant role of vitamin C. *Adv. in Free Radical Biology & Medicine* 2: 419.

Bennett, M., Medwadowski, B., Walberg, S., and Mitchell, A. 1965. Effect of vitamin E on serum levels of vitamins A and E and urinary creatine/creatinine ratio in fibrocystic children [Abstract]. *Fed. Proc.* 25: 546.

Bettger, W. J. and Olson, R. A. 1982. Effect of alpha-tocopherol and alpha-tocopheryl quinone on vitamin K dependent carboxylation in the rat. [Abstract]. *Fed. Proc.* 41: 344.

Brody, T., Shanc, B., and Stokstad, E. L. R. 1984. Folic acid. In *Handbook of Vitamins,* Machlin, L. J. (Ed.), p. 459. Marcel Dekker, Inc., New York.

Chen, L. H., 1981. An increase in vitamin E requirement induced by high supplementation of vitamin C in rats. *Amer. J. Clin. Nutr.* 34: 1036.

Clark, I. and Smith, M. R. 1964. Effects of hypervitaminosis A and D on skeletal metabolism. *J. Biol. Chem.* 239: 1266.

Coleman, M., Sobel, S., Bhagavan, H.N., Coursin, D., Marquardt, A., Guay, M., and Hunt, C. 1985. A double blind study of vitamin $B_6$ in Down's syndrome infants. Part 1- Clinical and biochemical results. *J. Ment. Defic. Res.* 29: 233.

Cooperman, J. M. and Lopez, R. , 1984. Riboflavin. In *Handbook of Vitamins.* Machlin, L. J. (Ed.), p. 299. Marcel Dekker, Inc., New York.

Corrigan, J. J. Jr. 1982. The effect of vitamin E on warfarin-induced vitamin K deficiency. *Ann. N.Y. Acad. Sci.* 393: 361.

Corrigan, J. J. Jr. and Marcus, F. I. 1974. Coagulopathy associated with vitamin E ingestion. J. Am. Med. Assoc. 230: 1300.

Driskell, J. A. 1984. Vitamin $B_6$. In *Handbook of Vitamins,* Machlin, L. J. (Ed.), p. 370. Marcel Dekker, Inc., New York.

Farrell, P. M. and Bieri, J. G. 1975. Megavitamin E supplementation in man. *Am. J. Clin. Nutr.* 28: 1381.

Fox, H. M. 1984. Pantothenic acid. In *Handbook of Vitamins*, Machlin, L. J. (Ed.), p. 437. Marcel Dekker, Inc., New York.

Frigg, M. and Broz, J. 1984. Relationships between vitamin A and vitamin E in the chick. *Int. J. Vit. Nutr. Res,* 54: 125.

Fujiwara, M. 1954. Influence of thiamin on riboflavin metabolism. *Vitamin* 7: 206.

Green, J. and Bunyan, J. 1969. Vitamin E and biological antioxidant theory. *Nutr. Abstr. Rev.* 39: 321.

Helson, L. 1984. The effect of intravenous vitamin E and menadiol sodium diphosphate on vitamin K dependent clotting factors. *Thromb. Res.* 35: 11.

Herbert, V. 1963a. Megaloblastic anemia. *N. Engl. J. Med.* 268: 201.

Herbert, V. 1963b. Megaloblastic anemia (concluded). *N. Engl. J. Med.* 268: 368.

Herbert, V. and Jacob, E. 1974. Destruction of vitamin $B_{12}$ by ascorbic acid. *J. Am. Med. Assoc.* 230: 241.

Herbert, V., Jacob, E., and Wong, K-T. J. 1977. Destruction of vitamin $B_{12}$ by vitamin C [Letter]. *Am. J. Clin. Nutr.* 30: 297.

Herbert, V., Jacob, E., Wong, K-T. J., Scott, J., and Pfeffer, R. D. 1978. Low serum vitamin $B_{12}$ levels in patients receiving ascorbic acid in mega-doses: studies concerning the effect of ascorbate on radioisotope vitamin $B_{12}$ assay. *Am. J. Clin. Nutr.* 31: 253.

Hines, J. D. 1975. Ascorbic acid and vitamin $B_{12}$ deficiency [Letter]. *J. Am. Med. Assoc.* 234: 24.

Hogenkamp, H. P. C. 1980. The interaction between vitamin $B_{12}$ and vitamin C. *Am. J. Clin. Nutr.* 33: 1.

Howard, L., Wagner, C., and Schenker, S. 1974. Malabsorption of thiamin in folate-deficient rats. *J. Nutr.* 104: 1024.

Hruba, F., Novakova, V., Ginter, E. 1982. The effect of chronic marginal vitamin C deficiency on the alpha-tocopherol content of the organs and plasma of guinea pigs. *Experientia* 38: 1454.

Inoue, K., Katsura, E., and Kariyone, S. 1956. Secondary riboflavin deficiency. *Vitamin* 10: 69.

Jaffe, G. M. 1984. Vitamin C. In *Handbook of Vitamins*, Machlin, L. J. (Ed.), p. 199. Marcel Dekker, Inc., New York.

Jenkins, M. Y. and Mitchell, G. V. 1975. Influence of excess vitamin E on vitamin A toxicity in rats. *J. Nutr.* 105: 1600.

Johnson, R. M. and Baumann, C. A. 1948. The effect of alpha-tocopherol on the utilization of carotene by the rat. *J. Biol. Chem.* 175: 811.

Kanazawa, K., Takeuchi, S., Hasegawa, R., Okada, M., Makiyama, I., Hirose, N., Toh, T., Hyon, C. S., and Kobayashi, M. 1948. Influence of ascorbic acid deficiency on the level of non-protein SH compounds and vitamin E in the blood of guinea pigs. *Nihon Univ. J. Med.* 23: 257.

Klopp, C. T., Abels, J. C., and Rhods, C. P. 1943. The relationship between riboflavin intake and thiamin excretion in man. *Am. J. Med. Sci.* 205: 852.

Kunert, K. J. and Tappel, A. L. 1983. The effect of vitamin C on in vivo lipid peroxidation in guinea pigs as measured by pentane and ethane production. *Lipids* 18: 271.

Lambelet, P., Saucy, F., and Löliger, J. 1985. Chemical evidence for interactions between vitamins E and C. *Experientia* 41: 1384.

Leung, H-W, Vang, M. J., and Mavis, R. D. 1981. The cooperataive interaction between vitamin E and vitamin C in suppression of peroxidation of membrane phospholipids. *Biochim. Biophys. Acta* 664: 266.

Machlin, L. J. and Gabriel, E. 1980. Interactions of vitamin E with vitamin C, vitamin $B_{12}$, and zinc. *Ann. N.Y. Acad. Sci.* 355: 98.

Maddock, C. L., Wolbach, S. B. and Jensen, D. 1948. Hypoprothrombincmia with hemorrhage as a cause of death in the rat with hypervitaminosis A. [Abstract]. *Fed. Proc.* 7: 275.

Marcus, M. 1981. Vitamin $B_{12}$: response to Dr. Herbert. [Letter]. *Am. J. Clin. Nutr.* 34: 1622.

Marcus, M., Prabhudesai, M., and Wassef, S. 1980. Stability of vitamin $B_{12}$ in the presence of ascorbic acid in food and serum: restoration by cyanide of apparent loss. *Am. J. Clin. Nutr.* 33: 137.

Mayfield, H. L. and Roehm, R. R. 1956. The influence of ascorbic acid and the source of the B vitamins on the utilization of carotene. *J. Nutr.* 58: 203.

McLaren, D. S. 1959. Influence of protein deficiency and sex on the development of ocular lesions and survival time of the vitamin A-deficient rat. *Br. J. Ophthal.* 43: 234.

Metz, A. L., Walser, M. M., and Olson, W. G. 1985. The interaction of dietary vitamin A and vitamin D related to skeletal development in the turkey poult. *J. Nutr.* 115: 929.

Morgan, A. F., Kimmul, L., and Hawkins, N. C. 1937. A comparison of the hypervitaminosis induced by irridated ergosterol and fish liver oil concentrates. *J. Biol. Chem.* 120: 85.

Newmark, H. L., Scheiner, J., Marcus, M., and Prabhudesai, M. 1976. Stability of vitamin $B_{12}$ in the presence of ascorbic acid. *Am. J. Clin. Nutr.* 29: 645.

Nishino, K. and Itokawa, Y. 1977. Thiamin metabolism in vitamin $B_6$ or vitamin $B_{12}$ deficient rats. *J. Nutr.* 107: 775.

Omaye, S. T. 1984. Safety of megavitamin therapy. In *Advances in Experimental Medicine and Biology.* Friedman, M. (Ed.), p. 169. Plenum Press, New York.

Pappu, A. S., Fatterpaker, R., and Sreenivasan, A., 1978. Possible interrelationship between vitamins E and $B_{12}$ in the disturbance in methylmalonate metabolism in vitamin E deficiency. *Biochem. J.* 172: 115.

Peifer, J. J. and Cleland, G. 1987. Multiple metabolic disturbances due to mild depletion of vitamin $B_{12}$ and a secondary deficiency of thiamin. [Abstract]. *Fed. Proc.* 46: 1158.

Pudelkiewicz, W. J., Webster, L., and Matterson, L. D. 1964. Effects of high levels of dietary vitamin A acetate on tissue tocopherol and some related analytical observations. *J. Nutr.* 84: 113.

Reddy, V. 1969. Vitamin A deficiency in children. *Indian J. Med. Res.* 57: 54.

Sadowski, J., Hood, S., Dallal, G., and Garry, P. 1987. Factors influencing the concentration of vitamin K in human plasma from young and old subjects. Fed. Proc. 46: 901.

Sauberlich, H. E. 1980. Interactions of thiamin, riboflavin, and other B-vitamins. *Ann. NY Acad. Sci.* 355: 80.

Sauberlich, H. E. 1985. Interaction of vitamin $B_6$ with other nutrients. In *Vitamin $B_6$: Its Role in Health and Disease,* Reynolds, R. D. and Leklum, J. E. (Ed.), p. 85. Alan R. Liss, Inc., New York.

Selivanova, V. M. 1960. Excretion of vitamin $B_6$ in the urine of a healthy individual. *Bull. Exp. Biol. Med.* 50: 37.

Shultz, T. D. and Leklem, J. E. 1982. Effect of high dose ascorbic acid on vitamin $B_6$ metabolism. *Am. J. Clin. Nutr.* 35: 1400.

Sklan, D. and Donoghue, S. 1982. Vitamin E response to high dietary vitamin A in the chick. *J. Nutr.* 112: 759.

Snyderman, S. E., Holt, L. E., Carretero, R., and Jacobs, K. 1953. Pyridoxine deficiency in the human infant. *J. Clin. Nutr.* 1: 200.

Suttie, J. W. 1984. Vitamin K. In *Handbook of Vitamins,* Machlin, L. J. (Ed.), p. 147. Marcel Dekker, Inc., New York.

Swaminathan, M. C., Susheela, T. P., and Thimmayamma, B. V. S. 1970. Field prophylactic trial with a single annual oral massive dose of vitamin A. *Am. J. Clin. Nutr.* 23: 119.

Taylor, T. G., Morris, K. M. L., and Kirley, J. 1968. Effect of dietary excesses of vitamin A and D on some constituents of the blood of chicks. *Br. J. Nutr.* 22: 713.

Tsai, A. C., Kelley, J. J., Peng, B., and Cook, N. 1978. Study on the effect of megavitamin E supplementation in man. *Am. J. Clin. Nutr.* 31: 831.

Zipursky, A., Milner, R. A., Blanchette, V S., and Johnston, M. A. 1980. Effect of vitamin E therapy on blood coagulation tests in newborn infants. *Pediatrics* 66: 547.

# 13

# Nutrient Interactions and the Toxic Elements Aluminum, Cadmium, and Lead

**M. R. Spivey Fox**

Food and Drug Administration
Washington, D.C.

## INTRODUCTION

### Toxic Elements of Concern

The focus of this chapter is on those elements for which there is limited or no evidence for essentiality and for which there is established or tentative evidence of an exposure hazard to one or more population groups in the United States. Aluminum, cadmium, and lead were chosen because they meet these criteria and because each interacts with several essential nutrients. Other toxic elements, such as arsenic, mercury, and tin, are of less practical concern in the United States today. The topics to be considered are based on an extensive literature. It is regretted that all relevant papers cannot be cited and that, in addition to recent publications, considerable reference to reviews has been necessary.

## Evaluation of Response

An adverse physiological or morphological effect of a toxic element is the primary concern. Other related measurements that change in relation to toxic element exposure can be useful indicators to predict and define adverse health effects, often more readily and precisely than physiological response measurements.

A model relationship between these two indices is shown in Fig. 13.1. Examples of indicator indices include concentrations of the toxic element in tissues and body fluids, certain blood enzymes, and blood and urinary metabolites.

To establish safe or tolerable levels of a toxic element in the food supply, cognizance must be taken of all relevant information in humans. This often includes the effects of industrial and environmental exposure (respiratory as well as oral), clinical exposure, and inadvertent poisonings (both acute and chronic). Retrospective evaluation is often difficult.

Experimental animals fed graded doses of the toxic element from background contamination to excessive levels can provide useful information regarding dose effects. Methods of relating animal data (for both essential and toxic dietary components) have been based on dietary concentration (Fox et al., 1981) and dietary content per unit energy intake (Nielsen, 1984).

In studies of toxic element–nutrient interactions it is desirable to maintain dietary levels of both the toxicant and the essential nutrient within ranges appropriate to human exposure and intake, respectively. Many of these interactions occur at the sites of intestinal absorption and subsequent transport, storage, and metabolism. The use of very high dietary concentrations may overwhelm one or more of these systems and yield data that have little meaning for humans. Probably the most difficult problem is the design of relatively short-term animal studies to obtain data applicable to the long-term chronic human toxicity situation.

## Factors Modifying Response

In general terms, interactions between nutrients and toxic elements involve bioavailability of each. Bioavailability has been defined (Fox et al., 1981) as follows: "Bioavailability is a quantitative measure of the utilization of a nutrient under specified conditions to support the organism's normal structure and physiological processes." The bioavailability of a toxic substance can be viewed in reverse terms as follows: Bioavailability is a quantitative measure of the utilization of a test dose of a toxic sub-

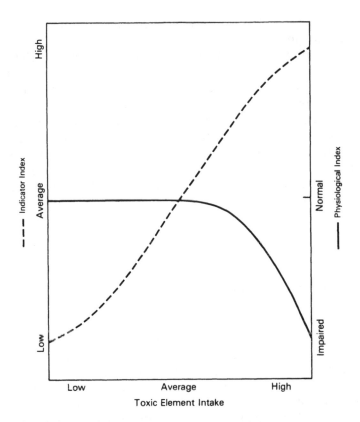

**FIG. 13.1**  Example of a model relationship between an indicator index and a physiological index (see text).

stance to modify normal structure and/or physiological processes of an organism under specified conditions.

The chemical species of either a toxic element or an essential element in a food can affect the individual's response to a given dose. This involves primarily the oxidation state of the element and/or the presence of organic molecules that may bind the element. Strength of binding and molecular size are important for the latter.

Population groups at risk to adverse effects of various stresses, including toxic elements, are generally composed of infants, young children, pregnant and lactating women, the elderly, and those persons suffering from some disease states. Individuals deficient in nutrients that interact with a toxic element are likely to be at significant risk. Typically, low intakes of the essential interacting nutrient exacerbate the adverse effect of the toxic element and, conversely, elevated intakes of the interacting nutrient may be protective.

The distribution of toxic elements usually varies among different types of foodstuffs and may be very high in a few foods. The toxic element content of a food is usually related to level of exposure to the element during growth; however, some species of both animals and plants and some tissues are efficient accumulators of toxic elements. Both nutritional status of an individual and nutrient content of a meal may affect absorption of a toxic element. Persons requiring total parenteral nutrition or dialysis have sometimes been exposed to toxic doses of specific elements. These diverse factors will be considered below with respect to each of the three elements.

## ALUMINUM

### Toxic Effects in Humans

Aluminum has long been considered to be relatively nontoxic to normal individuals because absorption from the gastrointestinal tract is low and urinary excretion of absorbed aluminum is rapid. Problems of aluminum toxicity have been identified in some treatments of clinical conditions. The possibility that long-term accumulation of aluminum in the brain might be involved in Alzheimer's disease has focused attention on sources and effects of aluminum in the general population. Reviews dealing with chemical, metabolic, and toxic aspects of aluminum include those by Kreuger et al. (1984), Kaiser and Schwartz (1985), Alfrey (1986), the Committee on Nutrition of the American Academy of Pediatrics (1986), Elinder and Sjögren (1986), Ganrot (1986), Nebeker and Coburn (1986), Anonymous (1987), and Greger and Lane (1987).

Much of our knowledge of aluminum toxicity in humans arose from treatments of clinical problems that led to aluminum exposure. Aluminum toxicity has been observed in patients with chronic renal failure who were given aluminum compounds to bind phosphate and decrease its intestinal absorption. Patients with liver disease have exhibited toxic effects from use of aluminum-containing antacids, apparently due to diminished excretion of aluminum in bile. In some cases, dialysis fluids and solutions used for total parenteral nutrition have been contaminated with aluminum.

The principal effect of aluminum under these conditions is the production of osteodystrophy, particularly osteomalacia, and sometimes encephalopathy. Anemia occurs in end stage renal disease, which involves both decreased production and increased destruction of red blood cells. Aluminum causes a microcytic, hypochromic anemia, which can adversely affect the preexisting anemia in these patients. The exact mechanisms are unknown.

Aluminum concentration in serum and bone is used to monitor aluminum status in relation to the toxic effects. Toxicity can be prevented by avoiding aluminum contamination and by using non-aluminum-containing phosphate binders, antacids, and other drugs. Children are at greater risk because they require proportionately much higher doses of aluminum-containing phosphate binders than do adults to control hyperphosphatemia. Low birthweight infants are particularly vulnerable to the adverse effects of aluminum exposure. The drug deferoxamine binds aluminum and can be used successfully to remove aluminum from the body and to ameliorate bone and brain injury.

The proposed relation of aluminum to Alzheimer's disease is still uncertain.

## Sources of Intake

Information on dietary sources of aluminum is presented in Table 13.1 for the infant, the toddler, and the adult male. These data are taken from the continuing Total Diet Study of the Food and Drug Administration. Presently 234 foods are analyzed, which permits calculation of intakes that represent the typical diet patterns of eight age-sex groups (Pennington, 1983). Types of foods that contributed the most aluminum overall to the eight diets were grain products (24–49%), dairy products (17–36%), desserts (9–26%), and beverages (5–10%) (Pennington et al., 1987).

Dietary intake of aluminum can vary widely depending on selection of foods with aluminum-containing additives, such as some leavening agents, anticaking agents, bleaches for some foods, and emulsifying agents in pasteurized processed cheeses. Occurrence of elevated aluminum concentrations in some infant formulas has been reviewed (Committee on Nutrition, 1986). Preparation of foods in contact with aluminum vessels or foil can also increase the aluminum content of the food. Acidity of the foods and condition of the aluminum surface affect the uptake of aluminum by the food.

**TABLE 13.1** Estimated Daily Dietary Intakes of Aluminum, Cadmium, and Lead and Provisional Tolerable Intakes of Cadmium and Lead for the Infant, Toddler, and Adult[a]

| Intake | Aluminum (mg/day) | Cadmium (μg/day) | Lead (μg/day) |
|---|---|---|---|
| *Infant* | | | |
| Estimated intake | 1.8[b] | 11[c] | 20[c] |
| Provisional tolerable intake | — | 8.2–9.8[d] | 30[e] |
| *Toddler* | | | |
| Estimated intake | 6.2 | 16 | 30 |
| Provisional tolerable intake | — | 13.7–16.4 | 49 |
| *Adult male* | | | |
| Estimated intake | 14 | 20[f] | 41[f] |
| Provisional tolerable intake | — | 70–84 | 497[d] |

[a]Age (yr) and body weight (kg), respectively, were 0.5–0.9 and 8.2 for the infant, 2 and 13.7 for the toddler, and 19+ and 70 for the adult male. Values for the adult male were based on an energy intake of 2850 kcal/day.
[b]Data for all age groups (Pennington et al., 1987; Pennington and Jones, 1988).
[c]Data for infant and toddler (Gartrell et al., 1986a).
[d]Values for cadmium: all age groups; value for lead: adult (FAO/WHO Expert Committee on Food Additives, 1972).
[e]Values for lead: infant and toddler (FAO/WHO Expert Committee on Food Additives, 1987).
[f]Gartrell et al. (1986b).

The amounts of aluminum in food are relatively trivial compared to the amounts in several over-the-counter drugs. Many antacids, buffered analgesics, antidiarrheals, and certain antiulcer drugs supply large amounts of aluminum, estimated to be as high as 0.1–0.5 g per day. These aspects have been reviewed (Greger and Lane, 1987).

The Joint FAO/WHO Expert Committee on Food Additives has not established a provisional tolerable intake of aluminum.

## Aluminum and Interacting Nutrients

Greger and Lane (1987) reviewed the effects of aluminum interactions with essential minerals. This section, therefore, is restricted to citation of controlled studies in humans and some recent reports of effects in animals. Table 13.2 summarizes the major effects of aluminum–nutrient interactions in humans, which are described in more detail below.

**TABLE 13.2**   Aluminum–Nutrient Interactions in Humans

| Nutrient | Effects | References |
|---|---|---|
| Ca and P | Low Ca intake: Al-antacids increased Ca and P in feces and Ca in urine. No effects with normal Ca diet | Spencer et al. (1982) |
|  | Al-antacids 12 yr: severe bone pain, osteomalacia, weakness. Corrected by good diet, no antacid | Insogna et al. (1980) |
| Vitamin D | Al-contaminated TPN: low serum 1,25-dihyroxyvitamin D (presumptive association) | Klein et al. (1981); Shike et al (1981); Klein et al. (1985) |
| F | Al-antacids (1.8 g Al/d): decreased F absorption with either typical or high F intakes | Spencer et al. (1985) |
| Zn | Al(OH)$_3$ (~0.5 g Al) with ZnSO$_4$ oral Zn tolerance test: reduced area under plasma Zn curve | Abu-Hamdan et al. (1986) |

**Calcium and phosphorus.**   The effect of ingesting large amounts of aluminum compounds in decreasing phosphorus absorption is well established. More recently it has been shown that with relatively small amounts of aluminum-containing antacids, fecal excretion of both phosphorus and calcium was increased when dietary calcium was low. Urinary calcium excretion increased, whereas that of phosphorus decreased (Spencer et al., 1982). With a normal calcium intake, there were no effects on calcium balance. A recent study in rats showed that feeding a low-calcium diet with 0.2% aluminum increased aluminum concentrations in bone and kidney (but not in liver and brain) when compared with rats fed a normal calcium diet plus aluminum (Burnatowska-Hledin et al., 1986).

Severe bone pain, osteomalacia, weakness, and hypophosphatemia occurred in a woman who had taken aluminum-containing antacids for 12 years (Insogna et al., 1980). Discontinuance of the antacids and consumption of an adequate diet markedly relieved all symptoms and other clinical abnormalities. Spencer and Kramer (1983) reviewed the significance of aluminum-containing antacids and calcium loss.

**Vitamin D.** It was noted above that aluminum contamination of total parenteral nutrition solutions has been a problem in some preparations. Low serum levels of 1,25-dihydroxyvitamin D were reported for some subjects receiving such products before the aluminum problem was recognized (Klein et al., 1981; Shike et al., 1981). In a subsequent study of total parenteral nutrition solutions without aluminum contamination, Klein et al. (1985) found that the serum levels of 1,25-dihydroxyvitamin D were normal or elevated. This is presumptive evidence that the aluminum was the cause of the original low values.

Adler and Berlyne (1985) measured aluminum absorption during in situ duodenal perfusion of vitamin D-deficient and -adequate rats. The nonsaturable component was identical for both groups; however, saturable absorption was lower in deficient rats. Aluminum in the perfusion medium decreased calcium absorption in controls, but not in deficient rats. Aluminum apparently competes with calcium by the saturable mechanism in the control rats. In rats receiving aluminum in the diet, injections of vitamin D increased aluminum in serum, bone, and kidney (Burnatowska-Hledin et al., 1986).

**Fluoride.** Three daily doses of aluminum hydroxide that supplied a total of 1.8 g aluminum per day caused a mean decrease of 57% in fluoride absorption in adult men consuming 4 mg fluoride per day (Spencer et al., 1985). A similar decrease in fluoride absorption was observed in osteoporotic patients receiving 50 mg fluoride per day as sodium fluoride. Plasma fluoride levels decreased by 41% and fluoride balances were either negative or less positive.

**Zinc.** When a 30-ml dose of aluminum hydroxide (~0.5 g aluminum) was administered with 25 mg zinc as the sulfate in an oral zinc tolerance test, the 2-hour area under the curve for plasma zinc concentration was decreased by 60% (Abu-Hamdan et al., 1986). This effect in decreasing apparent zinc absorption was observed in normal control subjects. A similar response was shown in patients with renal failure.

## Conclusions Regarding Aluminum

The present dietary aluminum intake does not pose a health hazard. Relatively little is known about the species of aluminum in foods and factors affecting bioavailability. Chronic use of pharmacologic doses of aluminum-containing drugs may adversely affect nutritional status with respect to calcium, phosphorus, and zinc in generally healthy individuals. Persons having low intakes of these nutrients may be at increased risk. Further studies are required to establish the practical significance of most aluminum–nutrient relationships and to establish adverse health effects of long-term accumulation of aluminum.

# CADMIUM

## Toxic Effects in Humans

Cadmium exposure has been known for many years to be a health hazard for industrially exposed workers. Approximately two decades ago, it was recognized that a severely disabling disease in the general population was caused by cadmium exposure in food (rice) and drinking water for approximately 20 years. This condition, known as Itai-itai disease, was characterized by bone demineralization with spontaneous fractures, anemia, enteropathy, and damage to the proximal renal tubule. It is now established that the tubular damage is the primary cadmium-induced lesion, which ultimately leads to the skeletal problems. Itai-itai disease has been observed only in multiparous, postmenopausal Japanese women. It is reasonably certain that during cadmium exposure they had low intakes of calcium, iron, protein, vitamin D, and fat. Bhattacharyya (1983), Friberg et al. (1985, 1986), and Kostial (1986) have reviewed the effects of cadmium in animals and humans.

Very little absorbed cadmium is excreted. Its biological half-life in the kidney has been estimated to be 10 to 30 years. When the cadmium concentration in the renal cortex reaches a critical concentration, approximately 200 $\mu$g/g, there is a loss of cadmium from the kidney via the urine and damage to the proximal renal tubule. It has been estimated that dietary intakes of 200 to 390 $\mu$g cadmium per day are required to attain the critical concentration of cadmium in the kidney by 50 years of age. There is no known means for accelerating cadmium excretion without causing kidney damage.

Most of the cadmium in the body is bound to a small cytoplasmic protein, metallothionein, which contains 30% cysteine and typically has a metal content of 5–10% w/w. The synthesis of metallothionein is inducible by cadmium, zinc, and copper, and its sequestration properties provide protection against toxicity of these metals.

## Sources of Intake

Negligible amounts of cadmium occur in the air and drinking water in the United States. Cigarette smoking significantly increases the cadmium intake. Otherwise, the primary source of cadmium exposure is from foods. In Table 13.1, data from the Food and Drug Administration Total Diet Study are presented for three age groups. For the infant and toddler, the estimated intakes are near the FAO/WHO provisional tolerable intakes, whereas the intake for the adult male is well below the FAO/WHO value.

Since the diet patterns varied considerably among the three age groups, the types of food that were significant sources of cadmium varied also. For the two younger age groups, milk was the major source of cadmium, but dairy products were also important for the adult male. Other food groups supplying significant cadmium for the adult male were grain and cereal products, potatoes, and vegetables. These food groups were also significant cadmium sources, but of lesser importance, for the younger age groups.

The amount of cadmium in a food depends on exposure of the organism to this element during growth. Principal sources of cadmium for plants are phosphate fertilizers, a few sites of natural high-cadmium soils, and industrial pollution. Urban sewage sludges, which have been proposed for fertilizing croplands, may be high in cadmium. Some tissues and some organisms markedly accumulate cadmium. Foods that typically contain elevated levels of cadmium include liver, kidney, and oysters and other marine bivalves. The Total Diet Study includes liver, but not the other high-cadmium foods. A serving of one of these foods can double the usual daily intake of cadmium.

## Cadmium and Interacting Nutrients in Humans

Although nutrition is thought to have played a pivotal role in the etiology of Itai-itai disease, there are relatively few studies in humans relating nutrients and health effects due to cadmium (Table 13.3).

**Iron.** Flanagan et al. (1978) showed that cadmium retention was markedly increased in both men and women when serum ferritin levels (an index of iron stores) were low. Shaikh and Smith (1980) reported confimatory data. Bunker et al. (1984) found a significant ($p < 0.001$) inverse correlation between the percentage of dietary cadmium absorbed and

**TABLE 13.3**  Cadmium–Nutrient Interactions in Humans

| Nutrient | Effects | References |
|---|---|---|
| Fe | Low Fe status: increased Cd absorption | Flanagan et al. (1978); Shaikh and Smith (1980); Bunker et al. (1984) |
| Cu | Cd-exposed women: increased urinary Cu | Nogawa et al. (1984) |
| Vitamin D | Cd-exposed men with kidney damage: decreased serum 1,25-dihydroxyvitamin D | Nogawa et al. (1987) |

body iron stores, based on serum iron, percentage iron saturation, and serum ferritin. The subjects (11 men and 12 women) were healthy non-smokers 69 to 85 years of age who consumed self-selected diets during the 5-day balance study.

**Copper.**  Nogawa et al. (1984) studied three groups of cadmium-exposed Japanese women: (a) Itai-itai patients, (b) patients with kidney damage but no bone disease, and (c) those with neither disorder. Group D was comprised of controls not exposed to cadmium. $\beta_2$-Microglobulin excretion in the urine, an index of renal proximal tubule damage, was significantly increased in groups A and B, as compared with controls, but was generally lower in group C subjects. Markedly increased urinary copper levels followed a similar pattern. For all subjects (69 exposed women and 31 controls), the correlation coefficient between $\beta_2$-microglobulin and copper concentrations in the urine was 0.95. There was no effect on urinary zinc excretion. Copper and zinc levels in the serum of cadium-exposed women were no different from those of the controls. The exposed women were apparently not markedly deficient in copper despite the urinary loss.

**Vitamin D.**  Serum 1,25-dihydroxyvitamin D concentrations were lower in cadmium-exposed men (but not women) with kidney damage than in nonexposed controls (Nogawa et al., 1987). A similar trend was shown for five Itai-itai patients but was not statistically significant due to wide variations among the small number of subjects. Compared with controls, the exposed women with kidney damage had elevated serum parathyroid hormone, alkaline phosphatase, and $\beta_2$-microglobulin levels. They also had decreased serum calcium and inorganic phosphorus. Exposed men and women had markedly elevated urinary cadmium and $\beta_2$-microglobulin values compared with their respective controls. The authors suggested that bone effects in cadmium-exposed subjects were related to disturbances in vitamin D and parathyroid hormone metabolism caused by the cadmium-induced renal damage. They also discussed the possible involvement of cadmium in decreasing calcium absorption and on bone collagen metabolism, such as through reduction of lysyl oxidase. Further studies are needed in this important area.

## Cadmium and Interacting Nutrients in Animals

Table 13.4 briefly summarizes the more recent literature related to cadmium–nutrient interactions in animals. Many of these effects have been the subject of earlier reviews (Fox, 1974, 1979; National Research Council, 1980; Whanger, 1985; Chmielnicka and Cherian, 1986). High levels of cadmium have been found to interfere with the absorption, metabolism, and/or function of most nutrients interacting with cadmium. When possible, reference is limited here to studies with lower levels of cadmium intake, which are thought to have greater relevance to humans. Since most of the cadmium–nutrient interactions seem to occur in the intestinal tract and at the site of absorption, studies with injected cadmium have been excluded.

**Zinc.**   Zinc and cadmium have been shown to be mutually antagonistic under many conditions. In young Japanese quail fed 62 µg cadmium/kg diet, a mild zinc deficiency caused increased cadmium in the liver (Fox et al., 1984). Adult rats made mildly zinc-deficient were fed cadmium at 12.5, 25, 50, 100, and 200 mg/kg diet (Waalkes, 1986). All levels of cadmium caused decreased zinc in the kidney, whereas increased cadmium in the kidney occurred with the three highest levels of cadmium. Zinc in the liver was not affected by zinc deficiency; however, cadmium in the liver was increased in the zinc-deficient rats by the two highest levels of cadmium. Hoadley and Cousins (1985) observed increased uptake of cadmium by the small intestine from the luminal perfusate in mildly zinc-deficient adult rats in a vascularly isolated small intestine system. There was no effect of zinc deficiency on transfer of cadmium from the tissue to the vascular perfusate. Fasting overnight did not affect the uptake of cadmium.

Providing zinc at 60 mg/kg diet, twice the requirement, caused a reduction of cadmium in the jejunum-ileum, liver, and kidney of young

Japanese quail receiving 145 μg cadmium/kg diet (Jacobs et al., 1983).

In young rats that received 75 mg cadmium/kg diet with or without 300 mg zinc/kg for 8, 16, 24, 36, and 48 weeks, cadmium accumulated in the femur and there was no effect of supplemental zinc (Bonner et al., 1980). At each time period, cadmium caused a reduction of zinc in the femur, which was overcome by the supplemental zinc, except at 48 weeks, when zinc was lower than that in the controls (no increase in dietary zinc or cadmium). Only at 48 weeks did bone alkaline phosphatase decrease with cadmium. Zinc plus cadmium or zinc alone increased the activity of the enzyme over that of the controls at most time periods. At least part of the bone lesions in humans exposed to cadmium might be due to a direct effect of cadmium in the bone involving zinc.

Lamphere et al. (1984) fed calves either 200 or 600 mg zinc/kg diet with 50 mg cadmium/kg for 60 days. The concentrations of cadmium in the liver and kidney were decreased by both levels of zinc compared with the control without added zinc. The decrease in cadmium by 600 mg zinc/kg diet was about 50% in both tissues, to 4.2 and 15.5 μg/g for liver and kidney cortex, respectively. Supplemental zinc also reduced cadmium in the muscle. The reduction in cadmium was not large enough to make these organ meats satisfactory for human consumption; however, supplemental zinc might be of practical use at lower levels of cadmium intake.

**Iron.** There are numerous reports in the literature on cadmium antagonism of iron. Recent animal studies provide additional information about these usually sensitive relationships. A mild iron deficiency in young Japaqnese quail fed a low-cadmium diet (62 μg/kg) caused increased concentrations of a tracer dose of cadmium-109 in the kidney but not in the liver (Fox et al., 1980, 1984). Iron deficiency in 28- and 42-day-old rats increased the uptake of cadmium in the jejunum during in situ perfusion (Leon and Johnson, 1985).

The substitution of an equal amount of $Fe^{2+}$ for the required level of 100 mg $Fe^{3+}$/kg diet of young Japanese quail reduced the tissue uptake of a single dose of cadmium-109 in a diet containing 62 μg cadmium/kg (Fox et al., 1980). The tracer was given at 9 days of age and the treatment effects were assessed at 16 days of age. The effect was greater when the iron substitution was made at 7 days of age rather than at hatching. Variations in iron absorption at the time the cadmium-109 dose was given may have been responsible for the differences.

Kostial et al. (1980) administered iron in milk to neonatal rats for several days and observed no change in whole body retention of a cadmium dose. Administration of milk and iron also had no effect on jejunal uptake of cadmium (Johnson et al., 1981). Injection of iron dextran or in-

**TABLE 13.4**  Cadmium–Nutrient Interactions in Animals

| Nutrient | Effects | References |
|---|---|---|
| Zn | Zn deficiency: increased tissue Cd; fasting, no effect | Fox et al. (1984); Waalkes (1986); Hoadley and Cousins (1985) |
|  | Zn supplements: decreased tissue Cd | Jacobs et al. (1983); Bonner et al. (1980); Lamphere et al. (1984) |
| Fe | Fe deficiency: increased tissue Cd | Fox et al. (1984); Fox et al. (1980); Leon and Johnson (1985) |
|  | Fe supplements: sometimes decreased tissue Cd | Leon and Johnson (1985) Kostial et al. (1980) Johnson et al. (1981) |
|  | High Cd: variable effects on Fe | Huebers et al. (1987); Webster (1979); Siewicki et al. (1986) |
| Cu | Cu deficiency: increased tissue Cd | Fox et al. (1984) |
| Ca | Low Ca intake: increased tissue Cd | Van Barneveld and Van den Hamer (1985) |
|  | Supplemental Ca: decreased Cd effects | Revis et al. (1980) |
|  | Ca–Cd in vitro interactions in the intestine: Cd decreased Ca absorption; Ca inhibited Cd uptake in the nonlabile compartment by the saturable mechanism | Chertok et al. (1981); Hoadley and Johnson (1987) |
| Se | Se deficiency: high Cd, increased liver glutathione peroxidase (GSH-Px) activity | Olsson (1985) |
|  | Adequate Se: low Cd, no effects on tissue Se or on GSH-Px activity; high Cd, variable effects on GSH-Px | Olsson (1985); Rahim et al. (1986); Flegal et al. (1980) |
|  | Supplemental Se: decreased tissue Cd with high Cd; corrected hypertension due to prior low Cd; high Cd, reduced tissue Se | Flegal et al. (1980); Perry et al. (1983) |

**TABLE 13.4** (*Continued*)

| Nutrient | Effects | References |
|---|---|---|
| Mn | Supplemental Mn: with low Cd, no effect on tissue Cd; with high Cd, depressed tissue Cd | Jacobs et al. (1983); Sarhan et al. (1986) |
| Ascorbic acid | Supplemental ascorbic acid: with low Cd, decreased tissue Cd | Fox et al. (1980) |
| Protein | Low protein: decreased tissue Cd | Revis (1981); Revis and Osborne (1984); Schäfer et al. (1986) |
| | Low protein plus cysteine: increased tissue Cd | Revis and Osborne (1984) |
| | High protein: increased tissue Cd | Revis (1981); Revis and Osborne (1984); Schäfer et al. (1986) |
| Fiber | Variable effects on tissue Cd, depending on type of fiber | Schäfer et al. (1986); Kiyozumi et al. (1982); Rose and Quarterman (1987) |

corporation of ferrous sulfate in the perfusion fluid decreased cadmium uptake by the jejunum (Leon and Johnson, 1985).

The effect of cadmium on iron absorption was studied by Huebers et al. (1987) in rats dosed intragastrically or by an in situ isolated upper jejunum loop. Cadmium reduced the absorption of iron salts by about half in both iron-deficient and control rats, whereas absorption of hemoglobin iron was enhanced by cadmium in iron-deficient rats. Absorption of transferrin iron was markedly decreased by cadmium in iron-deficient rats, but there was no effect of cadmium in controls. At the end of the 60-minute in situ studies, a high proportion of cadmium was bound to ferritin in control rats. The proportion of cadmium to iron in these studies was extremely high, far outside the dietary range of humans.

Webster (1979) compared the effects produced by feeding a purified iron-deficient diet with those obtained by administering 40 mg cadmium/ liter drinking water with a stock diet during pregnancy of mice. Both treatments caused anemia in the dams and the fetuses (day 19 of gestation), with a more pronounced effect in the latter. Anemic fetuses were growth-retarded. Adding 1000 mg $Fe^{2+}$/kg stock diet with cadmium partially prevented the anemia and the weight reduction in the fetuses. Anemia in the dams was entirely prevented by the $Fe^{2+}$ supplement.

When mildly anemic mice were fed 0.1 mg cadmium/kg diet, liver iron stores were reduced after 14 days by 100 mg zinc/kg diet from oyster tissue but not by zinc as the carbonate (Siewicki et al., 1986). In mice receiving 20 mg cadmium/kg diet in the form of cadmium chloride, liver and small intestine iron concentrations were decreased equally with 12 or 100 mg zinc/kg supplied by zinc carbonate or by 100 mg zinc/kg supplied by dried oyster tissue, as compared with mice receiving 12 mg zinc/kg and 0.1 mg cadmium/kg. Although zinc in oyster is considered to be as bioavailable as that in zinc salts, the effects on iron were not consistent under these conditions.

**Copper.** As with iron, a mild deficiency of copper increased cadmium levels in the kidney of young Japanese quail, with no change in liver cadmium (Fox et al., 1984). This was observed with 62 µg cadmium/kg diet. Also, all possible combinations of zinc, iron, and copper deficiencies had no greater effects on tissue cadmium accumulation than did single deficiencies.

**Calcium.** Numerous workers have shown increased cadmium toxicity and tissue uptake of cadmium with diets low in calcium. More recently Van Barneveld and Van den Hamer (1985) reported that feeding mice a low-calcium diet for 2 weeks increased the absorption of cadmium-115m from drinking water by 17%. Supplemental calcium had no effect.

Revis et al. (1980) found that the number and size of atherosclerotic plaques in male white carneau pigeons were increased after 6 months when the drinking water contained 0.6 mg cadmium/liter. When the drinking water also contained 100 mg calcium/liter, these effects of cadmium were reduced. Adding 30 mg magnesium/liter to the drinking water with cadmium had no effect.

Chertok et al. (1981) reported that under three in vitro experimental conditions, the absorption of calcium by everted rat gut sacs was decreased by cadmium. The conditions included oral pretreatment of the rats with 0, 0.05, 0.5, or 5 mg cadmium per day for 3 weeks as well as in vitro additions of cadmium.

Hoadley and Johnson (1987) investigated the in vitro processes of cadmium uptake and binding in everted duodenal, jejunal, and ileal segments of rat small intestine. By EDTA washing, it was possible to distinguish between labile and nonlabile cadmium compartments. Cadmium uptake in the nonlabile compartment was time-dependent (consistent with transport) and initial uptake rates showed saturable and first-order kinetics. Calcium inhibited only the saturable mechanism and the effect was greatest in the duodenum. There was no evidence of competitive inhibition, suggesting that calcium and cadmium did not share a common carrier-mediated uptake mechanism. The labile compartment was not time-dependent (consistent with adsorption) and cadmium binding was similar for all segments. Calcium had an apparent cooperative effect, increasing the second-order term of cadmium binding in the labile compartment. The authors discussed the possibility that adsorption of metals by the glycocalyx (the mucous covering of intestinal absorptive cells) may play a major role in metal absorption.

**Selenium.**    There is an extensive literature on the interactions between selenium and cadmium when the elements were either injected or administered orally at very high levels, as reviewed by Whanger (1985). These data are outside the scope of this review.

Addition of 50 mg cadmium/liter drinking water for selenium-deficient rats caused an increase in glutathione peroxidase (GSH-Px) activity in the liver as compared with values for selenium-deficient rats (Olsson, 1985). Cadmium also caused a decrease in the flavin-containing monooxygenase that performs N-oxygenation of N,N-dimethylanaline.

In rats fed 0.06 mg selenium/kg diet, near the requirement of 0.1 mg/kg, adding 1, 3, or 5 mg cadmium/kg diet had no effect on the GSH-Px activity in the liver or whole blood or in the uptake of a selenium-75 dose by seven tissues (Rahim et al., 1986). In rats receiving 0.2 mg selenium/liter drinking water, 50 mg cadmium/liter caused a decrease in GSH-Px activity in the liver and erythrocytes (Olsson, 1985). In pigs fed a diet containing 0.8 mg selenium/kg, the addition of 50 mg cadmium/kg doubled plasma GSH-Px activity (Flegal et al., 1980).

With 3 mg supplemental selenium/kg diet, 50 mg cadmium/kg caused a significant decrease in selenium in the blood and liver compared with pigs receiving supplemental selenium alone (Flegal et al., 1980). Compared with the lower dietary selenium level, the supplement of selenium caused a decrease in liver cadmium.

Rats receiving 1 mg cadmium/liter drinking water between 4 and 20 months of the study became consistently hypertensive after 6 months (Perry et al., 1983). When cadmium was removed from the water of another group at 8 months, the rats remained hypertensive to 16 months. Then, the addition of 0.3 mg selenium/liter drinking water between 16 and 20 months resulted in normal systolic pressures at 18 and 20 months.

From the above data, it appears that high levels of cadmium are required to affect selenium status; however, the metabolism of cadmium at low levels of intake may be sensitive to selenium intake, as indicated by the data on hypertension in rats. Further studies are needed.

**Manganese.**   There are relatively few recent data on the effects of manganese on cadmium. Jacobs et al. (1983) found that increasing the dietary intake of manganese from 12 to 24 mg/kg diet had no effect on the concentration of cadmium in the small intestine, liver or kidney of the young Japanese quail when the diet contained 145 µg cadmium/kg.

With higher levels of manganese and cadmium (56 and 112 mg/liter, respectively) added to the drinking water for 8 weeks, Sarhan et al. (1986) found significantly lower concentrations of cadmium in the blood, liver, whole kidney, kidney cortex, and brain cortex as compared with values for rats receiving cadmium alone. This level of manganese was not toxic and there were no increases in tissue concentrations of manganese. Thus, in an unusual case of unavoidable high cadmium intake, concomitant ingestion of supplemental manganese might be protective and safe for a limited period of time.

**Ascorbic acid.**   Fox et al. (1980) reported that a dietary supplement of 0.5% ascorbic acid reduced the concentrations of cadmium in the liver, kidney, and small intestine of young Japanese quail fed 60 to 80 µg cadmium/kg diet. This protective effect of ascorbic acid appears to be due to increased absorption of iron.

**Protein.**   Types of dietary protein have previously been shown to affect the toxicity of cadmium. Revis (1981) fed rats a purified casein diet with protein levels of 5.5, 21, and 67.5%. Cadmium at a level of 50 mg/liter was present in the drinking water. Both cadmium and metallothionein were significantly lower in the liver, kidney, and small intestine of rats fed the lower protein than in the controls. The levels of cadmium and metallothionein in these tissues of rats fed the high-protein diet were significantly higher than the controls. Proximal tubular necrosis was observed in the high-protein–fed rats, even at cadmium concentrations below 200 µg/g wet weight, generally considered to be the critical concentration for tubular damage.

These findings were confirmed in a later study (Revis and Osborn, 1984) with either 25 or 50 mg cadmium/liter drinking water. However, with 5 mg cadmium/liter the effect of high dietary protein level on increasing tissue cadmium concentration occurred in the kidney but not in the liver. When cysteine was added to the low-protein diet to equal that in the high-protein diet, the high accumulation of cadmium in the tissues was equal to that with the high-protein diet. These studies were carried out for 2 and 4 months.

Schäfer et al. (1986) also observed greater uptake of cadmium with higher protein diets. Mice were fed purified diets containing 10, 20, or 40% protein for 4 weeks. A single dose of 5 μmol $^{109}CdCl_2$/kg body weight was administered by stomach tube and measurements of whole body radioactivity were made after 1, 2, 3, 4, 7, and 10 days. There was a linear relationship between cadmium retention and dietary protein level, with a significant difference between 10 and 40% protein. The effects were apparent 4 days after the cadmium-109 dose.

**Fiber.**    Kiyozumi et al. (1982) fed rats a purified diet for 7 days containing 100 mg cadmium/kg with 5% cellulose, lignin, or sodium carboxymethyl cellulose. Compared with controls, lignin and sodium carboxymethyl cellulose reduced cadmium in the blood, liver, kidney, and testis, whereas cellulose had no effect.

In mice receiving 2, 7, or 15% fiber in a purified diet for 4 weeks, the whole body retention of a single oral dose of cadmium-109 was measured (Schäfer et al., 1986). Ten days after dosing there was a significant decrease in the cadmium retention with 15% wheat bran but no effect of cellulose.

Rose and Quarterman (1987) fed rats 4% fiber of several types in a purified diet containing 5 mg cadmium/kg for 4 weeks. Compared with controls receiving cadmium, pectin and agar increased the concentration of cadmium in the kidneys. A similar effect was observed with glucuronic acid fed at 0.67%, equivalent to the uronic acid in agar. Concentrations of cadmium in the kidneys and livers were not significantly increased with alginic acid, cellulose, indulin (a soluble form of lignin), or carrageenan. The authors discussed the factors that can complicate interpretation of these net effects on cadmium absorption.

## Cadmium in Foods and Its Bioavailability

Studies of cadmium-binding proteins in plants and animals show that most of the cadmium generally binds to low molecular weight proteins, many of which are similar to metallothionein. Some of these data have been reviewed (Fox, 1983; Chmielnicka and Cherian, 1986; Petering and Fowler, 1986). The bioavailability of cadmium in several foods has been investigated in experimental animals. In all cases, part of the cadmium in the food was absorbed, although generally less than that from control inorganic sources (Fox, 1983).

Seven male human volunteers consumed a single meal of crab hepato-pancreas intrinsically labeled with cadmium-115m and containing 24 to 166 μg cadmium (Newton et al., 1984). Cadmium retention after several weeks averaged 2.7%, similar to that observed by other investigators in humans who consumed a meal containing kidney extrinsically labeled with cadmium.

In another study, Schellmann et al. (1984) measured cadmium in the feces, urine, and blood for 10 days after humans consumed meals of wild (*agaricus*) mushrooms that supplied 315 to 908 μg cadmium. Concentrations of cadmium in the blood and urine did not change. There were large amounts of cadmium in the feces and it was concluded that there was no significant cadmium absorption from this high-cadmium food source, possibly due to its chitinous nature.

## Effects in Human Populations Consuming High-Cadmium Foods

The serious effects of consuming rice and drinking water contaminated with cadmium that led to Itai-itai disease were described above. It is now believed that these elderly Japanese women represented a special risk group due to initial poor skeletal status arising from multiple pregnancies and inadequate nutrient intakes during the period of exposure. Some of these nutrient deficiencies could have adversely impacted skeletal mineral as well as cadmium absorption. Itai-itai disease did not occur in all Japanese exposed to cadmium, nor has it been observed in other population groups that have consumed similarly high amounts of cadmium from other foods.

The most extensively studied of these recently recognized high cadmium intake populations is that of oyster consumers in Bluff, New Zealand (Sharma et al., 1983; McKenzie et al., 1985; McKenzie-Parnell and Eynon, 1988). The Bluff oyster *(Tiostrea lutaria),* which accumulates cadmium to an average of 5 mg/kg wet weight, is a highly regarded delicacy. Based on mean daily intakes of this oyster by 78 subjects during the 6-month oyster season, fecal cadmium excretion was closely related to estimated oyster cadmium intake. Half of the subjects consumed enough oysters to exceed the FAO/WHO provisional tolerable intake for cadmium during 6 months of each year. There were no effects on urinary cadmium (an index of body stores), small increases in blood cadmium (an index of recent cadmium intake), and there was an approximately four-fold increase in maternal milk cadmium concentration of the one woman studied (Eynon et al., 1985). The cadmium intakes of 5 to 15% of the subjects were high enough to expect renal tubular proteinuria; however, $\beta_2$-

microglobulin in the urine was within the normal range, even in those who had ingested oysters regularly for more than 30 years. No subject had hypertension or a medically treated skeletal condition.

Results of a preliminary autopsy study showed increased cadmium in the renal cortex with age (McKenzie-Parnell and Eynon, 1988). The higher concentrations were about half those observed in Itai-itai patients. Smoking cigarettes had a greater effect on blood and tissue cadmium concentrations than did consumption of the high-cadmium oysters. Estimated body burdens of cadmium revealed no differences between men and women. The results on the effects of smoking and sex of the subjects agree with those of a 3-year global pilot study sponsored by UNEP/WHO (Friberg and Vahter, 1983).

The initial studies of 78 Bluff oyster consumers have recently been validated by 14-week fecal cadmium excretion studies in 3 nonsmoking women who consumed 5 Bluff oysters per day, approximately 132 to 147 µg cadmium (McKenzie-Parnell and Eynon, 1988). Fecal cadmium was markedly elevated and returned to normal low levels 1 to 2 weeks after oyster consumption was discontinued. At the end of the study, whole blood cadmium concentrations were increased by one fourth, onefold, or threefold over the subject's initial value. The highest increase was similar to that of earlier nonsmoking subjects who consumed the most oysters.

In another 72-day fecal cadmium excretion study, two women consumed 176 g cooked Japanese brown rice that provided 300 µg cadmium, equally divided between breakfast and lunch on day 15. On day 43, each woman consumed 96 g oyster puree that provided 300 µg cadmium, equally divided between breakfast and lunch. Otherwise, they consumed their regular diet with exclusions of shellfish, fish, liver, and kidney. With each food, fecal cadmium excretion increased markedly and similarly. Twenty-eight days after the oyster consumption, fecal cadmium excretion increased markedly and similarly. Twenty-eight days after the oyster consumption, fecal cadmium values had not quite returned to baseline levels. For precise determination of absorption, studies are needed with foods intrinsically labeled with cadmium.

Adequate nutritional quality of the diet, as indicated by survey questionnaires, may have prevented any adverse health effects due to high cadmium intake from the oysters. The possibility cannot be excluded, however, that the chemical form of cadmium in the oysters was responsible for the results, at least to some degree.

Hansen et al. (1985) reported that the blood cadmium concentrations of an East Greenland population group were no different from those of East Greenlanders residing in Copenhagen. The daily cadmium intake from seal meat and other marine foods in East Greenland had been estimated to be approximately 109 to 710 µg, whereas that in Copenhagen was approximately 32 µg. All blood cadmium levels were similar to those in populations not exposed to cadmium and, as in other groups, smoking significantly increased the values.

In Shipham, England, soil contamination with cadmium from a nearby zinc mine increased the cadmium content of homegrown vegetables to more than 1 mg/kg. The total cadmium intake was estimated to be about twice that for the general UK population (Sherlock, 1984). There were reports of elevated blood cadmium levels for 22 of 31 individuals and elevated $\beta_2$-microglobulin levels in the urine of five persons (Carruthers and Smith, 1979) and elevated liver cadmium levels (Harvey et al., 1979).

Inskip et al. (1982) compared mortality statistics for the 1939 to 1979 period for Shipham residents and those from a nearby town with no cadmium contamination. There were small excesses of death due to hypertensive, cerebrovascular, and genitourinary disease in Shipham as compared with the control town. There were low death rates in both towns from respiratory disease and cancer, which resulted in overall mortality rates well below the national average.

A preliminary report of an extensive clinical study of 548 Shipham residents matched by age and sex with nonexposed residents of a nearby town revealed no adverse health effects attributable to cadmium intake (Barltrop and Strehlow, 1982). Although cadmium in the soil at Shipham was higher than that associated with Itai-itai disease in Japan, the consumption of vegetables grown in these soils was low so that cadmium intakes seldom exceeded the FAO/WHO provisional tolerable intake. These data also indicate that bioavailability of cadmium from the vegetables was not uniquely high in this population group.

## Conclusions Regarding Cadmium

Deficiencies of several nutrients permit increased tissue uptake of cadmium and increased toxicity at higher levels of cadmium intake by humans and/or animals. The nutrients that appear to be of greatest practical importance include zinc, iron, copper, and calcium, but more information is needed on others, such as selenium. More needs to be learned about the influence of the chemical forms of cadmium in foods upon its bioavailability.

The Japanese women who developed Itai-itai disease appear to be a uniquely susceptible population due to multiple nutrient deficiencies and multiple pregnancies. Other well-nourished population groups exposed to high cadmium intake have not developed Itai-itai disesase.

The cadmium content of the diet in the United States is presently within a safe range. Due to the long-term nature of cadmium retention in the body and uncertainties regarding its effects, it is desirable to avoid increases of cadmium in the food supply.

## LEAD

### Toxic Effects in Humans

Lead toxicity, which dates to antiquity, is a serious contemporary problem. Infants and young children are much more sensitive to lead and are subject to greater exposure than are most adults. Great scientific and public attention has focused on lead in the last two decades and significant progress has been made in defining and dealing with the problem. The literature has been comprehensively treated in recent reviews (Mahaffey, 1985; Quarterman, 1986; Tsuchiya, 1986).

Lead toxicity can adversely affect the nervous system, the vascular system, and the kidney, as manifested by behavioral abnormalities, cognitive impairment, peripheral neuropathy, anemia, and nephropathy. Lead interferes in several steps of heme and globin synthesis, including the suppression of the enzymes δ-aminolevulinic acid dehydrase and ferrochelatase. Erythrocyte protoporphyrin (EP), which is elevated in both iron deficiency and with high blood lead levels, is a useful marker to augment the traditional determination of blood lead.

The most comprehensive attempt to determine the extent of lead toxicity in the United States was part of the second National Health and Nutition Examination Survey (NHANES II), conducted between 1976 and 1980. An evaluation of those blood lead values by Mahaffey et al. (1982a) showed that among children 6 months to 5 years of age, 4% had significantly elevated values. The prevalence was six times higher in black than in white children and was associated with low income. From adolescence onward, males had higher blood level values than females.

There has been marked progress in establishing the adverse effects of lead on cognitive development of children (reviewed in Mahaffey, 1985), including prenatal lead exposure (Bellinger et al., 1987). Although the changes caused by lead are not great, Needleman et al. (1982) pointed out that children with high blood lead values were three times more likely to have a verbal IQ below 80. The effects on neurophysiological development are irreversible.

As a result of these findings in young children, the Centers for Disease Control (1985) established more restrictive criteria for identifying children at risk from lead exposure. Lead toxicity is now defined as a whole blood lead concentration of 25 µg/dl or greater (formerly 30 µg/dl or greater) in conjunction with an EP level of 35 µg/dl or greater (formerly 50 µg/dl or greater). The Committee on Environmental Hazards and the Committee on Accident and Poison Prevention of the American Academy of Pediatrics (1987) have endorsed these criteria.

Persons with lead poisoning are commonly treated with calcium disodium ethylenediaminetetraacetate to remove lead from the body. This drug is not a specific chelator for lead and treatment can cause loss of essential minerals, such as zinc (Thomas and Chisolm, 1986).

## Sources of Intake

Table 13.1 summarizes information on daily lead intake of the infant, toddler, and adult male based on estimates from the Total Diet Study of the Food and Drug Administration. For each age group, the intake values are below the provisional tolerable intake levels established by the FAO/WHO (1972, 1987). The food groups that supplied the most lead to the diets of infants and toddlers were milk; grain and cereal products; meat, fish, and poultry; and vegetables (Gartrell et al., 1986a). Milk was more important for the infant than the toddler. For the adult male, the major sources of lead were grain and cereal products, potatoes, dairy products, and vegetables (Gartrell et al., 1986b). Progress has been made by the food industry in reducing lead contamination, particularly from lead-soldered seams of cans.

For many children, the diet is not as great a source of lead intake as is the environment. Those sources include ambient air, household dust, lead-based paints, and soil in play areas. Separation of the child from high-contamination sources of lead is essential to avoid toxicity.

## Lead and Interacting Nutrients in Humans

The effects of specific nutrients on lead uptake and toxicity have been well reviewed (DeMichele, 1984; Mahaffey, 1985; Quarterman, 1986). There are clearcut effects in humans for many of these lead–nutrient interactions. This section, therefore, will be limited primarily to the major effects and more recent studies in humans (Table 13.5).

**TABLE 13.5**   Lead–Nutrient Interactions in Humans

| Nutrient | Effects | References |
|---|---|---|
| Ca and P | Inverse relation between Pb uptake and Ca and P intake with food or Pb dose | Ziegler et al. (1978); Heard and Chamberlain (1982); Blake and Mann (1983); Heard et al. (1983) |
| Fe | Decreased Fe status: elevated blood Pb | Yip et al. (1981); Mahaffey and Annest (1986) |
|  | Variable relationships in absorption/retention between Fe and Pb | Watson et al. (1980a); Flanagan et al. (1982); Watson et al. (1986) |
| Vitamin D | High blood Pb: decreased serum 1,25-dihyroxyvitamin D | Rosen et al. (1980); Mahaffey et al. (1982b) |
|  | High blood Pb: decreased serum 25-hydroxyvitamin D | Box et al. (1981) |
| Zn | Supplemental Zn: no effect on Pb status of exposed workers | Lauwerys et al. (1983) |
| Ascorbic acid | Ascorbic acid supplements: no effect on Pb status of exposed workers | Lauwerys et al. (1983) |
| Fasting/foods | Markedly greater retention of Pb dose during fasting than with foods | Heard et al. (1983); Flanagan et al. (1982); Rabinowitz et al. (1980); Blake et al. (1983) |
|  | Minerals in meal: reduced Pb uptake | Blake et al. (1983) Heard and Chamberlain (1982) |
|  | Intrinsic vs extrinsic $^{203}$Pb in liver and kidney in a meal: no difference in Pb retention |  |

**Calcium and phosphorus.**   An inverse relationship between lead retention and dietary calcium and phosphorus levels was shown by balance studies in formula-fed infants (Ziegler et al., 1978). Similar effects of calcium and phosphorus on retention of lead-203 were produced in adults given aqueous doses of lead with and without $CaCO_3$ and $Na_2HPO_4$ (Heard and Chamberlain, 1982; Blake and Mann, 1983; Heard et al., 1983). Each element reduced lead retention, but the combined effect was greater. Addition of milk to the test dose markedly reduced lead-203 retention; however, the decrease was not as great as that with inorganic sources of calcium and phosphorus (Blake and Mann, 1983).

Pounds (1984) reviewed the biochemical and physiological aspects of calcium–lead interactions.

**Iron.** A variety of data link high lead uptake with compromised iron status or intake. Yip et al. (1981) examined 2150 children for blood level concentration, EP, and hemoglobin, with additional measurements of iron status in those with low EP values. The prevalence of iron deficiency increased with increasing severity of lead poisoning. Mahaffey and Annest (1986) reported that a higher percentage of all subjects studied in the NHANES II with high EP and high blood level values had low iron status (based on percentage transferrin saturation and total iron-binding capacity) as compared with normal iron status. These data raise concerns about increased lead absorption in individuals with low iron status and emphasize the need for determination of both blood lead and EP to ensure recognition and treatment of all persons with lead poisoning.

Using radiotracer techniques, Watson et al. (1980a) reported a correlation between iron and lead retention following oral administration of test doses to 10 fasting subjects (6 men and 4 women). Uptakes of both elements were highest in 5 subjects with low serum ferritin and in one with normal serum ferritin. Important aspects of methodology, such as accuracy of lead-203 measurements in the presence of iron-59 and practical significance of the molar ratios of the iron and lead doses have been discussed (Flanagan, 1980; Watson et al., 1980b).

The retention of lead-203 from an aqueous solution (containing ascorbic acid and acetic aid) was approximately 60% in 85 fasting adults (Flanagan et al., 1982). Lead retention was proportional to dose up to 400 μg lead. The retention of lead was not influenced by iron stores, capacity of the subject to absorb iron, or the simultaneous administration of a 10-fold molar excess of iron. Deletion of ascorbic acid from the test dose increased lead retention, whereas addition of ethylenediaminetetraacetic acid decreased lead retention.

In a subsequent study, Watson et al. (1986) investigated the whole body retention of lead-203 and iron-59 incorporated into three successive meals. Of the subjects with high iron absorption, only half had high lead retention. Two subjects with hemochromatosis had high retention of iron but normal retention of lead.

These studies leave many unanswered questions about the effects of iron status on lead uptake and toxicity. The findings on lead retention by adults may not be relevant to young children, the major risk population group.

**Vitamin D.** Children with increased blood lead levels had reduced serum concentrations of 1,25-dihydroxyvitamin D (Rosen et al., 1980). When these children were treated with calcium disodium ethylenediaminetet-raacetate to remove lead, serum 1,25-dihydroxyvitamin D levels returned to normal; however, there was no change in serum 25-hydroxyvitamin D concentrations.

In a subsequent study, there was a negative correlation between serum concentrations of 1,25-dihydroxyvitamin D and blood lead in 177 human subjects aged 1 to 16 years (Mahaffey et al., 1982b). The negative association applied to the range of blood lead values between 12 and 120 µg/dl.

Box et al. (1981) reported a negative correlation ($p < 0.001$) between plasma 25-hydroxyvitamin D and blood lead concentrations.

**Zinc and ascorbic acid.** Lauwerys et al. (1983) supplied lead-exposed male workers with a placebo, 1 g ascorbic acid, or 60 mg zinc per day, five times per week for 20 weeks in a single-blind study. The workers had moderate exposure to inorganic lead and they were well-nourished. The supplements had no beneficial effect on either blood lead concentrations or excretion of lead in the urine.

**Fasting and ingestion of foods.** Fasting overnight has been shown by numerous workers to permit high uptake of lead from an aqueous dose, as reviewed by Heard et al. (1983). Rabinowitz et al. (1980) reported retention values of 37, 31, and 30% for lead tracer doses of lead nitrate, lead sulfide, and lead cysteine, respectively. The aqueous doses of lead were administered in the ninth hour of a 16-hour fast. When the same doses were taken with a breakfast of juice, eggs, toast, and coffee, the absorption values were 9, 8, and 6.4%, respectively.

Flanagan et al. (1982) observed 60% retention of lead from an aqueous test dose (containing ascorbic acid and citric acid) that was administered after an overnight fast. Lead retention decreased to 42% when it was given with coffee and a doughnut. Blake et al. (1983) reported retentions of 65 and 4% when the lead-203 dose was given to fasting subjects in water alone or with a purified diet meal, respectively.

In one subject, $^{203}PbCl_2$ that was taken 10 minutes before, during, and 10 minutes after lunch resulted in 36, 7, and 16% retention values, respectively (Heard et al., 1983). The mean uptake in two fasting subjects was 44%. Lead ingested with either coffee or tea in the early morning, mid-morning, or midafternoon gave an overall lead uptake value of 14%. There were no clearcut effects of beverage type or time of ingestion. The uptake of lead in beer consumed before lunch was 19%, which was not different from the values with tea and coffee.

Blake et al. (1983) deleted either minerals, protein, fiber, fat, or vitamins from an adequate purified meal containing lead-203. The retention of lead was not affected by any deletions except by that of minerals, which increased lead retention from 4% for the test meal to 35%. Flanagan et al. (1982) found that giving lipid (10% soybean oil, 1.2% egg yolk phospholipids, and 2.25% glycerin) alone significantly increased lead retention. Single additions of lactose, hard water, cobalt, or zinc had no effects on lead retention.

Physiological incorporation of lead-203 into the liver and kidney of a lamb resulted in lead retention values approximating 3% when the meats were cooked and consumed with a meal of bread, peas, yogurt, and cheese (Heard and Chamberlin, 1982). There were no differences in lead retention between the two types of meat and between intrinsic and extrinsic labels. Intrinsically labeled spinach that was cooked and consumed in a meal resulted in 5.5% retention of lead-203.

## Lead and Interacting Nutrients in Animals

From the studies summarized in the above section, it is well established in humans that several nutrients and dietary components profoundly influence the bioavailability of lead. Data in animals have been summarized by Quarterman (1986). In addition to the nutrient–food–lead effects described for humans, there are data in animals to show that copper, manganese, sulfur, selenium, protein, vitamin E, thiamine, nicotinic acid, pyridoxine, phytic acid, pectin, and alginate have at least sometimes modified lead uptake and/or toxicity. It is likely that some of these effects are of importance to humans.

## Conclusions Regarding Lead

Lead concentrations in the United States diet are currently within safe ranges; however, lead toxicity is a health problem for many infants and children in the United States due to environmental lead exposure. Lead toxicity can be exacerbated by low intakes of certain nutrients, such as calcium and iron. Also, lead toxicity in children interferes with vitamin D metabolism. The uptake of lead from aqueous solutions and from beverages is far greater after a short fast or even when consumed a few minutes before a meal. The effect of fasting disappears when lead is consumed with food. Oral intake of environmental lead, such as that in dust and paint chips, is common in children at times apart from meals. This probably results in greater risk than equivalent lead in the diet. Monitoring the lead status and treatment, if necessary, of pregnant women, infants, and children who are exposed to lead is necessary to avoid the risk of irreversible learning deficits associated with lead toxicity in infants and children.

## GENERAL CONCLUSIONS

High levels of toxic elements can interfere with the bioavailability of essential nutrients. Deficient intakes of certain essential nutrients can increase the adverse effects of the toxic elements due to fairly specific interactions, often in the intestinal lumen or at the site of intestinal absorption. Metabolic interactions also occur after absorption, which can markedly alter nutrient function as well as the adverse effects of the toxic element. Most experimental studies have addressed the effects of single nutrient–single toxic element interactions. This is, however, simpler than the real-life situation in humans. Many studies designed to yield data applicable to humans remain to be done.

To avoid increased risk from toxic elements, consumption of an adequate and varied diet is essential. Many of the foods that supply significant amounts of a toxic element are also good sources of one or more nutrients that reduce the absorption/toxicity of the toxic element. Increased intakes above requirement of one or a limited number of nutrients to protect against toxic elements could result in imbalances among essential nutrients themselves and cause adverse health effects.

## REFERENCES

Abu-Hamdan, D. K., Mahajam, S. K., Migdal, S. D., Prasad, A. S., and McDonald, F. D. 1986. Zinc tolerance tests in uremia. Effect of ferrous sulfate and aluminum hydroxide. *Ann. Int. Med.* 104: 50.

Adler, A. J. and Berlyne, G. M. 1985. Duodenal aluminum absorption in the rat: Effect of vitamin D. *Am. J. Physiol.* 249 (*Gastrointest. Liver Physiol.* 12): G209.

Alfrey, A. C. 1986. Aluminum. Ch. 9. In *Trace Elements in Human and Animal Nutrition,* 5th ed., Vol. 2. Mertz, W. (Ed.), p. 399. Academic Press, New York.

Anonymous. 1987. Toxicological consequences of oral aluminum. *Nutr. Rev.* 45: 73.

Barltrop, D. and Strehlow, C. D. 1982. Cadmium and health in Shipham. *Lancet* 2: 1394.

Bellinger, D., Leviton, A., Waternaux, C., Needleman, H., and Rabinowitz, M. 1987. Longitudinal analysis of prenatal and postnatal lead exposure and early cognitive development. *N. Engl. J. Med.* 316: 1037.

Bhattacharyya, M. H. 1983. Bioavailability of orally administered cadmium and lead to the mother, fetus, and neonate during pregnancy and lactation: An overview. *Sci. Total Environ.* 28: 327.

Blake, K. C. H., Barbezat, G. O., and Mann, M. 1983. Effect of dietary constituents on the gastrointestinal absorption of [203]Pb in man. *Environ. Res.* 30: 182.

Blake, K. C. H. and Mann, M. 1983. Effect of calcium and phosphorus on the gastronintestinal absorption of [203]Pb in man. *Environ. Res.* 30: 188.

Bonner, F. W., King, L. J., and Parke, D. V. 1980. Cadmium-induced reduction of bone alkaline phosphatase and its prevention by zinc. *Chem. Biol. Interact.* 29: 369.

Box, V., Cherry, N., Waldron, H. A., Dattani, J., Griffiths, K. D., and Hill, F. G. H. 1981. Plasma vitamin D and blood lead concentrations in Asian children. *Lancet* 2: 373.

Bunker, V. W., Lawson, M. S., Delves, H. T., and Clayton, B. W. 1984. The intake and excretion of lead and cadmium by the elderly. *Am. J. Clin. Nutr.* 39: 803.

Burnatowska-Hledin, M. A., Doyle, T. M., Eadie, M. J., and Mayor, G. H. 1986. 1,25-Dihydroxyvitamin $D_3$ increases serum and tissue accumulation of aluminum in rats. *J. Lab. Clin. Med.* 108: 96.

Carruthers, M. and Smith, B. 1979. Evidence of cadmium toxicity in a population living in a zinc-mining area. Pilot survey of Shipham residents. *Lancet* 2: 845.

Centers for Disease Control. 1985. Preventing lead poisoning in young children - United States. *Morbid. Mortal. Weekly Rep.* 34: 66.

Chertok, R. J., Sasser, L. B., Callahan, M. F., and Jarboe, G. E. 1981. Influence of cadmium on the intestinal uptake and absorption of calcium in the rat. *J. Nutr.* 111: 631.

Chmielnicka, J. and Cherian, M. G. 1986. Environmental exposure to cadmium and factors affecting trace-element metabolism and metal toxicity. *Biol. Trace Element Res.* 10: 243.

Committee on Environmental Hazards and Committee on Accident and Poison Prevention. 1987. Statement on childhood lead poisoning. *Pediatrics,* 79: 457.

Committee on Nutrition. 1986. Aluminum toxicity in infants and children. *Pediatrics* 78: 1150.

DeMichele, S. J. 1984. Nutrition and lead. *Comp. Biochem. Physiol.* 78A: 401.

Elinder, C. -G. and Sjögren, B. 1986. Aluminum. Ch. 1. In *Handbook on the Toxicology of Metals, Vol. II: Specific Metals,* 2nd ed. Friberg, L., Nordberg, G. F., and Vouk, V. B. (Ed.), p. 1. Elsevier, New York.

Eynon, G. R., McKenzie-Parnell, J. M., Robinson, M. F., and Wilson, P. D. 1985. Cadmium in non-smoking and smoking New Zealand women immediately following childbirth. *Proc. Univ. Otago Med. Sch.* 63: 38.

FAO/WHO Expert Committee on Food Additives. 1972. Evaluation of certain food additives and the contaminants mercurey, lead, and cadmium. *WHO Tech. Rep. Ser.* No. 505: 16.

FAO/WHO Expert Committee on Food Additives. 1987. Report of the thirtieth meeting. *WHO Tech. Rep. Ser.* 751: 35.

Flanagan, P. R. 1980. Oral absorption of lead and iron. *Lancet* 2: 699.

Flanagan, P. R., Chamberlain, M. J., and Valberg, L. S. 1982. The relationship between iron and lead absorption in humans. *Am. J. Clin. Nutr.* 36: 823.

Flanagan, P. R., McLellan, J. S., Haist, J., Cherian, G., Chamberlain, M. J., and Valberg, L. S. 1978. Increased dietary cadmium absorption in mice and human subjects with iron deficiency. *Gastroenterology* 74: 841.

Flegal, K. M., Cary, E. E., Pond, W. G., and Krook, L. P. 1980. Dietary selenium and cadmium interrelationships in growing swine. *J. Nutr.* 110: 1255.

Fox, M. R. S. 1974. Effect of essential minerals on cadmium toxicity. A review. *J. Food Sci.,* 39: 321.

Fox, M. R. S. 1979. Nutritional influences on metal toxicity: Cadmium as a model toxic element. *Environ. Health Perspect.* 29: 95.

Fox, M. R. S. 1983. Cadmium bioavailability. *Fed. Proc.* 42: 1726.

Fox, M. R. S., Jacobs, R. M., Jones, A. O. L., Fry, B. E., Jr., Rakowska, M., Hamilton, R. P., Harland, B. F., Stone, C. L., and Tao, S. -H. 1981. Animal models for assessing essential and toxic elements. *Cereal Chem.* 58: 6.

Fox, M. R. S., Jacobs, R. M., Jones, A. O. L., Fry, B. E. Jr., and Stone, C. L. 1980. Effects of vitamin C and iron on cadmium metabolism. *Ann. N. Y. Acad. Sci.* 355: 249.

Fox, M. R. S., Tao, S. -H., Stone, C. L., and Fry, B. E., Jr. 1984. Effects of zinc, iron and copper deficiencies on cadmium in tissues of Japanese quail. *Environ. Health Perspect.* 54: 57.

Friberg, L., Elinder, C. -G., Kjellström, T., and Nordberg, G. F. 1985. *Cadmium and Health: A Toxicological Appraisal, Vol. I. Exposure, Dose and Metabolism.* CRC Press, Boca Raton, FL.

Friberg, L., Elinder, C. -G., Kjellström, T., and Nordberg, G. F. 1986. *Cadmium and Health: A Toxicological Appraisal, Vol. II. Effects and Response.* CRC Press, Boca Raton, FL.

Friberg, L. and Vahter, M. 1983. Assessment of exposure to lead and cadmium through biological monitoring: Results of a UNEP/WHO global study. *Environ. Res.* 30: 95.

Ganrot, P. O. 1986. Metabolism and possible health effects of aluminum. *Environ. Health Perspect.* 65: 441.

Gartrell, M. J., Craun, J. C., Podrebarac, D. S., and Gunderson, E. L. 1986a. Pesticides, selected elements, and other chemicals in infant and toddler total diet samples, October 1980–March 1982. *J. Assoc. Off. Anal. Chem.* 69: 123.

Gartrell, M. J., Craun, J. C., Podrebarac, D. S., and Gunderson, E. L. 1986b. Pesticides, selected elements, and other chemicals in adult total diet samples, October 1980–March 1982. *J. Assoc. Off. Anal. Chem.* 69: 146.

Greger, J. L. and Lane, H. W. 1987. The toxicology of dietary tin, aluminum, and selenium. Ch. 9. In *Nutritional Toxicology,* Vol. II, Hathcock, J. N. (Ed.) p. 223. Academic Press, New York.

Hansen, J. C., Wulf, H. C., Kromann, N., and Albøge, K. 1985. Cadmium concentrations in blood samples from an East Greenlandic population. *Dan. Med. Bull.* 35: 277.

Harvey, T. C., Chettle, D. R., Fremlin, J. H., Al Haddad, I. K., and Downey, S. P. M. 1979. Cadmium in Shipham. *Lancet* 1: 551.

Heard, M. J. and Chamberlain, A. C. 1982. Effect of minerals and food on uptake of lead from the gastrointestinal tract in humans. *Human Toxicol.* 1: 411.

Heard, M. J., Chamberlain, A. C., and Sherlock, J. C. 1983. Uptake of lead by humans and effect of minerals and food. *Sci. Total Environ.* 30: 245.

Hoadley, J. E. and Cousins, R. J. 1985. Effects of dietary zinc depletion and food restriction on intestinal transport of cadmium in the rat. *Proc. Soc. Exp. Biol. Med.* 180: 296.

Hoadley, J. E. and Johnson, D. R. 1987. Effects of calcium on cadmium uptake and binding in the rat intestine. *Fundam. Appl. Toxicol.* 9: 1.

Huebers, H. A., Huebers, E., Csiba, E., Rummel, W., and Finch, C. A. 1987. The cadmium effect on iron absorption. *Am. J. Clin. Nutr.* 45: 1007.

Inskip, H., Beral, V., and McDowall, M. 1982. Mortality of Shipham residents: 40-year follow-up. *Lancet* 1: 896.

Insogna, K. L., Bordley, D. R., Caro, J. F., and Lockwood, D. H. 1980. Osteomalacia and weakness from excessive antacid ingestion. *J. Am. Med. Assoc.* 244: 2544.

Jacobs, R. M., Lee, A. O. L., Fox, M. R. S., and Lener, J. 1983. Effects of dietary zinc, manganese, and copper on tissue accumulation of cadmium by Japanese quail. *Proc. Soc. Exp. Biol. Med.* 172: 34.

Johnson, D. R., Foulkes, E. C., and Leon, L. 1981. Intestinal transport of cadmium in newborn rats. *Fed. Proc.* 40: 1073.

Kaiser, L. and Schwartz, K. A. 1985. Aluminum-induced anemia. *Am. J. Kidney Dis.* 6: 348.

Kiyozumi, M., Mishima, M., Noda, S., Miyata, K., Takahashi, Y., Mizunaga, F., Nakagawa, M., and Kojima, S. 1982. Studies on poisonous metals. IX. Effects of dietary fibers on absorption of cadmium in rats. *Chem. Pharm. Bull.* 30: 4494.

Klein, G. L., Horst, R. L., Alfrey, A. C., and Slatopolsky, E. 1985. Serum levels of 1,25-dihydroxyvitamin D in children receiving parenteral nutrition with reduced aluminum content. *J. Pediatr. Gastroenterol. Nutr.* 4: 93.

Klein, G. L., Horst, R. L., Norman, A. W., Ament, M. E., Slatopolsky, E., and Coburn, J. W. 1981. Reduced serum levels of 1-alpha,25-dihydroxyvitamin D during long-term total parenteral nutrition. *Ann. Int. Med.* 94: 638.

Kostial, K. 1986. Cadmium. Ch. 5. In *Trace Elements in Human and Animal Nutrition,* 5th ed., Mertz, W. (Ed.), p. 319. Academic Press, New York.

Kostial, K., Rabar, I., Blanusa, M., and Simonovic, I. 1980. The effect of iron additive to milk on cadmium, mercury and manganese absorption in rats. *Environ. Res.* 22: 40.

Kreuger, G. L., Morris, T. K., Suskind, R. R., and Widner, E. M. 1984. The health effects of aluminum compounds in mammals. *CRC Crit. Rev. Toxicol.* 13: 1.

Lamphere, D. N., Dorn, C. R., Reddy, C. S., and Meyer, A. W. 1984. Reduced cadmium body burden in cadmium-exposed calves fed supplemental zinc. *Environ. Res.* 33: 119.

Lauwerys, R., Roels, H., Buchet, J. -P., Bernard, A. A., Verhoeven, L., and Konings, J. 1983. The influence of orally-administered vitamin C or zinc on the absorption of and the biological response to lead. *J. Occup. Med.* 25: 668.

Leon, L. and Johnson, D. R. 1985. Role of iron in jejunal uptake of cadmium in the newborn rat. *J. Toxicol. Environ. Health,* 15: 687.

Mahaffey, K. R. (Ed.). 1985. *Dietary and Environmental Lead: Human Health Effects.* Elsevier, New York.

Mahaffey, K. R. and Annest, J. L. 1986. Association of erythrocyte protoporphyrin with blood lead level and iron status in the second National Health and Nutrition Examination Survey, 1976–1980. *Environ. Res.* 41: 327.

Mahaffey, K. R., Annest, J. L., Roberts, J., and Murphey, R. S. 1982a. National estimates of blood lead levels: United States, 1976–1980. Association with selected demographic and socioeconomic factors. *N. Eng. J. Med.* 307: 573.

Mahaffey, K. R., Rosen, J. F., Chesney, R. W., Peeler, J. T., Smith, C. M., and DeLuca, H. F. 1982b. Association between age, blood lead concentration, and serum 1,25-dihydroxycholecalciferol levels in children. *Am. J. Clin. Nutr.* 35: 1327.

McKenzie, J. M., Kjellström, T. E., and Sharma, R. P. 1985. Cadmium status in a New Zealand population group ingesting high daily intakes of cadmium. In *Trace Elements in Man and Animals,* Mills, C. F., Bremner, I., and Chesters, J. K. (Ed.), p. 526. Commonwealth Agricultural Bureaux, Farnham Royal, Slough, UK.

McKenzie-Parnell, J. M. and Eynon, G. 1988. Effect on New Zealand adults consuming large amounts of cadmium in oysters. *Trace Sub. Environ. Health,* 21: In Press.

National Research Council. 1980. *Mineral Tolerance of Domestic Animals,* p. 93. National Academy of Sciences, Washington, DC.

Nebeker, H. G. and Coburn, J. W. 1986. Aluminum and renal osteodystrophy. *Ann. Rev. Med.* 37: 79.

Needleman, H. L., Leviton, A., and Bellinger, D. 1982. Lead-associated intellectual deficit. *N. Engl. J. Med.* 306: 367.

Newton, D., Johnson, P., Lally, A. E., Pentreath, R. J., and Swift, D. J. 1984. The uptake by man of cadmium ingested in crab meat. *Human Toxicol.* 3: 23.

Nielsen, F. H. 1984. Ultratrace elements in nutrition. *Ann. Rev. Nutr.* 4: 21.

Nogawa, K., Tsuritani, I., Kido, T., Honda, R., Yamada, Y., and Ishizaki, M. 1987. Mechanism for bone disease found in inhabitants environmentally exposed to cadmium: Decreased serum 1α-25-dihydroxyvitamin D level. *Arch. Environ. Occup. Health* 59: 21.

Nogawa, K., Yamada, Y., Honda, R., Tsuritani, I., Kobayashi, E., and Ishizaki, M. 1984. Copper and zinc levels in serum and urine of cadmium-exposed people with special reference to renal tubular damage. *Environ. Res.* 33: 29.

Olsson, U. 1985. Selenium deficiency and detoxication functions in the rat: Effect of chronic dietary cadmium. *Drug Nutr. Interact.* 3: 129.

Pennington, J. A. T. 1983. Revision of the Total Diet Study food list and diets. *J. Am. Diet. Assoc.* 82: 166.

Pennington, J. A. T. and Jones, J. W. 1988. Aluminum in American diets. In *Aluminum in Health, A Critical Review,* Gitleman, H. J. (Ed.), Marcel Dekker, New York. In Press.

Pennington, J. A. T., Jones, J. W., and Vanderveen, J. E. 1987. Aluminum in total diet study foods and diets. *Fed. Proc.* 46: 1002.

Perry, H. M., Erlanger, M. W., and Perry, E. F. 1983. Effect of a second metal on cadmium-induced hypertension. *Arch. Environ. Health* 38: 80.

Petering, D. H. and Fowler, B. A. 1986. Discussion summary, roles of metallothionein and related proteins in metal metabolism and toxicity problems and perspectives. *Environ. Health Perspect.* 65: 217.

Pounds, J. G. 1984. Effect of lead intoxication on calcium homeostasis and calcium-mediated cell function. A review. *NeuroToxicology* 5: 295.

Quarterman, J. 1986. Lead. Ch. 4. In *Trace Elements in Human and Animal Nutrition,* 5th ed., Mertz, W. (Ed.), p. 281. Academic Press, New York.

Rabinowitz, M. B., Kopple, J. D., and Wetherill, G. W. 1980. Effect of food intake and fasting on gastrointestinal lead absorption in humans. *Am. J. Clin. Nutr.* 33: 1784.

Rahim, A. G. A., Arthur, J. R., and Mills, C. F. 1986. Effects of dietary copper, cadmium, iron, molybdenum and manganese on selenium utilization by the rat. *J. Nutr.* 116: 403.

Revis, N. W. 1981. The relationship of dietary protein to metallothionein and cadmium-induced renal damage. *Toxicology* 20: 323.

Revis, N. W., Major, T. C., and Horton, C. Y. 1980. The effects of calcium, magnesium, lead or cadmium on lipoprotein metabolism and atherosclerosis in the pigeon. *J. Environ. Pathol. Toxicol.* 4-2, 3: 293.

Revis, N. W. and Osborne, T. R. 1984. Dietary protein effects on cadmium and metallothionein accumulation in the liver and kidney of rats. *Environ. Health Perspect.* 54: 83.

Rose, H. E. and Quarterman, J. 1987. Dietary fibers and heavy metal retention in the rat. *Environ. Res.* 42: 166.

Rosen, J. F., Chesney, R. W., Hamstra, A., DeLuca, H. F., and Mahaffey, K. R. 1980. Reduction in 1,25-dihydroxyvitamin D in children with increased lead absorption. *N. Engl. J. Med.* 302: 1128.

Sarhan, M. J., Roels, H., Lauwerys, R., Reyners, H., and Gianfelici de Reyners, E. 1986. Influence of manganese on the gastrointestinal absorption of cadmium in rats. *J. Appl. Toxicol.* 6: 313.

Schäfer, L., Anderson, O. and Nielsen, J. B. 1986. Effects of dietary factors on G. I. Cd absorption in mice. *Acta Pharmacol. Toxicol.* (Copenhagen) 59, Suppl. 7: 549.

Schellmann, B., Rohmer, E., Schaller, K. -H., and Weltle, D. 1984. Cadmium- und Kupferkonzentrationen in Stuhl, Urin und Blut nach Aufnahme wildwachsender Champignons. *Z. Lebensm. Unters. Forsch.* 178: 445.

Shaikh, Z. A. and Smith, J. C. 1980. Metabolism of orally ingested cadmium in humans. In *Mechanisms of Toxicity and Hazard Evaluation,* Holmstedt, B., Lauwerys, R., Mercier, M., and Roberfroid, M. (Ed.), p. 569. Elsevier/North-Holland Biomedical Press, Amsterdam, The Netherlands.

Sharma, R. P., Kjellström, T. E., and McKenzie, J. M. 1983. Cadmium in blood and urine among smokers and non-smokers with high cadmium intake via food. *Toxicology* 29: 163.

Sherlock, J. C. 1984. Cadmium in foods and the diet. *Experientia* 40: 152.

Shike, M., Sturtridge, W. C., Tam, C. S., Harrison, J. E., Jones, G., Murray, T. M., Husdan, H., Whitwell, J., Wilson, D. R., and Jeejeebhoy, K. N. 1981. A possible role of vitamin D in the genesis of parenteral nutrition-induced metabolic bone disease. *Ann. Int. Med.* 95: 560.

Siewicki, T. C., Sydlowski, J. S., Van Dolah, F. M., and Balthrop, J. E., Jr. 1986. Influence of dietary zinc and cadmium on iron bioavailability in mice and rats: Oyster versus salt sources. *J. Nutr.* 116: 281.

Spencer, H. and Kramer, L. 1983. Antacid-induced calcium loss. *Arch. Int. Med.* 143: 657.

Spencer, H., Kramer, L., Norris, C., and Osis, D. 1982. Effect of small doses of aluminum-containing antacids on calcium and phosphorus metabolism. *Am. J. Clin. Nutr.* 36: 32.

Spencer, H., Kramer, L., Osis, D., and Wiatrowski, E. 1985. Effects of aluminum hydroxide on fluoride and calcium metabolism. *J. Environ. Pathol. Toxicol. Oncol.* 6: 33.

Thomas, D. J. and Chisolm, J. J. 1986. Lead, zinc and copper decorpora-
tion during calcium disodium ethylenediamine tetraacetate treatment
of lead-poisoned children. *J. Pharmacol. Exp. Ther.* 239: 829.

Tsuchiya, K. 1986. Lead. Ch. 14. In *Handbook on the Toxicology of Metals.
Vol. II: Specific Metals,* 2nd ed., Friberg, L., Nordberg, G. F., and Vouk, V.
B. (Ed.), p. 298. Elsevier, New York.

Van Barneveld, A. A. and Van den Hamer, J. A. 1985. Influence of Ca and
Mg on the uptake and deposition of Pb and Cd in mice. *Toxicol. Appl.
Pharmacol.* 79: 1.

Waalkes, M. P. 1986. Effect of dietary zinc deficiency on the accumulation
of cadmium and metallothionein in selected tissues of the rat. *J. Toxicol.
Environ. Health* 18: 301.

Watson, W. S., Hume, R., and Moore, M. R. 1980a. Oral absorption of lead
and iron. *Lancet* 2: 236.

Watson, W. S., Moore, M. R., and Hume, R. 1980b. Oral absorption of lead
and iron. *Lancet* 2: 699.

Watson, W. S., Morrison, J., Bethel, M. I. F., Baldwin, N. M., Lyon, D. T. B.,
Dobson, H., Moore, M. R., and Hume, R. 1986. Food iron and lead absorption in humans. *Am. J. Clin. Nutr.* 44: 248.

Webster, W. S. 1979. Iron deficiency and its role in cadmium-induced fetal
growth retardation. *J. Nutr.* 109: 1640.

Whanger, P. D. 1985. Metabolic interactions of selenium with cadmium,
mercury, and silver. *Adv. Nutr. Res.* 7: 221.

Yip, R., Norris, T. N., and Anderson, A. S. 1981. Iron status of children
with elevated blood lead concentrations. *J. Pediatr.* 98: 922.

Ziegler, E. E., Edwards, B. B., Jensen, R. L., Mahaffey, K. R., and Fomon, S.
J. 1978. Absorption and retention of lead by infants. *Pediatr. Res.* 12:
29.

# 14

## Interactions of Food Additives and Nutrients

**John E. Vanderveen**

Food and Drug Administration
Washington, D.C.

## INTRODUCTION

The Food, Drug and Cosmetic Act defines a food additive as "any substance the intended use of which results or may reasonably be expected to result, directly or indirectly, in its becoming a component or otherwise affecting the characteristics of any food (including any substance intended for use in producing, manufacturing, packing, processing, preparing, treating, packing, transporting, or holding food; and including any source of radiation intended for any such use" (Anonymous, 1985).

The definition excludes pesticides associated with raw agricultural commodities, color additives, substances used in meat and poultry prior to January 1, 1958, and new animal drugs. The law requires that food additives must provide some function ("effect") in the foods to which they are added. Often this function is to change the physical or chemical properties in some desirable way. Since the major constituents in all foods are nutrients, it is conceivable that interactions between food additives and nutrients in the foods will occur. Some of these interactions are desirable and may even be intended. Other interactions have no effect on the stability, digestibility, or metabolism of the nutrients and therefore are of no biological consequence. Still other interactions have a negative impact on the stability and/or bioavailability of nutrients.

It is important to nutritionists and food technologists that knowledge of these interactions be understood and used to maintain and improve the nutrition quality of the national food supply.

It is the intent of this chapter to provide a general review of the types of interactions between food additives and nutrients that have been identified through scientific research. Although nutrients added to foods are also food additives by definition, such interactions will not be included because they are discussed, in part, in other chapters. This paper will also address the critieria by which food additives are to be reviewed for approval.

## TYPES OF FOOD ADDITIVE–NUTRIENT INTERACTIONS

The interactions between food additives and nutrients fall into at least seven categories. These are interactions that 1) result in decreased nutrient absorption; 2) result in enhanced nutrient absorption; 3) alter nutrient metabolism; 4) alter nutrient excretion; 5) cause nutrient destruction; 6) foster nutrient synthesis; and 7) protect nutrients. In some situations, two or more of these processes may occur simultaneously.

### Decreased Nutrient Absorption

Among the substances permitted to be added to foods are such complex carbohydrates as cellulose, carrageenan, pectin, and gums. These carbohydrates are part of the category of substances currently referred to as dietary fiber. Several scientific reviews have examined the data on the interactions between dietary fiber and essential trace elements and mineral elements. These data demonstrate that the addition of dietary fiber to the diets of experimental animals decreased the absorption of trace elements, including zinc, manganese, magnesium, and iron (Kelsay, 1978). Data from human studies are less definitive. Kelsay (1984) concluded from data collected in human studies that diets that contained less than 25 g of neutral detergent fiber had no effect on mineral absorption. Other studies indicate that levels of dietary fiber above 30 g per day may interfere with the absorption of calcium, magnesium, iron, zinc, and copper. These studies were reviewed and summarized by an expert panel convened by the Life Sciences Research Office (LSRO) of the Federation of American Societies for Experimental Biology (Pilch, 1987). As indicated in that review, the results from these studies were highly variable, and there appeared to be some adaptation in the level of mineral absorption with time. Since fiber–mineral interactions are the subject of another chapter, no further discussion is warranted here.

Complex carbohydrates have also been shown to interfere with the absorption of proteins and lipids. Nilson and Schaller (1941) investigated the effect of adding carrageenan to the diets of rats at levels up to 20% by weight and reported that the digestibilty of protein and fat was reduced with increased carrageenan in the diet. Several investigators have since duplicated these findings: Nilson and Wagner (1959) in rats; Rhee et al. (1981) in rats; and Vohra and Kratzer (1964) in chickens. It should be noted that the level of carrageenan necessary to observe the decreases in absorption of protein and fat is far above that required to provide the technological functions for which carrageenan is used (Stanley, 1982). In addition, other complex carbohydrates have the same effect, as shown by Rhee et al. (1981).

The impact of dietary fiber on lipid absorption/metabolism has been investigated in more than 65 clinical studies in the last 10 years. A review of these studies by the LRSO Expert Panel on Dietary Fiber (Pilch, 1987) provides an assessment of the overall importance of this interaction. It is now clear that dietary fiber lowers serum cholesterol levels, but there appears to be little or no effect on serum triglycerides. At one time, observed changes in serum lipid levels induced by increased dietary fiber were considered evidence of direct or indirect interaction between the fiber and lipids. One of the primary hypotheses was that dietary fiber sequestered bile acids, which had two effects on serum lipid levels (Story, 1984). The binding of bile acids prevented reabsorption and, thus, tended to increase the body's use of cholesterol for synthesis of primary acids. In addition, it was reasoned that the binding of bile acids also reduced the emulsification of lipids and interfered with the micelle formation and, hence, cholesterol absorption. However, as discussed later, soluble forms of dietary fiber have the largest effect on serum lipids and the action appears to be achieved through a metabolic rather than an absorptive mechanism.

The impact of dietary fiber on water-soluble vitamins is less certain. In 1983 Lindberg et al. reported that wheat bran decreased the availability of vitamin $B_6$ to human subjects; however, no other data support these findings. Several studies showed an increased excretion of folic acid in the feces when insoluble fiber was added to the diet of rats. Originally this increase in folic acid excretion was thought to be an indication of decreased absorption (Keagy and Oace, 1984). However, as discussed later, this may not be a true indication of decreased absorption. In a similar manner, Keltz (1978) observed that the addition of pectin to a diet decreased urinary excretion of ascorbic acid and concluded that this decreased absorption reflected a decrease in the level of vitamin C absorption. Finally, Cullen and Oace (1978) reported that pectin and cellulose reduced vitamin $B_{12}$ utilization in the rat.

The addition to foods of lipids with high melting points, such as hydrogenated palm kernel oil, has been shown to significantly reduce the digestibility of protein in the diets of young men (Vanderveen et al., 1966). Fat with a high melting point was added to bite-size foods designed for the United States Space Program. The fat was selected to provide integrity to the foods and keep them from forming crumbs. The decrease in protein digestibility may have been partly due to a decreased transit time since the subjects had several nonformed stools each day.The same effects were observed with the astronauts in space, and the use of the high-melting-point fats was hastily withdrawn. Ironically, Crockett and Deuel (1947) had reported similar results in studies with rats many years before, but the food technologists in the space program were unaware of these data.

Crockett and Deuel did not study the impact of the high-melting-point fats on the absorption of fat-soluble vitamins. Such measurements are important in view of data obtained on experimental substances tested for lowering cholesterol absorption. One such substance, known as sucrose polyester, when given in large amounts has been shown to lower the levels of vitamin E in plasma in human subjects (Mellies et al., 1985). The effects of sucrose polyester on the absorption of other fat-soluble vitamins is less certain. Earlier studies by Mellies and other investigators show decreased absorption of vitamins A and K with ingestion of sucrose polyester, but the most recent data show no significant change (Fallat et al., 1976; Crouse and Grundy, 1979; Glueck et al., 1979; Jandacek et al., 1980; Glueck et al., 1983; Mellies et al., 1983). Such losses can be compensated for by increasing the levels of fat-soluble vitamins in the diet.

The addition of chelating agents in foods can also affect absorption. High levels of ethylenediaminetetraacetic acid (EDTA) have been shown to bind mineral elements sufficiently to lower absorption. However, as discussed in the next section, when added to foods at levels permitted by regulation, EDTA has a greater role in enhancing mineral absorption. Chelation also is known to occur between caffeine and riboflavin, thereby rendering the vitamin less available to animals (Jusko and Levy, 1975).

The clinical significance of this reaction and other such reactions on the availability of riboflavin is not known.

## Enhanced Nutrient Absorption

In the late 1950s it was discovered that if EDTA was added to a diet, the bioavailability of zinc increased (Kratzer et al., 1959). Later research showed that EDTA mobilizes zinc, manganese, and iron from other constituents in the diet such as dietary fiber and some proteins (Layrisse et al., 1976; Viteri et al., 1978; Martinez-Torres et al., 1979). These chelated elements were then more available to the molecules that transported them across the mucosal cells of the gastrointestinal tract. El Guindi et al. (1988) showed that the addition of EDTA to breads baked at very high temperatures increased the bioavailability of iron to human subjects.

The bioavailability of iron has also been shown to be enhanced by the addition of ascorbic acid and ethanol to the diet. Recently there have been claims that the bioavailability of calcium is enhanced significantly by the addition of citric and/or malic acids. These claims are based on the observation that calcium excretion in the urine is increased when these acids are added to the diets of rats. This conclusion seems reasonable since the addition of these acids would increase the solubility of calcium, and calcium must be in solution to be absorbed. However, additional data will be needed to prove that the increased urinary excretion is not caused by a lower than normal level of excretion of calcium through the gastrointestinal tract. In a related example, it has long been known that the treatment of corn with alkali and the steaming or roasting of sweet corn release the niacin from a bound form (Kodicek et al., 1974; Carter and Carpenter, 1982).

## Altered Nutrient Metabolism

There has been considerable interest in the impact of food additives on metabolism of nutrients ever since it was discovered that compounds such as butylated hydroxytoluene (BHT) are metabolized by the cytochrome P-450 enzyme system. In a recent publication by Leo et al. (1987), it was reported that the addition of BHT to a diet marginal in vitamin A content resulted in depletion of liver vitamin A stores and low serum vitamin A levels in rats. Suzuki et al. (1983) showed that the addition of BHT to the diets of rats deficient in vitamin K resulted in decreased levels of vitamin K in liver tissue and increased excretion of vitamin K in the feces. It appears from the data in this report that BHT either inhibited vitamin K absorption or enhanced excretion of vitamin K into the gastrointestinal tract.

An interacting relationship between vitamin $B_6$ and monosodium L-glutamate (MSG) was reported by Folkers et al. (1984). Their research indicated that the reaction to MSG, commonly called Chinese Restaurant Syndrome, is associated with a deficiency of vitamin $B_6$. These authors proposed that MSG might be used to test for vitamin $B_6$ deficiency.

Gaunt et al. (1977) found that caramel (prepared by the ammonia process) added to the diets of rats caused the development of lymphocytopenia. For a time these results caused the FAO/WHO Expert Committee on Food Additives to temporarily eliminate the acceptable daily intake for caramel. Several investigators found that the condition was corrected by the addition of pyridoxine (vitamin $B_6$) to the diet of the rats. Spector and Huntoon (1982) reported that caramel produced by the ammonia process contains low weight competitive inhibitor(s) of mammalian pyridoxal kinase activity. These authors demonstrated that the contaminant(s) could be effectively removed from caramel by treatment with charcoal or by dialysis.

As indicated above, it now appears that the addition of soluble fiber to the diet alters the metabolism of lipids. The current hypothesis is that these carbohydrates, which are not digested by human enzymes, are metabolized by bacteria in the lower gastrointestinal tract and that volatile fatty acids are synthesized. Indications are that when these volatile fatty acids are absorbed from the colon, they inhibit the synthesis of cholesterol.

Finally, Alexander (1969) described the impact of cobalt as a food additive on metabolism of pyruvate and fatty acids in cardiac tissue. In the early 1960s, cobalt was added to beer to promote desired foam characteristics. This was safe for most individuals. However, when an individual consumed large quantities of beer with added cobalt and at the same time was on a protein-deficient diet, the cobalt blocked oxidation of pyruvate and fatty acids. A number of individuals in the United States, Canada, and Belgium died before this condition was identified.

## Altered Nutrient Excretion

Changes in urinary excretion levels of nutrients are often caused by nutrient–nutrient interactions; however, fewer instances are apparent with food additive–nutrient interactions. One observation of the latter is the enhanced excretion of vitamin C in urine following the consumption of hemicellulose by humans (Keltz, 1978). The mechanism for this change in excretion is unknown. Another situation known to affect nutrient excretion is the presence of chelating agents in the body. Chelation is used as therapy for the removal of some toxic elements, such as lead, from the body. Approximately 5% of EDTA in the diet has been shown to be absorbed from the diet and is excreted most frequently bound to calcium. However, the amounts of EDTA permitted in foods in the United States are not sufficient to alter significantly the rate of calcium excreted in the urine. Finally, an animal food additive, roxarsone, which is an organic form of arsenic, causes the reduction of tissue copper in chickens (Czarnecki and Baker, 1984).

## Destruction of Nutrients

The impact of sulfiting agents on the thiamine content of foods has been known for more than 40 years (Mallette et al., 1946). The reaction between sulfite and thiamine results in the formation of pyrimidine and thiazole.

Because this reaction is sufficiently complete, the FDA prohibits the use of any sulfiting agents on foods that are a significant source of thiamine. Changes in pH during thermal processing can destroy some vitamins. The addition of pH-lowering substances causes the destruction of the polyglutamate forms of folacin, whereas the addition of pH-raising substances increases the destruction of thiamine (Sauberlich, 1985).

The addition of nitrates to foods containing ascorbic acid results in the oxidation of ascorbic acid to dehydroascorbic acid (Basu et al., 1984). This reaction, which proceeds without activation, has been used to protect against the formation of nitrosamines in products preserved with nitrites.

At first glance, irradiation of food appears to be an unusual food additive. However, under the Food, Drug and Cosmetic Act this process is subject to the food additives provisions of the Act. Since the process of irradiation is defined as a food additive, losses of nutrients during this process fit within this discussion. A review by Murray (1983) concludes that irradiation causes losses of some vitamins, including vitamin C, thiamine, and vitamin A as beta-carotene. These losses are modest and vary with the type of food and level of radiation. However, it is important to point out that any process of preservation results in some loss of nutrients and that the losses associated with the irradiation process are comparable to losses observed with other preservation processes.

## Enhancement of Nutrient Formation

Studies by Keagy and Oace (1984) have shown that certain fiber substances in the diet cause a rise in folic acid content of feces of rats. Furthermore, such increases also occurred in liver tissues of rats consuming some types of fiber. It is postulated that the fiber supports the growth of bacteria that produce the folic acid, some of which is absorbed from the lower gastrointestinal tract. In a similar manner, rats fed a biotin-free diet showed substantial enhancement in growth when cellulose was added to the diet (Rader et al., 1985). The growth enhancement was comparable to that obtained from adding biotin to the diet.

## Interactions That Protect Nutrients

A paper on food additive–nutrient interactions would not be complete without consideration of the protective role of food additives on nutrients. The protective effect of butylated hydroxytoluene (BHT) and butylated hydroxyanisole (BHA) on the oxidation of lipids in foods is well known. Stohs et al. (1984) reported that BHA also protected against microsomal lipid peroxidation caused by contaminants formed in the manufacture of the herbicide 2,4,5-trichlorophenoxyacetic acid. These antioxidants are widely recognized as having a key role in preserving essential fatty acid integrity in formulated foods. Despite the potential for complex carbohydrates to bind essential elements, complex carbohydrates serve to protect essential fatty acids from oxidation. Allen et al. (1982) demonstrated that polyanionic hydrocolloids were capable of preventing cupric ions from catalyzing the oxidation of linoleic acid. This property of such polyanionic hydrocolloids as alkali-modified kappa-carrageenans is critical to the integrity of infant formulas and other formula foods.

Finally, any discussion on the protection of nutrients by food additives would not be complete without consideration of the process of encapsulation. Two nutrients for which encapsulation is widely used are vitamin A and D (Borenstein, 1987, personal communication). Use of encapsulation occurs less often for some of the B vitamins (Borenstein, 1987, personal communication) and vitamin C (Werner, 1980). The food additives used for encapsulation of nutrients vary widely with the food and environmental conditions. Much of this technology is protected by trade secrets, and basic mechanisms of protection are not well understood. Nevertheless, the role of food additives in protecting these nutrients through encapsulation is very important in maintaining the quality of the national food supply.

## REGULATORY CONCERNS

In reviewing the area of food additive–nutrient interactions, it is apparent that the impact of such interactions is small but perceivable. Nevertheless, such data must be considered in regulation of the nation's food supply. As part of the premarket approval process for new food additives, the FDA requires that animal studies be conducted to demonstrate that the use of a food additive will not have harmful effects on consumers. Such studies should be designed to adequately address estimated maximum exposures that might be experienced by any population segment (Anonymous, 1983). Guidelines for such data requirements are available from the Agency (Anonymous, 1982). In some instances where it is considered difficult to extrapolate from animal data to anticipated human experience, clinical studies are needed. In all of these situations, impact on nutrition is a critical element.

In a similar manner, the cyclic review of Generally Regarded as Safe (GRAS) substances must consider data on nutrition as well as toxicological effects of the substance being evaluated. The Agency has also implemented a formal program in postmarket surveillance, which will seek to relate any reports of adverse food reactions to specific nutritional or toxicological conditions. If warranted, the Agency will initiate experimental research to further understand these relationships.

In summary, the database on approved food additive–nutrient interactions demonstrates that no major impact on the nutritional quality of the national food supply exists. Furthermore, mechanisms for approving the use of food additives take into consideration the impact of the nutritional quality of the foods to which they are added. Nutritionists and food technologists have a responsibility to use available scientific data on food additive–nutrient interactions to maintain or improve the nutrition quality of the food supply.

## REFERENCES

Alexander, C. S. 1969. Cobalt and the heart. *Ann. Intern. Med.* 70: 411.

Allen, J. C., Gardner, R. S., Parry, D. E., and Wedlock, D. J. 1982. The influence of polyanionic hydrocolloids on the $Cu^{2+}$-catalysed oxidation of linoleic acid. *Prog. Food Nutr. Sci.* 6: 191.

Anonymous. 1982. Toxicological principles for the safety assessment of direct food additives and color additives used in food. U. S. Food and Drug Administration, Bureau of Foods, Washington, DC.

Anonymous. 1983. Risk Assessment in the Federal Government: Managing the Process. National Academy Press, Washington, DC.

Anonymous. 1985. Federal Food, Drug and Cosmetic Act. As amended. United States Code under Title 21 Section 201(s), 403, and 409.

Basu, T. K., Weiser, T., and Dempster, J. F. 1984. An in vitro effect of ascorbate on the spontaneous reduction of sodium nitrite concentration in a reaction mixture. *Int. J. Vitam. Nutr. Res.* 54: 233.

Carter, E. G. and Carpenter, K. J. 1982. Available niacin values of foods for rats and their relation to analytical values. *J. Nutr.* 112: 2091.

Crockett, M. E. and Deuel, H. J. 1947. Fat digestibility and absorption. *J. Nutr.* 33: 187.

Crouse, J. R. and Grundy, S. M. 1979. Effects of sucrose polyester on cholesterol metabolism in man. *Metabolism* 28: 994.

Cullen, R. W. and Oace, S. M. 1978. Methylmalonic acid and vitamin B-12 excretion of rats consuming diets varying in cellulose and pectin. *J. Nutr.* 108: 640.

Czarnecki, G. L. and Baker, D. H. 1984. Feed additive interactions in the chicken: Reduction of tissue copper deposition by dietary roxarsone in healthy and in Eimeria acevulina-infected or Eimeria tenella-infected chicks. *Poult. Sci.* 63: 1412.

El Guindi, M., Lynch, S. R., and Cook, J. D. 1988. Iron absorption from fortified flat bread. *Br. J. Nutr.,* in press.

Fallat, R. W., Glueck, C. J., Lutmer, R., and Mattson, F. H. 1976. Short-term study of sucrose polyester a nonabsorbable fat-like material as a dietary agent for lowering plasma cholesterol. Am. J. *Clin. Nutr.* 29: 1204.

Folkers, K., Shizukuishi, S., Willis, R., Scudder, S. L., Takemura, K., and Longenecker, J. B. 1984. The biochemistry of vitamin B6 is basic to the cause of Chinese restaurant syndrome. *Hoppe-Seyler's Z. Physiol. Chem.* 365: 405.

Gaunt, I. F., Lloyd, A. G., Grasso, P., Gangolli, S. D., and Butterworth, K. R. 1977. Short-term study in the rat on two caramels produced by variations of the 'ammonia process'. *Food Cosmet. Toxicol.* 15: 509.

Glueck, C. J., Mattson, F. H., and Jandacek, R. J. 1979. The lowering of plasma cholesterol by sucrose polyester in subjects consuming diets with 800, 300, or less than 50 mg of cholesterol per day. *Am. J. Clin. Nutr.* 32: 1636.

Glueck, C. J., Jandacek, R., Hogg, E., Allen, C., Baehler, L., and Tewksbury, M. 1983. Sucrose polyester: Substitution for dietary fat in hypocaloric diets in the treatment of familial hypercholesterolemia. *Am. J. Clin. Nutr.* 37: 347.

Jandacek, R. J., Mattson, F. H., McNeely, S., Gallon, L., Yunker, R., and Glueck, C. J. 1980. Effects of sucrose polyester on fecal steroid excretion by 24 normal men. *Am. J. Clin. Nutr.* 33: 251.

Jusko, W. J. and Levy, G. 1975. Absorption, protein binding and elimination of riboflavin. In *Riboflavin*, R. S. Rivilin (Ed.), p. 99. Plenum Press. New York.

Keagy, P. A. and Oace, S. M. 1984. Folic acid utilization from high fiber diets in rats. *J. Nutr.* 114: 1252.

Kelsay, J. L. 1978. A review of research on effects of fiber intake on man. *Am. J. Clin. Nutr.* 31: 142.

Kelsay, J. L. 1984. Update on fiber and mineral availability. In *Dietary Fiber*, G. V. Vahouny and D. Kritchevsky (Eds.), p. 361. Plenum Press, New York.

Keltz, P. R. 1978. Urinary ascorbic acid excretion in the human as affected by dietary fiber and zinc. *Am. J. Clin. Nutr.* 31: 1167.

Kodicek, E., Ashby, D. R., Muller, M., and Carpenter, K. J. 1974. The conversion of bound nicotinic acid to free nicotinamide on roasting sweet corn. *Proc. Nutr. Soc.* 33: 105A.

Kratzer, F. H., Allred, J. B., Davis, P. N., Marshall, B. J., and Vohra, P. 1959. The effects of autoclaving soybean protein and the addition of EDTA on biological availability of dietary zinc for turkey poults.*J. Nutr.* 68: 313.

Layrisse, M., Martinez-Torres, C., Renzi, M., Velez, F., and Gonzalez, M. 1976. Sugar as a vehicle for iron fortification. *Am. J. Clin. Nutr.* 29: 8.

Leo, M. A., Lowe, N., and Lieber, C. S. 1987. Potentiation of ethanol-induced hepatic vitamin A depletion by phenobarbital and butylated hydroxytoluene. *J. Nutr.* 117: 70.

Lindberg, A. S., Leklem, J. E., and Miller, L. T. 1983. The effects of wheat bran on the bioavailability of vitamin B-6 in young men. *J. Nutr.* 113: 2578.

Mallette, M. F., Dawson, C. R., Nelson, W. L., and Gortner, W. A. 1946. Commercially dehydrated vegetables. Oxidative enzymes, vitamin content and other factors. *Ind. Eng. Chem.* 38: 437.

Martinez-Torres, C., Romano, E. L., Renzi, M., and Layrisse, M. 1979. Fe III-EDTA complex as iron fortification: Further studies. *Am. J. Clin. Nutr.* 32: 809.

Mellies, M. J., Jandacek, R. J., Taulbee, J. D., Tewesbury, M. B., Lamkin,

G., Baehler, L., King, P., Boggs, D., Goldman, S., Gauge, A., Tsang, R., and Glueck, C. J. 1983. A double blind, placebo-controlled study of sucrose polyester in hypercholesterolemic outpatients. *Am. J. Clin. Nutr.* 37: 339.

Mellies, M. J., Vitale, C., Jandacek, R. J., Lamkin, G. E., and Glueck, C. J. 1985. The substitution of sucrose polyester for dietary fat in obese, hypercholesterolemic outpatients. *Am. J. Clin. Nutr.* 41: 1.

Murray, T. K. 1983. Nutritional aspects of food irradiation. In *Recent Advances in Food Irradiation,* P. S. Elias and A. J. Cohen (Eds.), p. 203. Elsevier Biomedical Press, New York.

Nilson, H. W. and Schaller, J. S. 1941. Nutritive value of agar and Irish moss. *Food Res.* 6: 461.

Nilson, H. W. and Wagner, J. A. 1959. Feeding test with carrageenin. *Food Res.* 24: 235.

Pilch, S. M. (Ed.). 1987. Physiological effects and health consequences of dietary fiber. In "Report of Expert Panel of Life Sciences Research Office." Federation of American Societies for Experimental Biology, Bethesda, MD.

Rader, J. I., Gaston, C. M., Wolnik, K. A., Fricke, F. L., and Fox, M. R. S. 1985. Growth and tissue minerals in weanling rats fed purified biotin-free or fiber-free diets. *Ann. N. Y. Acad. Sci.* 447: 417.

Rhee, M., Pittz, E. P., and Abraham, R. 1981. Effect of combination of *Irideae* carrageenan and cellulose on the absorption of some nutrients from the alimentary tract of rats. *Ecotoxicol. Environ. Saf.* 5: 1.

Sauberlich, H. E. 1985. Bioavailability of vitamins. *Prog. Food Nutr. Sci.* 9: 1.

Spector, R. and Huntoon, S. 1982. Effects of caramel color (ammonia process) on mammalian vitamin B6 metabolism. *Toxicol. Appl. Pharmacol.* 62: 172.

Stanley, N. F. 1982. The effects of carrageenan on peptic and tryptic digestion of casein. *Prog. Food Nutr. Sci.* 6: 161.

Stohs, S. J., Hassan, M. Q., and Murray, W. J. 1984. Effects of BHA d-α-tocopherol and retinol acetate on TCDD-mediated changes in lipid peroxidation glutathione peroxidase activity and survival. *Xenobiotica* 14: 533.

Story, J. A. 1984. Modification of steroid excretion in response to dietary fiber. In *Dietary Fiber,* G. V. Vahouny and D. Kritchevsky, (Eds.), p. 253. Plenum Press, New York.

Suzuki, H., Nakao, T., and Hiraga, K. 1983. Vitamin K content of liver and feces from vitamin K-deficient and butylated hydroxytoluene (BHT) treated male rats. *Toxicol. Appl. Pharmacol.* 67: 152.

Vanderveen, J. E., Heidelbaugh, N. D., and O'Hara, M. J. 1966. Study of man during a 56-day exposure to an oxygen-helium atmosphere at 258 mm. Hg total pressure. IX. Nutritional evaluation of feeding bite-size foods. *Aerosp. Med.* 37: 591.

Viteri, F. E., Garcia-Ibanez, R., and Torun, B. 1978. Sodium iron - NaFe-EDTA as an iron fortification compound in Central America: Absorption studies. *Am. J. Clin. Nutr.* 31: 961.

Vohra, P. and Kratzer, F. H. 1964. Growth inhibitory effect of certain polysaccharides for chickens. *Poult. Sci.* 43: 1164.

Werner, L. E. 1980. Encapsulation makes vitamin C more efficient, more available. *Food Eng.* 52: 116.

# 15

# A Comparison Of Drug–Nutrient And Nutrient–Nutrient Interactions

**Daphne A. Roe**

Cornell University
Ithaca, New York

## INTRODUCTION

Drugs and nutrients share certain attributes, including physicochemical characteristics, capacity to modulate physiological processes, and dose-related toxicity. Mechanisms by which drugs may react with nutrients are similar to those by which nutrients can react with nonnutrient components of food. These mechanisms include physicochemical, physiological, and toxicological interactions.

Sites of drug and nutrient interactions include the drug formula, the gut, blood, and tissues. Similarly, sites of interaction between nutrients and nonnutrient components of food include the food itself as well as the gut and postabsorptive locations in the body. Outcomes of drug and nutrient interactions may be classified into those that are beneficial and those that are adverse in effect. Within drug formulations or nutrient supplements, interactions can promote or decrease the stability of the ingredients. In the gut, biopharmaceutical interactions can enhance or reduce the absorption of drugs or nutrients. Also within the gut, physiological interactions can alter the rate or extent of drug or nutrient metabolism or absorption. In the blood, interactions between drugs and nutrients can affect their plasma transport as well as the uptake of these substances into red blood cells. In the liver, interactions between drugs and nutrients can affect drug metabolism and nutrient storage or release. In the kidney, these interactions can affect the rate of drug clearance and the rate of clearance of end products of nutrient metabolism. The disposition of drugs and nutrients is affected by body composition (Roe, 1985a).

Toxic effects of drugs and nutrients include cytotoxicity and metabolic aberrations. Evidence from studies in laboratory animals and clinical reports provide ample evidence that these pathophysiological events are influenced by nutritional status (Krishnaswamy, 1985).

Practical concerns relative to drug and nutrient interactions include their effects on drug efficacy, nutritional status, and health status. Interactions between nutrients and nonnutrient components of food are of concern because of loss of product stability and because the nonnutrient component of the food (including naturally occurring substances, intentional additives, and contaminants) may reduce the nutritional value of the food or create the risk of food toxicity. The aim of this review is to make food scientists at academic institutions and in industry more aware of the importance of drug and nutrient interactions. Currently it is essential that drug and nutrient interactions be understood by those engaged in the development of new food products, by those producing liquid diets, including vitamin and mineral mixtures as well as infant, enteral, and parenteral formulas, by those concerned with intentional food additives, and by those whose research is on means to provide the public with foods that meet national guidelines for health promotion and disease prevention.

Food scientists need to be concerned with drug and nutrient interactions for the following reasons:

1. Components of new foods (e.g., fiber) affect the rate and the extent of absorption of drugs and nutrients (Roe et al., 1980; Toothaker and Welling, 1980).
2. The composition of nutrient supplements affects the disposition of therapeutic drugs (Roe, 1984a).
3. Intentional food additives of natural or synthetic origin can speed up the rate of drug metabolism and can impair nutritional status (Hathcock, 1985).

4. Proposed changes in the American diet which will reduce its fat content will affect the extent of drug absorption but also may necessitate rethinking of appropriate drug dosages (Somogyi et al., 1986).

In order to clarify these issues, drugs will be compared with nutrients with respect to their absorption, transport, metabolism, and excretion. The beneficial and adverse effects of drug and nutrient interactions will be presented.

## DRUG VERSUS NUTRIENT DISPOSITION

### Absorption

It is commonly stated that a major difference between drugs and nutrients is that drugs are absorbed by passive diffusion and that nutrients are absorbed by active transport mechanisms (Gibaldi, 1978). However, such differentiation of drug and nutrient disposition only holds true for physiological levels of nutrients. When the absorption of pharmacological doses of nutrients are in question, then the nutrient (e.g., a B vitamin such as riboflavin) is absorbed in a manner similar to a drug. Further, when pharmacological or megadoses of nutrients are absorbed, factors within the gut that influence the rate and efficiency of their absorption are similar to those affecting the absorption of drugs. A summary of the factors affecting drug and nutrient absorption is given in Table 15.1. Absorp-

**TABLE 15.1** Factors Affecting the Absorption of Drugs and Nutrients

Physical and chemical properties
Constituents
Formulations
Dose
Frequency
Route of administration
Food vs. drug times
Patient type
Pathology present

tion of drugs that are similar in chemical structure to nutrients (for example, methyl dopa, an antihypertensive drug, which has a structure similar to an amino acid) is competitive with absorption of those nutrients (Myhre et al., 1982).

Examples of foods that affect the absorption of drugs and nutrients by similar mechanisms are shown in Table 15.2.

## Transport

Drugs and nutrients are transported in the plasma in the free state or bound to plasma proteins. The binding of drugs and nutrients to plasma proteins is competitive, and one can displace the other. A schematic representation of nutrient and drug interactions in the plasma is shown in Fig. 15.1. The competitive binding of drugs and nutrients to plasma proteins has been well described (Gayte-Sorbier and Airaudo, 1984).

Drugs and nutrients are also transported into erythrocytes competitively. Branda and Nelson (1981) have summarized their semi–in vitro studies in which they demonstrated that several drugs known to inhibit red blood cell anion transport also reduce the folate permeability of these cells. Drugs in this category include several that are important therapeutically, including the loop diuretics, ethacrynic acid, and furosemide.

## Metabolism

Drugs and nutrients are metabolized in the gut by drug-metabolizing enzymes present in the intestinal mucosa as well as by gut bacteria. Examples of nutrients and drugs that are metabolized in the gut include beta-carotene and sulfasalazine, which is used in the treatment of ulcerative colitis. Food factors that can influence the metabolism of drugs and nutrients metabolized in the gut include changes in protein and fiber intake.

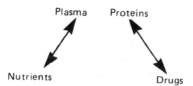

**FIG. 15.1** Schematic representation of nutrient and drug interactions in plasma.

**TABLE 15.2** Comparison of the Absorption Characteristics of Drugs and Vitamins and the Effects of Food on Their Bioavailability

| Absorption mechanism | Drug | Vitamin | Food effect on drug and nutrient absorption |
|---|---|---|---|
| Lipid-soluble | Griseofulvin | β-Carotene | Fat promotes |
| Site-specific saturable process | Chlorothiazide | Riboflavin | Fiber promotes due to slowed gastric emptying |

**TABLE 15.3** Mechanisms Underlying Food Effects on the Metabolism of Drugs and Nutrients

| Intestinal | Hepatic |
|---|---|
| Altered microflora | Altered flood flow |
| Change in activity of drug metabolizing enzyme | |

This can affect the activity of the drug-metabolizing enzymes. The fiber content of the food not only affects the activity of the drug-metabolizing enzymes in the mucosa, but also the gut microflora (Roe, 1984b).

The hepatic metabolism of drugs and nutrients can also be affected by changes in the diet, including its protein content, as well as the presence in the food of nonnutrient substances in herbs and spices. The method of cooking (e.g., the broiling of meat) also affects the rate of metabolism of drugs (Alvares, 1984). The mechanisms responsible for effects of food on the metabolism of drugs and nutrients are shown in Table 15.3.

## Excretion

The excretion of drugs and nutrient metabolites may be competitive in the sense that both compete for the anion transport system. Further, dietary changes can affect the excretion of both drugs and nutrients. However, while it is known from clinical experience that concurrent use of low potassium diets and diuretics is likely to lead to potassium depletion, the effects of other dietary restrictions, such as subsistence on very low calorie diets, on drug and nutrient excretion are only known from a few studies of children and adults with protein–energy malnutrition. Protein-energy malnutrition can lead to decreased excretion of certain drugs because of an impairment in glomerular filtration (Mehta et al., 1981).

## OUTCOMES OF DRUG AND NUTRIENT INTERACTIONS

### Beneficial Effects

Beneficial effects of drug and nutrient interactions include infectious disease control, control of cancer, prevention of thromboses, treatment of metabolic disease, and prevention of acute drug toxicity.

**TABLE 15.4** Beneficial Effects of Drug and Nutrient Interactions

| Drug | Nutrient | Effect |
|---|---|---|
| Coumarin anticoagulant | Vitamin K | Prevents thromboses |
| Isoniazid | Vitamin $B_6$ | Prevents convulsions |
| Methotrexate | Folic acid | Controls cancer |
| Colestipol | Niacin | Controls hypercholesterolemia |

Antibacterial drugs as well as cancer chemotherapeutic drugs include many that are vitamin antagonists. Examples are sulfa drugs, which inhibit folate utilization by bacteria, and methotrexate, which inhibits folate utilization by cancer cells. Anticoagulants in common use are vitamin K antagonists. Drugs in these groups, in fact, owe their therapeutic efficacy to their antinutrient effects (Roe, 1985b). Nutrients may also be used for their pharmacological effects in the control of inborn errors of metabolism. An example is the use of high doses of niacin in familial hypercholesterolemia (Havel and Kane, 1982).

Convulsions due to an overdose of the antituberculosis drug isoniazid can be prevented by giving vitamin $B_6$, because the drug is a vitamin $B_6$ antagonist and this effect of the drug is overcome by giving the vitamin at an antidotal dose level.

Beneficial outcomes of drug and nutrient interactions are in certain instances of great concern and interest to food scientists. For example, when a vitamin antagonist such as a coumarin anticoagulant is in use, it is important that the desired antivitamin K effect of the drug not be overcome by a high intake of vitamin K by the patient. While it is unlikely that the patient will overcome the effect of the drug by eating green vegetables, which are good natural sources of vitamin K, the drug effect may be obviated by concurrent intake of a formula food containing a high level of vitamin K. Beneficial effects of drug and nutrient interactions are listed with examples in Table 15.4.

## Adverse Effects

Adverse effects of drug and nutrient interactions include loss of drug efficacy, drug-induced nutritional deficiencies, toxic reactions, and blocked feeding tubes.

Loss of drug efficacy may be due to administration of a high dose of a vitamin, which overcomes a drug's desired effect. Examples include loss of effectiveness of anticoagulant due to high intake of vitamin K. Indeed, this could be used beneficially to reduce the magnitude of the anticoagulant effect. Administration of pharmacological doses of folic acid (>5 mg/day) will actually lessen the anticonvulsant effect of phenytoin, which is still commonly used for the treatment of seizure disorders (Baylis et al., 1971).

The duration of the therapeutic effect of a drug may also be reduced when the drug's metabolism is speeded up due to the effect of a high protein diet. Asthmatics on theophylline will start to wheeze sooner if they increase their intake of dietary protein without changing the drug dose (Kappas et al., 1976).

Drug-induced nutritional deficiencies may be due to the anorectic effect of the drug, to impairment of digestion by the drug, to drug-induced malabsorption, to loss of ability for nutrient utilization, or to mineral depletion (Roe, 1985a).

Toxic reactions may occur either because of the build-up of toxic levels of a drug in the body or because a drug triggers the release of an endogeneous substance which causes a reaction. An example of the former situation brought about by a drug and nutrient interaction is lithium toxicity, which may be induced by sodium depletion possibly due to prescription of a low sodium diet or administration of a formula food with a very low sodium content. Examples of toxic reactions due to release of endogeneous metabolites include tyramine reactions, which may occur when patients taking antidepressants or other drugs that are monoamine oxidase inhibitors, such as procarbazine, eat foods such as aged cheese, which is high in tyramine. When the tyramine cannot be metabolized due to the effect of the drug, there is release of catecholamines, which elevates blood pressure and an acute hypertension that can lead to a fatal stroke (Roe, 1985c). Foods, special diets, and active components of foods that may cause adverse drug reactions are shown in Table 15.5.

Blocking of nasogastric feeding tubes employed for the infusion of enteral formulas is commonly due to the administration of acidic liquid formulations of drugs concurrently with the nutrient formula. The blockage may not depend on a complexation of the drug with the nutrient formula ingredients, but rather is attributable to the low pH of the drug suspension (Cutie et al., 1983). Drug formulations that are incompatible with enteral formulas are shown in Table 15.6.

**TABLE 15.5** Foods, Diets, and Food Components that Can Cause Adverse Drug Reactions

| Food | Diets | Food component | Drug | Drug reactions |
|------|-------|----------------|------|----------------|
| Aged cheese | — | Tyramine | Phenelzine (other MAOIs) | Acute hypertension |
| Tuna | — | Histamine | Isoniazid | Flush, itching, headache |
| Rum cake | — | Alcohol | Chlorpropamide | Flush |
| — | Sodium-restricted | — | Lithium | Thirst, confusion, blurred vision |

**TABLE 15.6**   Drug Formulations Incompatible with Enteral Formulas[a]

| Drug group | Formulation | Product |
|---|---|---|
| Antihistamine | Elixir | Dimetane (A.H. Robins) |
|  | Elixir | Dimetapp (A.H. Robins) |
| Urinary antiseptic | Suspension | Mandelamine (Parke-Davis) |
| Iron | Elixir | Feosol (Menley and James) |
| Cardiac glycoside | Elixir | Lanoxin (Burroughs-Wellcome) |
| Phenothiazine tranquilizers | Oral solution | Mellaril (Sandoz) |

[a]Compatibility was tested with Ensure, Ensure Plus, and Osmolite (Ross Laboratories).

## CRITIQUE OF METHODS OF INVESTIGATING DRUG AND NUTRIENT INTERACTIONS

### In Vitro

Physicochemical interactions have been investigated by in vitro studies. However, great caution should be exercised in the interpretation of findings because methods have commonly failed to simulate the interaction of drugs with digestion products of food. Other shortcomings of such studies include nonphysiological pH and temperature.

### Animal Models

Studies of drug and nutrient interactions in animal models are essential when histopathological or biochemical studies of tissues are required. Problems are related to species specificity of outcomes of interactions (e.g., flush reactions), as well as differences in the nutritional economy of rats and human beings, which are related to the coprophagous habit of the rodents.

### Human Studies

Pharmacokinetic studies are essential in the investigation of effects of foods and nutrients on drug absorption and metabolism. However, most investigations are single drug dose studies, and therefore they do not mimic the conditions pertaining when a patient is on a daily multiple dose regimen.

Studies of the effects of drugs on nutritional status are mainly observations of patients receiving particular drugs with related laboratory findings relative to their nutritional status. Unfortunately, these studies are notably lacking in information about the patients' diets, and it is therefore difficult to assess the extent to which dietary deficiency might explain, at least in part, evidences of nutritional deficiency attributed to a drug. To examine the effect of a drug on nutritional status, it is clearly necessary to minimize intersubject variability with respect to nutrient intake and health status (Roe, 1981). It is noteworthy that the criticisms which have been leveled against design of human and animal studies of drug and nutrient interactions may equally be leveled against studies of nutrient–nutrient interactions.

## SUMMARY

Drug and food interactions may be physicochemical, may be explained by effects of drugs or food on gastrointestinal function, or may be due to toxic effects of a drug on the gut or liver. Incompatibility of drugs and food may be related to the release of vasoactive substance by a nonnutrient substance within a food which is not metabolized in the normal manner due to the effects of a drug.

Outcomes of drug and nutrient interactions, or of interactions between drugs and nonnutrient substances in food, may be beneficial or adverse. Beneficial effects include infectious disease control, control of cancer, prevention of thromboses, treatment of metabolic disease, and prevention of drug toxicity. Adverse outcomes include loss of drug efficacy, drug-induced nutritional deficiencies, toxic reactions, and blocked nasogastric tubes. These interactions can be studied by in vitro experiments, by studies in animal models, or by human experiments. Problems related to the interpretation of the findings from these investigations include failure to simulate physiological conditions in vitro, inability to extrapolate findings in animals to men, women, and children, and lack of dietary control of human studies. Shortcomings of the research design in investigations of drug and food or drug and nutrient interactions are similar to those which apply to investigations of nutrient–nutrient interactions.

## REFERENCES

Alvares, A. P. 1984. Environmental influences on drug biotransformation in humans. *World Rev. Nutr. Dietet.* 43: 45.

Baylis, E. M., Crowley, J. M., Preece, J. M., et al. 1971. Influence of folic acid on blood phenytoin levels. *Lancet* 1: 62

Branda, R. F. and Nelson, N. L. 1981. Interactions of drugs and folate compounds at the cellular level. In *Nutrition in Health and Disease and International Development,* p. 773. Symposia from the XII Internat. Congress of Nutrition. Alan R. Liss, Inc., New York.

Cutie, A. J., Altman, E., and Lenkel, L. 1983. Compatibility of enteral products with commonly employed drug additives. *J. Parent. Ent. Nutr.* 7: 186.

Gayte-Sorbier, A. and Airaudo, C. B. 1984. Nutrition and drug binding. *World Rev. Nutr. Diet.* 43: 95.

Gibaldi, M. 1978. *Biopharmaceutics and Clinical Pharmacokinetics,* 2nd ed., p. 15. Lea and Febiger, Philadelphia, PA.

Hathcock, J. N. 1985. Nutrient and non-nutrient effects on drug metabolism. *Drug-Nutrient Interact.* 4: 217.

Havel, R. J. and Kane, J. P. 1982. Therapy of hyperlipidemic states. *Ann. Rev. Med.* 33: 417.

Kappas, A., Anderson, D. E., Conney, A. H., and Alvares, A. P. 1976. Influence of dietary protein and carbohydrate on antipyrine and theophylline metabolism in man. *Clin. Pharmacol. Ther.* 20: 643.

Krishnaswamy, K. 1985. Nutrients/nutrients and drug metabolism. *Drug-Nutrient Interact.* 4: 235.

Mehta, S., Nain, C. K., Sharma, B., and Mathur, V. S. 1981. Drug metabolism in malnourished children. In *Nutrition in Health and Disease and International Development,* p. 739. Symposia from the XII Internat. Congress of Nutrition. Alan R. Liss, Inc., New York.

Myhre, E., Rugstad, H. E., and Hansen, T. 1982. Clinical pharmacokinetics of methyldopa. *Clin. Pharmacokinet.* 1: 221.

Roe, D. A., Wrick, K., McLain, D., and Van Soest, P. 1980. Effects of dietary fiber sources on riboflavin absorption. *Fed. Proc.* 37: 756.

Roe, D. A. 1981. Intergroup and intragroup variables affecting interpretation of studies of drug effects on nutritional status. In *Nutrition in Health and Disease and International Development,* p. 757. Symposia from the XII Internat. Congress of Nutrition. Alan R. Liss, Inc., New York.

Roe, D. A. 1984a. Vitamin use and abuse among the elderly. In *Drugs and Nutrition in the Geriatiric Patient,* Roe, D. A. (Ed.), p. 135. Churchill Livingstone, New York.

Roe, D. A. 1984b. Food, formula and drug effects on the disposition of nutrients. *World Rev. Nutr. Dietet.* 43: 80.

Roe, D. A. 1985a. Drug effects on nutrient absorption, transport and metabolism. *Drug-Nutrient Interact.* 4: 117.

Roe, D. A. 1985b. *Drug Induced Nutritional Deficiencies,* p. 183. AVI Publ. Co., Westport, CT.

Roe, D. A. 1985c. Therapeutic effects of drug-nutrient interactions in the elderly. *J. Amer. Dietet. Assoc.* 85: 174.

Somogyi, J. C., Hotzel, D., and Fujiwara, M. 1986. Biological Interactions and Nutrition: A Workshop Report. *Proc. XIII Internat. Congress of Nutrition, 1985,* Taylor, T. G. and Jenkins, N. K. (Ed.), p. 480. John Libbey, London.

Toothaker, R. D. and Welling, P. G. 1980. The effect of food on drug bioavailability. *Ann. Rev. Pharmacol. Toxicol.* 20: 173.

# Index